Natural Computing Series

Muddassar Farooq

Bee-Inspired Protocol Engineering

From Nature to Networks

With 128 Figures and 61 Tables

 Springer

Author

Dr. Muddassar Farooq
Director
Next Generation Intelligent Networks
Research Center (nexGIN RC)
National University of Computer and
Emerging Sciences (NUCES-FAST)
A.K. Brohi Road, Sector H-11/4
Islamabad, Pakistan
and
Informatik III
Technical University of Dortmund
Germany
muddassar.farooq@nu.edu.pk
muddassar.farooq@cs.uni-dortmund.de

Series Editors

G. Rozenberg (Managing Editor)
rozenber@liacs.nl

Th. Bäck, J.N. Kok, H.P. Spaink
Leiden Center for Natural Computing
Leiden University
Niels Bohrweg 1
2333 CA Leiden, The Netherlands

A.E. Eiben
Vrije Universiteit Amsterdam
The Netherlands

ISBN 978-3-642-09946-5   e-ISBN 978-3-540-85954-3

DOI 10.1007/978-3-540-85954-3

Natural Computing Series ISSN 1619-7127

ACM Computing Classification (1998): C.2, I.2.11

© 2009 Springer-Verlag Berlin Heidelberg
Softcover reprint of the hardcover 1st edition 2009

Cover design: KünkelLopka, Heidelberg

Printed on acid-free paper

9 8 7 6 5 4 3 2 1

springer.com

This book is dedicated to my father Barkat Ali Chaudry and my mother Asmat Begum.

Foreword

The beginning of the computer era was accompanied by a couple of exciting interdisciplinary concepts. Norbert Wiener established the discipline cybernetics, which emphasizes (self)-regulation as a principle in natural and artificial systems. McCulloch's and Pitts' artificial neuron, Rosenblatt's perceptron, and Steinbuch's 'Lernmatrix' as means for pattern recognition and classification raised hopes for brain-like machines. Not much later, Jack Steele coined the term bionics (later also called biomimetics) for all kinds of efforts to learn from living systems in order to create technical devices or processes for solving tasks in innovative ways. Three (at least) groups of people, at the same time but at different locations (San Diego and Ann Arbor in the US and Berlin in Germany) began to mimic mutation, recombination, and natural selection as search principles for many kinds of amelioration, if not approximate optimization, tasks that sometimes resist traditional approaches.

Since the mid-1990s, several of these interdisciplinary endeavors have come together under the umbrella "Computational Intelligence," including artificial neural nets, fuzzy systems, and evolutionary computation, and/or under the umbrella "Natural Computing," including ever more approaches gleaned from nature. There are, for example, DNA and quantum computing, and a couple of successors to evolutionary algorithms like artificial immune systems and simulated particle swarms.

One of these new problem-solving aids uses the bee hive as metaphor to create a novel routing strategy in telecommunication networks. As always with bio-inspired computing procedures, it is important to choose an appropriate level of abstraction. If this level is too low, rigorous analysis becomes impossible; if it is too high, the algorithm may lose its efficacy. The author of this unique and innovative work has found an admirable route between Scylla and Charybdis.

Be curious!

Hans-Paul Schwefel
Dortmund, September 2007

Preface

The constant improvement in communication technologies and the related dramatic increase in user demand to be connected anytime and anywhere, to both the wealth of information accessible through the Internet and other users and communities, have boosted the pervasive deployment of wireless and wired networked systems.[1] These systems are characterized by the fact of their being *large* or *very large*, highly *heterogeneous* in terms of communication technologies, protocols, and services, and very *dynamic*, due to continual changes in topology, traffic patterns, and number of active users and services.

Intelligent and *autonomic* management, control, and service provisioning in these complex networks, and in the future networks resulting from their integration and evolution, require the definition of novel protocols and techniques for all the architectural components of the network.

In this book we focus on the *routing* component, which is at the very core of the functioning of every network since it implements the strategies used by network nodes to discover and use paths to forward data/information from sources to destinations. An effective design of the routing protocol can provide the basic support to unleash the intrinsic power of the highly pervasive, heterogeneous, and dynamic complex networks of the next generation. In this perspective, the routing path selection must be realized in a fully *automatic* and *distributed* way, and it must be *dynamic* and *adaptive*, to take into account the constant evolution of the network state, which is defined by multiple concurrent factors such as topology, traffic flows, available services, etc.

The literature in the domain of routing is very extensive. Routing research has fully accompanied the evolution of networking to constantly adapt the routing protocols to the different novel communication technologies and to the changes in user demand. In this book we review routing protocols and algorithms which have been specifically designed taking inspiration from, and reverse engineering the characteristics of, processes observed in nature in general and in *insect societies* in particular.

---

[1] The author would like to sincerely thank Gianni Di Caro for his time and effort in co-authoring this preface.

This class of routing protocols is indeed relatively large. The first notable examples date back to the beginning of the second half of the 1990s, and a number of further implementations rapidly followed the first ones and gained the attention of the scientific community.

The fact that insect societies have and, in general, nature has served as a major source of inspiration for the design of novel routing algorithms can be understood by noticing that these biological systems are characterized by the presence of a set of distributed, autonomous, minimalist units that, through local interactions, self-organize to produce system-level behaviors which show life-long adaptivity to changes and perturbations in the external environment. Moreover, these systems are usually resilient to minor internal failures and losses of units, and scale quite well by virtue of their modular and fully distributed design. All these characteristics, both in terms of system organization and resulting properties, meet most of the necessary and desired properties of routing protocols for next-generation networks. This fact makes it potentially very attractive to look at insect societies to draw inspiration for the design of novel routing protocols featuring *autonomy, distributedness, adaptivity, robustness*, and *scalability*. These are desirable properties, not only in the domain of network routing, but also in a number of other domains. As a matter of fact, in the last 20 years, collective behaviors of insect societies related to operations such as *foraging, labor division, nest building and maintenance, cemetery formation*, etc. have provided the impetus for a growing body of scientific work, mostly in the fields of telecommunications, distributed systems, operations research, and robotics. Behaviors observed in colonies of *ants* and of *termites* have fueled the large majority of this work. In this book, however, we focus our attention on *bee colonies* that since the beginning of our research have been attracting a growing interest.

All the algorithms that will be discussed in the book are characterized by the fact of their being composed by a potentially very large number of *autonomous* and *fully distributed* controllers, and of having been designed according to a *bottom-up* approach, relying on basic *self-organizing* abilities of the system. These characteristics, together with the biological inspiration from behaviors of *insect societies*, are the very fingerprints of the *Swarm Intelligence* (SI) paradigm.

These peculiar design guidelines contrast with those of the more common *top-down* approach followed for the design of the majority of "classical" routing protocols. In typical top-down design a centralized algorithm with well-known properties is implemented in a distributed system. Clearly, this requires us to modify the original algorithm to cope with the intrinsic limitations of a distributed architecture in terms of full state observability and delays in the propagation of the information. The main effect of these modifications is that several properties of the original algorithm do not hold anymore if the network dynamics is non-stationary, which is the most common case. Still, it is relatively easy to assert some general formal properties of the system.

On the other hand, with the bottom-up approach, the design starts with the definition of the behavior and interaction modalities of the individual node with the objective of obtaining the wanted global behavior as the result of the joint actions of all nodes interacting with one another and with the environment at the local level. It is, in general, "easier" to follow a bottom-up approach, and the resulting algorithm is usually more flexible, scalable, and capable of adapting to a variety of different situations. This is precisely the case for SI protocols and our bee-inspired routing protocols that we will discuss in this book. The negative aspect of this way of proceeding is that it is usually hard to state the formal properties and the expected behavior of the system.

In this book we follow a presentation style that nurtures the cognitive faculties of a reader in such a manner that he becomes a curious traveler in an adventurous journey that takes him from nature to *networks*. We expect him to ask himself questions during this adventure: (1) What is the correlation between nature and *networks*? (2) How do *bees* in nature provide inspiration for *bee agents*? (3) What are the peculiar characteristics of *bee agents*? (4) Can we utilize tools of mathematics to model behavior of *bee agents*? (5) How do we develop testing theaters to appreciate the role of *bee agents* in different acts? (6) How can we engineer nature to develop systems that can be deployed in the real world? We feel most of these questions will be answered sooner or later in the book. We believe that the book will also reveal unconventional design philosophies to classical networking researchers and engineers, who will appreciate the importance of cross-fertilization of concepts from nature for *engineering*. We call this discipline *Natural Engineering*, in which nature and its principles are used as a driving impulse to raise the awareness and the consciousness of a designer. This principle is also at the center of *Bionics* research.

Acknowledgements

First, I would like to emphasize that the dedication to my father should not be considered as a traditional dedication because my father is not a person but an institution. He retired as a senior bank executive. The financial experts could imagine the stress related to such a job. He used to teach me at least for two to three hours daily in my primary school after coming home from his tiring job routine. I still remember that once when he was posted in a rural town of Saudi Arabia, I was unable to go to any school for two years because of the unavailability of any English or Urdu medium school. However, I had the honor of being educated by my father. He taught me everything from science to mathematics and from drawing to literature during these two years. I just used to go to the Dhahran province at the end of the academic year to take my final examination in an Urdu medium school. Some of you might be surprised to know that I stood second in both grades 5 and 6 and missed the top position by only a couple of marks. I think that without his tremendous hard work I would not have been successful in my life. I believe that the world would be a better place for many children if their fathers could give them only 20% of the time that my father gave to me. I thank you and salute you my teacher, tutor and father. This book is in fact your book and this success is of course your success. My mother is a housewife and she gave me all that a mother could give to her child. Without her strong encouragement and prayers, I would not have achieved this success in my life. I am thankful to God that He gave me parents like mine.

After my parents, I thank Prof. Dr. Horst F. Wedde (LS III, TU Dortmund), who showed his confidence in me by allowing me to tread on a labyrinthine research path many other professors would have not even dared to. He always encouraged me and remained patient while I was reading the two masterpieces: *The Dance Language and Orientation of Bees* and *The Wisdom of the Hive*. Finally, his patience and confidence was generously rewarded once our paper won the best paper award at the ANTS conference in Brussels in 2004. Currently, we are working on two projects that are inspired by the bee behavior: *BeeHive* deals with routing in fixed networks and *BeeAdHoc* deals with routing in Mobile Ad Hoc Networks (MANETs). The projects have received enormous attention from nature-inspired routing algorithm groups around the world. Moreover, my special gratitude goes to Prof. Wedde for the way he thoroughly read the draft version of this manuscript. Last but not least, he pushed a lazy person like me to limits to finish the writing of this manuscript in time. I would also like to thank Prof. Dr. Heiko Krumm and Dr. Thomas Bartz-Beielstein for their valuable comments and suggestions on an earlier version of the book. These helped in improving the quality of the book.

My stay of five years at Lehrstuhl III of the Technical University of Dortmund is a story of dedicated friendship. I consider this friendship an even bigger achievement than *BeeHive* or *BeeAdHoc*. Frank-Thorsten Breuer and his parents accepted my wife, my son and me like family members. Every couple of months they invited us for a dinner or a party at their home. Arnim Wedig took care of me with his nice tea and cookies. He also assisted me in the procurement of expensive computational resources for the bee-inspired projects. Mario Lischka helped me quickly learn La-TeX. I must not dare to forget Mrs. Düsenberg, who is the heart of our department. She is reputed to be our de facto psychotherapist. She gave me useful tips on how to be a successful husband.

BeeHive would have never been realized inside the network stack of the Linux kernel without the dedicated work of my students Yue Zhang and Alexander Harsch. I find myself lucky that I had the opportunity to supervise them for their Master's theses. Constantin Timm deserves my special indebtedness for developing a plotter utility that automated the process of reading the data files and then plotting the important performance values. Later he also became my student and helped me in realizing security frameworks for *BeeHive*. Then I moved from TU Dortmund, Germany to the National University of Computer and Emerging Sciences (NUCES), Pakistan. I again found myself lucky that I had students like Saira Zahid and Muhammad Shahzad who helped in developing the formal model for *BeeHive*. Finally, Mohammad Saleem started working on developing *BeeSensor* for Wireless Sensor Networks (WSNs). I would also like to thank Gianni Di Caro at IDSIA, Switzerland. We extensively exchanged emails and our discussions resulted in identifying the important directions for our *BeeHive* and *BeeAdHoc* projects. He also helped in auditing the source code of our *AntNet* implementation in the OMNeT++ simulator. His suggestions were useful in reproducing the desired behavior of *AntNet*.

Both projects would not have been successful without two special persons: my wife Saadi (Dua) and my son Yousouf. Saadi is my friend, and my love. She has sacrificed her career in order to enable me to quickly finish my projects and the current manuscript. She is a gynecologist and I wish that a day would come when I could do something for her as well. Yousouf kept me busy in everything except my *BeeHive* and *BeeAdHoc* projects. He showed me that there are more important things in life than *BeeHive*, e.g., *Teletubbies* and *Barney*. I now remember their names by heart (Tinky Winky, Dipsy, Laa-Laa and Po) because we saw them almost daily during the time period when I was writing the first half of the book. In the meantime, God has blessed us with a daughter, Hajra. Her cute smiles were the best source of stress therapy for me, when I was writing the second phase of the book that consists of Chapters 6 to 8.

Finally, I would like to thank Prof. Dr. Hans-Paul Schwefel for his valuable time writing an informative foreword for my book.

Muddassar Farooq
Islamabad, March 2008

Contents

1

Introduction

During recent years, telecommunication networks have become a special focus of research, both in academia and industry [91, 93, 205]. This is certainly due to the unprecedented growth of the Internet during the last decade of the previous century as it developed into a nerve center of the communication infrastructure [168]. One important reason for the success of the Internet is its connectionless packet-switching technology (no connection is established between a sender and a receiver). Such a paradigm results in a simple, flexible, scalable and robust network layer architecture [135, 19, 189]. This is in contrast to traditional connection-oriented telecommunication networks in which a circuit is reserved for a connection between a sender and a receiver [91, 93, 205].

The Internet's success motivated researchers to realize the dream of *Ubiquitous Computing*, including the concept of "one person–many computers" [276, 278, 277, 279]. Research and development in *Ubiquitous Computing* resulted in an exponential growth of smart handheld computing devices, which have to be interconnected and connected to the Internet to satisfy highly demanding users. In turn, these requirements resulted in a phenomenal growth in wireless telecommunication networks and their supporting Internet Protocol (IP) (the standard protocol for the network layer of the Internet) on wireless networks. However, these wireless networks require an infrastructure (base station) for providing connectivity to mobile terminals. As a result, work on Mobile Ad Hoc Networks (MANETs) has become a vigorous effort. Here mobile terminals communicate with one another without the need for a communication infrastructure. These networks have turned useful or even indispensable in search and rescue operations, disaster relief management, and military command and control.

Ubiquitous Computing has created a demanding community of users, who are utilizing its potential in novel applications like the World Wide Web (WWW), Computer Supported Collaborative Work (CSCW), e-commerce, tele-medicine, and e-learning. An essential feature of most of these applications is the ability to transmit audio and video streams to the participants under some Quality of Service (QoS) constraints. The users want all of these services on their desktops as well as on their mobile terminals. Such challenging requirements can only be met if a network's resources are utilized in an efficient manner.

The efficient utilization of limited network resources and infrastructures by enhancing/optimizing the performance of operational IP networks is defined as *Traffic Engineering* [15, 167]. Its goals are accomplished by devising efficient and reliable routing strategies. The important features and characterizations of such routing protocols are: *loadbalancing, constraint-based routing, multi-path routing, fast rerouting, protection switching, faulttolerance and intelligent route management.* Currently, the Internet community employs multi-path routing algorithms like MPLS (Multi-Protocol Label Switching) [181], which is based on managing virtual circuits on top of the IP layer, and hence lacks scalability and robustness. Another approach avoids completely the use of virtual circuits and manages the resources of each session by per-flow fair scheduling of the links. Nevertheless, flows are set up along the shortest paths determined by the underlying routing protocols. The reservation of flows are managed by the Resource Reservation Protocol (RSVP) [297, 260]. However, the deterministic service guarantees are provided to real-time applications using the Interserv architecture [28, 297]. In large networks, this per-flow mechanism does not scale (they can have hundreds of thousands of flows); therefore, in [102], RSVP has been extended by replacing the per-flow routing state with per-source/destination routing state. This results in a state size that grows only quadratically with the number of nodes. Both of these protocols suffer from serious performance bottlenecks because they utilize the single-path routing algorithm *Open Shortest Path First* (OSPF) at the IP layer. Consequently, the bandwidth of the single path is quickly consumed, which results in a high call-blocking probability [260].

The major challenge in *traffic engineering* in a nutshell is *to design multi-path routing protocols for IP networks in which multiple/alternative paths are efficiently discovered and maintained between source and destination pairs.* Such routing protocols will provide solutions to existing technical challenges by using the connectionless paradigm of the IP layer.

1.1 Motivation of the Work

We believe that a complete reengineering of the network layer is the logical solution not only to the *traffic engineering* problem but also to network management. The growth of the Internet demands design and development of novel and intelligent routing protocols that result in an intelligent and knowledgeable network layer. Currently, the network layer is relegated to just switching data packets to the next hop based on the information in the routing tables collected by non-intelligent control packets. The new protocols, however, have to be designed with a careful engineering vision in order to reduce their communication, processing, and router's resource costs.

The research in agent-based routing systems has resulted in our developing many novel networking systems [250, 51, 107, 164]. The algorithms utilize software agents which have the following properties [303]:

- *Autonomous:* the capability of performing autonomous actions.
- *Proactive:* the capability of exhibiting opportunistic and goal-oriented behavior and taking initiative where appropriate.
- *Responsiveness:* the capability of perceiving the environment and responding in a timely fashion to the changes that occur in it.
- *Social:* the capability of interaction with other artificial agents and humans when appropriate in order to achieve their own objectives and to help others in their activities.

This design paradigm, therefore, focuses on robust and intelligent agent behavior. In [281], White blames the Artificial Intelligence (AI) community for this. The AI community has been strongly influenced by *Symbol Hypothesis* [176] and first-order predicate logic. The symbols and theorem proving are classical tools, based on the *Resolution Principle* [196]. Consequently, such systems coordinate their activities by exchanging symbolic information and theorem proving. In addition, all properties of a system could not be inferred by representing knowledge in a symbol formula and then manipulating it using first-order predicate logic [204, 281]. Another shortcoming is the *Frame* problem, which results from the need to specify states and state transitions. The measured data obtained from real-world systems has to be represented in symbols, which leads to the *sensor fusion* problem. Connectionist systems and artificial neural networks try to overcome these problems. However, their black box nature makes it difficult to synthesize and utilize them in distributed network systems [281].

The real-world networks represent a dynamic environment in which good routing decisions need to be taken in real time under a number of performance and cost constraints; therefore, applying such complex paradigms to achieve intelligence in the network layer is not feasible. The processing complexity and communication cost of launching such complex agents will be overwhelming, and they will also consume significant amounts of a router's resources, especially in large networks.

The above-mentioned problems in traditional agent-based approaches could be easily solved if we followed a dramatically novel paradigm for designing the agents: *agents need not be rational in order to solve complex problems* [281]. This conjecture, at first, appears to completely boggle the mind because it suggests that *intelligence could result from simple non-intelligent agents.* However, systems based on this design paradigm are rigorously studied in Swarm Intelligence [21]. It takes inspiration from self-organization in natural colony systems, e.g., ants' or bees' [33], and utilizes their principles as a metaphor to design simple agents that take decisions based on local information without the need of a central complex controller. However, such agents are situated in their environment and they utilize either a direct agent-agent communication paradigm or an agent-group paradigm in which they indirectly communicate through the environment. In [33, 24], the authors have defined the basic ingredients of self-organization, which are the following:

1. The *positive feedback* in the system amplifies the good solutions that the agents have discovered. Consequently, other agents are recruited to exploit these good solutions.

2. The *negative feedback* in the system helps in counterbalancing the positive feedback; as a result, good solutions cannot dominate forever.
3. *Amplification of random solutions* helps in discovering and exploring new solutions.
4. *Multiple interactions* help in enabling individuals to use the results of their own activities as well as of others' activities.

In this way a colony is able to achieve a complex and intelligent behavior at the colony level that is well beyond the intelligence and capabilities of an individual in the colony. We believe that self-organization systems have all the features that we could wish for in large network systems.

1.2 Problem Statement

We believe that the complexity of the manifold task of endowing intelligence and knowledge to the network layer through self-organizing agents, which are inspired by the communicative and evaluative principles of a honeybee colony, is overwhelmingly phenomenal. Therefore, in our research, we take a cardinal first step to achieve this objective. Our problem statement could be outlined as: *efficient, scalable, robust, fault-tolerant, dynamic, decentralized and distributed solutions to traffic engineering could be provided within the existing connectionless model of IP through a nature-inspired population of agents, which have simple behavior. The agents explore multiple paths between all source/destination pairs and then distribute the network traffic on them. This approach could significantly enhance the network performance.*
Our routing protocol should be able to meet the following challenging requirements:

1. The *agents* must not require existing Multi-Agent System (MAS) software for their realization. Rather, their behavior and learning algorithm should be simple enough to be implemented directly in the network layer by utilizing semantics of C/C++ languages.
2. The processing complexity of *agents* must be kept at a minimum level and the time a router spends in processing them should only be a fraction of the time that it spends in switching data packets. This requirement is necessary because the performance of a router could significantly degrade if *agent processing* steals most of its time [295].
3. The *agents* must explore the network in an asynchronous manner.
4. The protocol must be robust against loss of *agents*.
5. The size of agents must be such that they could fit into the payload of an IP packet. This requirement will significantly reduce communication-related costs.
6. The protocol must be able to scale to large networks.
7. It must be designed with a vision to install it on real-world routers. Therefore, the simulation model must be realizable inside the network stack of a Linux router.
8. It must be realizable in real-world routers without the need for additional resources in both hardware and software. This requirement would simplify its installation, though in a cost-effective manner, on existing routers.

9. It must not require synchronization of clocks in the network.
10. It must not require that the routing tables of different routers should be in a consistent state for taking correct routing decisions.

1.2.1 Hypotheses

The study of honeybees has revealed a remarkable sophistication of their communication capabilities. Nobel laureate Karl von Frisch deciphered and structured these into a language in his book *The Dance Language and Orientation of Bees* [259]. Upon their return from a foraging trip, bees communicate the distance, direction, and quality of a flower site to their fellow foragers by waggle dances on a dance floor inside the hive. By dancing zealously for a good foraging site they recruit foragers for it. In this way a good flower site is exploited, and the number of foragers at this site are reinforced. A honeybee colony has many features that are desirable in networks:

- efficient allocation of foraging force to multiple food sources;
- different types of foragers for each commodity;
- foragers evaluate the quality of food sources visited and then recruit the optimum number of foragers for their food source by dancing on a dance floor inside the hive;
- no central control;
- foragers try to optimize the energetic efficiency of nectar collection and make decisions without any global knowledge of the environment.

In our work we use the following hypotheses

(a) **H1:** If a honeybee colony is able to adapt to countless changes inside the hive or outside in the environment through simple individuals without any central control, then an agent system based on similar principles should be able to adapt itself to an ever-changing network environment in a decentralized fashion with the help of simple agents who rely only on local information. This system should be dynamic, simple, efficient, robust, flexible, reliable, and scalable because its natural counterpart has all these features.

(b) **H2:** If designed with a careful engineering vision, nature-inspired solutions are simple enough to be installed on real-world systems. Therefore, their benefit-to-cost ratio should be better than that of existing real-world solutions.

We believe that all of these objectives can be achieved by contemplating novel paradigms for developing agents. The research, however, is of multidisciplinary nature because it involves cross-fertilization of ideas from biology, AI, agent technology, network management, and network engineering. Therefore, we developed a *Natural Engineering* approach[1] to successfully accomplish our objectives in a given time frame.

---

[1] The focus of our work is on following an engineering approach for nature-inspired routing protocols. However, the engineering approach itself is general enough and complements the existing approaches of Bionik [175, 199] and CI (Computational Intelligence) [3].

1.3 An Engineering Approach to Nature-Inspired Routing Protocols

In this section we will introduce our engineering approach[2], which we followed in the design and development of a routing protocol inspired by a natural system (a honeybee colony).

Definition 1 (Natural Engineering) *Natural Engineering is an emerging engineering discipline that enables scientists and engineers in search of efficient or optimal solutions for real-world problems under resource constraints to take inspiration and utilize observations from organizational principles of natural systems, and to transform them into structural principles of software organization of algorithms or industrial product design.*

The above-mentioned concept emphasizes six aspects:

1. Understanding the working principles of natural systems.
2. Developing algorithmic models of the organizational principles of natural systems.
3. Understanding the operational environment of target systems.
4. Mapping concepts from the natural system to the technical system.
5. Adapting the algorithmic model to the operational environment of a technical system.
6. Following a testing and evaluating feedback loop in search of optimum solutions under the resource constraints (time, space, computation, money, labor, etc.).

There is no clear-cut way to achieve a perfect match between structures and principles in natural life organizations and working principles in technical systems. The most important challenge, therefore, is to identify a natural system of which the working principles could be appropriately abstracted for deriving suitable principles to work in a given technical system. Instead of adding numerous non-biological features to a natural system, we believe that it is more advisable to look to other natural systems for inspiration. In our case we chose honeybee colonies because the foraging behavior of bees could be transformed into different types of agents performing different routing tasks in telecommunication networks. Both systems have to maximize the amount of a commodity (nectar delivered to hives and data delivered to nodes respectively) as quickly as possible, under a permanently and even unpredictably changing operating environment.

The major focus of research is to design and develop cost-efficient bio/nature-inspired business solutions for highly competitive markets. Therefore, the development of a nature-inspired routing algorithm must follow a feedback-oriented engi-

---

[2] This section is reproduced by permission of the publisher, Chapman & Hall/CRC Computer and Information Science, from our Chapter 21:*BeeHive: New Ideas for Developing Routing Algorithms Inspired by Honey Bee Behavior* (pages 321–339), published in Handbook of Bioinspired Algorithms and Applications, Albert Zomaya and Stephan Olariu, editors, 2005.

neering approach (see Figure 1.1) that incorporates most of the features discussed above.

First, we considered the ensemble of constraints under which the envisioned routing protocol is supposed to operate:

- Nonavailability of a global clock for trip time calculation.
- Routers and links could crash.
- Routers have limited queue capacity.
- Links have a BER (bit error rate) associated with them.
- The requirements from the Linux kernel routing framework needed to support the protocol.
- The requirements of the IP protocol, which is currently used in the network layer of the Internet.

At the same time we decided that the *bee agents* should explore the network, collect important parameters, and make the routing decisions in a decentralized fashion (in the style in which real scouts/foragers make decisions while collecting nectar from flowers). *Bee agents* should measure the quality of a route and then communicate it to other *bee agents* like foragers do in nature. The structure of the routing tables should provide the functionality of a dance floor for exchanging information among *bee agents* as well as among *bee agents* and data packets. Moreover, we must be able to realize it in a real kernel of the Linux operating system later on.

We implemented our ideas in a simulation environment and then refined our algorithmic mapping through the feedback channel 1 (see Figure 1.1). During this phase we did not use any simulation-specific features that were not available inside the Linux kernel, e.g., vector, stack, or similar data structures. Once we reached a relative optimum of our protocol in a simulator, we started to develop an engineering model of the algorithm. The engineering model can be easily transported to the Linux kernel routing framework. We tested it in the real network of Linux routers and refined our engineering model through the feedback channel 2 (see Figure 1.1). We evaluated our conceptual approach in two prototype projects: *BeeHive* [273], which deals with the design and development of a routing algorithm for fixed networks, and *BeeAdHoc*, the goal of which is to design and develop an energy-efficient routing algorithm for Mobile Ad Hoc Networks (MANETs) [269, 270, 271].

1.4 The Scientific Contributions of the Work

In this section we will list the general scientific contributions achieved during our research in the past six years. The reader will appreciate the overwhelming complexity of the work due to the diverse nature of accomplishments achieved in the *BeeHive* and *BeeAdHoc* projects. Some of the information might be duplicated here, but we believe that it is important to make the section self-contained.

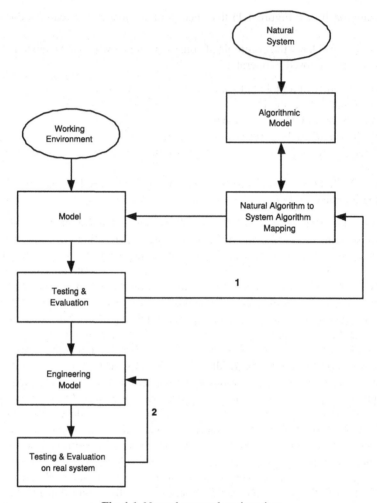

Fig. 1.1. Natural protocol engineering

1.4.1 A Simple, Distributed, Decentralized Multi-Agent System

We have developed a simple and distributed multi-agent system in which a population of agents collectively achieves an objective. The agents are simple entities with limited processing and memory capabilities and they make their decisions based on their local view of the network state. The state is determined by local information, which is collected in a small region around their launching node. Such a simple agent model is the result of borrowing communication principles from the wisdom of the hive. The agents try to undertake the daunting task of optimizing a number of competing performance values like throughput, packet delay, etc. under different cost constraints.

1.4.2 A Comprehensive Routing System

The multi-agent system, as described above, was instrumental in designing and developing a multi-path routing protocol, *BeeHive*, which is dynamic, simple, efficient, robust, flexible, and scalable. As demonstrated by our results, the algorithm achieves a similar or better performance than the existing state-of-the-art algorithms. *BeeHive*, however, achieves this objective with significantly lesser costs in terms of processing, communication, and router's resources. The algorithm does not require access to the complete network topology; rather, it works with a local view of the network. The agents take their decisions in an autonomous and decentralized fashion.

1.4.3 An Empirical Comprehensive Performance Evaluation Framework

The other major contribution of the work is a comprehensive performance evaluation framework, which calculates a number of important performance values and the associated costs of a routing algorithm. The framework can also vary a number of network configurations from traffic patterns to network topology. As a result, the developer of a routing protocol can study the behavior of an algorithm on a wide operational landscape with a focus on its benefit-to-cost ratio in an unbiased manner. The framework proved to be useful in identifying reasons behind the anomalous behavior of *BeeHive* in different scenarios. Subsequently, we were able to improve our algorithm through the feedback channel 1 as shown in Figure 1.1.

1.4.4 A Scalability Framework for (Nature-Inspired) Agent-Based Routing Protocols

We developed a comprehensive framework that facilitates the study of the scalability of agent-based distributed systems in general and of routing protocols in particular. The framework provides a formal model and a set of empirical tools to protocol developers that are useful in investigating the scalability of their protocols at an early stage of development. To our knowledge, this is the first model that provides an unbiased way of studying the scalability of (nature-inspired) agent-based routing protocols.

1.4.5 Protocol Engineering of Nature-Inspired Routing Protocols

One of the most important contributions of our work is the vision of *Natural Engineering*, introduced in the last section. We believe that developing a nature-inspired system, which can be installed or utilized in real-world systems, is a challenging task. The nature-inspired community, at times, lacks vision about the real operational environments. As a result, most of the proposed solutions are never realized in the intended real-world systems. Our work, according to our knowledge, is an important step from "Swarm Intelligence" to "Natural Engineering." We believe that the work will stimulate other researchers to adopt a similar approach for their projects as well.

1.4.6 A Nature-Inspired Linux Router

Our *Natural Engineering* approach significantly helped us in developing an algorithmic model in the simulation environment that is mostly independent of the underlying features of a simulation system. It rather utilizes only those components in a simulation environment which are available in real-world Linux routers. This approach showed its benefits once we started developing an engineering model in the form of a nature-inspired Linux router because we were able to make a quantum leap with significantly limited man power and computing resources.

1.4.7 The Protocol Validation Framework

Another important contribution of the work is a comprehensive validation framework in which we implemented the same traffic generators in the simulation and in an application layer of a Linux network stack. We also utilized the same network topology in both simulation and the real network of Linux routers. Our validation principle is: if we generate the same traffic patterns in identical topologies in both simulation and the real network, then the performance values of the algorithms should be traceable from one environment to another with acceptable deviations. To the best of our knowledge, *BeeHive* is the first nature-inspired algorithm that has been implemented in real networks and shown substantial performance benefits for existing real-world applications.

1.4.8 The Formal Framework for Nature-Inspired Protocols

An important contribution of our research is a formal framework developed by utilizing probabilistic recursive functions and formal concepts of M/M/1 queuing theory. By utilizing the model, we were able to model both agents' and data traffic flows passing through a node in a network running our *BeeHive* protocol. We then used this traffic flow model to formally represent relevant performance parameters. We believe that the framework is generic and will help protocol designers and developers to model the behavior of their nature-inspired routing protocols by utilizing its relevant concepts. In line with our *Natural Engineering* approach we validated our formal model by comparing its estimated values with the values obtained from the OMNeT++ network simulator. The performance metrics estimated by the formal model approximately map to the metrics obtained from the simulator.

1.4.9 A Simple, Efficient, and Scalable Nature-Inspired Security Framework

Another important contribution of the work is the conducting of a pilot study for the vulnerabilities of our *BeeHive* protocol that malicious nodes can exploit to disrupt normal network operations. To the best of our knowledge, this is the first detailed pilot study within nature-inspired community. Consequently we developed an immune-inspired simple, efficient, and scalable nature-inspired security framework for *BeeHive* that provides the same security level as that of signature-based cryptographic solutions but at significantly smaller processing and communication costs. Our results show that our enhanced framework can counter a number of threats launched by malicious nodes in the network.

1.4.10 Emerging Mobile and Wireless Sensors Ad Hoc Networks

Another important contribution of our work is demonstrating that bee-inspired protocol engineering is not limited to just fixed telecommunication networks. We show that by taking inspiration from the wisdom of the hive, we can also develop an energy-efficient routing protocol, *BeeAdHoc*, for Mobile Ad Hoc Networks (MANETs), and *BeeSensor*, for Wireless Sensor Networks (WSNs). Both protocols take inspiration from the energy conservation behavior of a bee colony. Following our *Natural Engineering* approach, we implemented *BeeAdHoc* on mobile laptops running Linux and tested our protocol in a real-world MANET. We also designed a novel testing methodology in which we gradually move from a simulator-only environment to real MANET. This work shows the potential of nature-inspired protocols in next-generation networks.

1.5 Organization of the Book

The work presented in this book is organized into nine chapters. Each chapter, except the first and the last, will provide a comprehensive review of the research conducted in a particular phase of our *Natural Engineering* cycle, from our conceiving the ideas from the working principles of a natural system, to our developing an algorithmic model from them, to our realizing the algorithmic model both in a simulation environment and in a real network of Linux routers. The realization phase, both in simulation and real networks, is complemented by extensive testing, analysis, evaluation, and feedback channels.

Chapter 2: A Comprehensive Survey of Nature-inspired Routing Protocols

The chapter presents the true challenges that a routing protocol is expected to meet in complex networks of the new millennium. We provide classifications of the algorithms based either on their characteristics or on their design philosophy. The basic objective of the survey is to understand the design doctrine of different communities involved in the design and development of routing algorithms. This will motivate researchers to develop state-of-the-art routing algorithms through a process of cross-fertilization of useful features and characteristics of different design doctrines. We classify the communities into three categories: Networking, Artificial Intelligence (AI), and Natural Computing (NC). The focus of the survey presented in Chapter 2 is on the algorithms developed by the Natural Computing community. We provide a detailed survey of routing algorithms inspired from the pheromone-laying principles of ant colonies. The algorithms are based on the Ant Colony Optimization (ACO) metaheuristic. We also provide a comprehensive review of routing algorithms based on the principle of evolution in natural systems. Later in the chapter, we introduce routing algorithms based on the principles of *Reinforcement Learning*. These routing algorithms are developed by the Artificial Intelligence community. Finally, we briefly summarize the routing algorithms recently developed by the networking community. The comprehensive survey proved helpful in identifying the merits and deficiencies of existing state-of-the-art routing protocols developed by different communities. Most of the chapter has been reproduced from the following paper with the kind permission of Elsevier.

- H. F. Wedde and M. Farooq. A comprehensive review of nature inspired routing algorithms for fixed telecommunication networks. *Journal of System Architecture*, 52(8-9):461-484, 2006.

Chapter 3: From *the Wisdom of the Hive* to Routing in Telecommunication Networks

The chapter describes the most important steps in our *Natural Engineering* approach. It starts with a brief introduction to the foraging principles of a honeybee colony. We present the biological concepts in such a manner that the reader conveniently conceives a honeybee colony as a population-based multi-agent system, in which simple agents coordinate their activities to solve the complex problem of the allocation of labor to multiple forage sites in dynamic environments. The agents achieve this objective in a decentralized fashion with the help of local information they acquire while foraging. We argue that an efficient, reliable, adaptive, and fault-tolerant routing algorithm has to also deal with similar daunting issues.

We then provide the mapping of concepts from a natural honeybee colony to an artificial multi-agent system, which can be utilized for routing in telecommunication networks. The mapping of concepts appears to be a crucial step in developing an algorithmic model of an agent-based routing system. We emphasize the motivation behind important design principles of our *BeeHive* routing algorithm. We provide a comprehensive description of our *bee agent* model by emphasizing the communication paradigm utilized by the *bee agents*, which is instrumental in reducing the costs associated with a routing algorithm: communication, processing, and router's resources. Later in the chapter, we introduce our comprehensive empirical performance evaluation framework that calculates a number of preliminary and auxiliary performance values. These values provide an in depth insight into the behavior of a routing algorithm under a variety of challenging network configurations.

Finally, we introduce our extensive experimental framework in a simulation environment. The experiments are designed through extensive brainstorming exercises in order to meticulously analyze the behavior of a routing protocol under diversified network operations. The results obtained from our performance evaluation framework are discussed. We compare *BeeHive* with a state-of-the-art ACO routing algorithm, *AntNet*, a state-of-the-art evolutionary routing algorithm *Distributed Genetic Algorithm (DGA)*, *OSPF*, and *Daemon*. *Daemon* is an ideal algorithm that can instantly access the complete network topology and size of the queues in all routers to make an optimum routing decision. The algorithm, though, is not realizable in real networks due to the associated costs; but, nevertheless, it serves as an important benchmark for different algorithms.

The results of the experiments unequivocally suggest that *BeeHive* is able to achieve similar or better performance under congested loads compared with *AntNet* and is able to achieve similar or better performance under normal static loads as compared with *OSPF*. However, this excellent performance of *BeeHive* is achieved with significantly smaller communication and processing costs, and routing tables, which have the order of the size as in *OSPF*. The chapter contains extracts from our following published papers, reproduced [273, 268] with the kind permission of Springer Verlag and Chapman & Hall/CRC Computer and Information Science:

1. Horst F. Wedde, Muddassar Farooq, and Yue Zhang. BeeHive: An Efficient Fault Tolerant Routing Algorithm under High Loads Inspired by Honey Bee Behavior. In Marco Dorigo, M. Birattari, C. Blum, L. M. Gambardella, F. Mondada, and T. Stützle, editors, Proceedings of the Fourth International Workshop on Ant Colony and Swarm Intelligence (ANTS 2004), volume 3172 of Lecture Notes in Computer Science, pages 83–94, Brussels, Belgium, September 2004. Springer Verlag. (Winner of the Best Paper Award ANTS 2004.)
2. Horst F. Wedde and Muddassar Farooq. A Performance Evaluation Framework for Nature Inspired Routing Algorithms. In Franz Rothlauf et al., editors, Applications of Evolutionary Computing – Proceedings of EvoWorkshops 2005, volume 3449 of Lecture Notes in Computer Science, pages 136–146, Lausanne, Switzerland, March/April 2005.
3. Horst F. Wedde and Muddassar Farooq. *BeeHive: Routing Algorithms Inspired by Honey Bee Behavior.* Künstliche Intelligenz. Schwerpunkt: Swarm Intelligence, pages 18–24, Nov 2005.
4. Horst F. Wedde and Muddassar Farooq. *BeeHive: New Ideas for Developing Routing Algorithms Inspired by Honey Bee Behavior.* In Handbook of Bioinspired Algorithms and Applications, Albert Zomaya and Stephan Olariu, Ed. Chapman & Hall/CRC Computer and Information Science, Chapter 21, pages 321–339, 2005.

Chapter 4: A Scalability Framework for Nature-inspired Routing Algorithms

The chapter presents a new scalability framework that designers and developers of the routing algorithms in general, and of nature-inspired routing protocols in particular, can utilize to analyze the scalability of their routing protocols. We believe that our new framework will enable the designers to establish the scalability of their routing protocols in an early stage of *protocol engineering* [140]. Such a framework will be instrumental in practicing the principles of *Software Performance Engineering* (SPE), which also emphasizes the consideration of performance and scalability issues early in the design and architectural phase in order to rectify the deficiencies in a simulation environment. This will not only obviate the risk of a disaster once the algorithm is deployed on large-scale networks, but also avert the cost overruns due to tuning or redesign of the algorithm later in the protocol engineering cycle. Consequently, such a pragmatic protocol engineering cycle will be capable of reducing the time to market of a new protocol.

Our scalability model defines power and productivity metrics for a routing protocol. The productivity metric provides insight into the benefit-to-cost ratio of a routing protocol. The cost model includes the communication, processing, and memory costs related to a routing algorithm. We believe that the productivity of a routing algorithm is an important performance value which can be used for an unbiased investigation of a routing protocol. Later we define a scalability metric, which is a ratio of productivity values of two network configurations, and its value should be ideally 1 if the algorithm is perfectly scalable from one network configuration to the other.

The framework is general enough to act as a guideline for analyzing the scalability of any agent-based network system. However, in our work, we restricted our analysis to only three protocols due to lack of high performance simulation platforms. We studied the scalability behavior of *BeeHive*, *AntNet*, and *OSPF* in six topologies which vary in their degree of complexity and connectivity. The size of the topologies is gradually increased from eight nodes to 1,050 nodes. According to our knowledge, this is the first extensive effort to empirically study the scalability of nature-inspired routing protocols.

The results demonstrate that *BeeHive* is able to deliver superior performance under both high and low network traffic loads in all topologies. We believe that an engineering vision during the design and development phase, in which we emphasized the scalability as an important metric, has significantly helped *BeeHive* in achieving better scalability metrics for the majority of the network configurations compared with *AntNet* and *OSPF*. It took more than six months to extensively evaluate the algorithms under a variety of network configurations.

Chapter 5: *BeeHive* in Real Networks of Linux Routers

This chapter describes the second phase of our *Natural Engineering* approach: the realization of an engineering model of *BeeHive* inside the network stack of the Linux kernel, and then the comparison of its performance values with *OSPF* in a real network of eight Linux routers. The work presented in the chapter is novel in the sense that, to our knowledge, *BeeHive* is the first nature-inspired routing algorithm which has been realized and tested in real networks.

The chapter begins by illustrating different design options that are available for realizing a nature-inspired routing algorithm in a Linux router. We then describe the motivation behind our *engineering model* that we realized in a Linux router. Subsequently, we define the software architecture of our Nature-inspired Linux router. Here, we emphasize the challenges we encountered because of the unique features of the *BeeHive* algorithm.

We also migrated our *performance evaluation framework* to the application level of the Linux network stack. The motivation behind this significant step is to follow the protocol verification principle: *if we generate the same traffic patterns through the same traffic generators in both simulation and real networks and utilize the same performance evaluation framework in both simulation and real networks then the performance values obtained from the simulation environment should be traceable to the ones obtained from the real Linux network with minor deviations, provided our simulation environment depicts a somewhat realistic picture of a real network.* We believe that this verification principle will help in tracking the performance values in simulation with those of their counterparts in real networks. If the values are similar, then this would strengthen our thesis: *Nature-inspired routing protocols, if engineered properly, could manifest their merits in real networks.*

Finally, we discuss the results obtained from extensive experiments both in simulation and in a real network. We feel satisfied because the performance values obtained from the simulation are consistent with the values in the real network, with an acceptable degree of deviation. This, according to our knowledge, is the first substantive work which shows the benefits of utilizing nature-inspired routing protocols in real networks running real-world applications, e.g., File Transfer Protocol (FTP) and Voice over IP (VoIP). The success in this phase satisfyingly concludes our last phase in the protocol development cycle of our *Natural Engineering* approach.

Chapter 6: A Formal Framework for Analyzing the Behavior of *BeeHive*

In this chapter we report our formal framework to analyze the behavior of our *Bee-Hive* protocol. The motivation for such a formal framework comes from the fact that most researchers in Natural Computing follow a well-known protocol engineering philosophy: inspire, abstract, design, develop, and validate. Consequently, researchers even today have little understanding of the reasons behind the superior performance of nature-inspired routing protocols. We argue that formal understanding about the merits of *BeeHive* is important in order to get an in depth understanding about its behavior.

We revisit in this chapter our *BeeHive* protocol, which is introduced in Chapter 3, to understand the merits of different design options with the help of our formal framework. We show why different quality formulas provide the same performance. We used probabilistic recursive functions for analyzing online the stochastic packet-switching behavior of the algorithm. The queuing delays experienced due to the congestion have been analyzed using the formal concepts of M/M/1 queuing theory. With the help of this framework we model bee traffic and data traffic on the links of a given node. Using this traffic model we derive formulas for throughput (bits correctly delivered at the destination in unit time) and end-to-end delay of a packet.

Towards the end of the chapter we describe our empirical verification framework in OMNeT++ to validate the correctness of our formal model. We validated our formal model on two topologies and compared its results with the results obtained from the simulations. The estimated performance values of the model have only a small deviation from the real values measured in the network simulator. We believe that this formal treatment will add an important phase of *formal modeling* to protocol engineering of nature-inspired protocols. This formal treatment is the key to widespread acceptability of such protocols in the networking community. The chapter has been reproduced from our following paper with the kind permission of IEEE.

- S. Zahid, M. Shehzad, S. Usman Ali, and M. Farooq. A comprehensive formal framework for analyzing the behavior of nature inspired routing protocols. In Proceedings of Congress on Evolutionary Computing (CEC), pages 180–187. IEEE, Singapore, September 2007.

Chapter 7: An Efficient Nature-Inspired Security Framework for *BeeHive*

In this chapter we investigate the vulnerabilities and related security threats that malicious nodes in a network can exploit to seriously disrupt the networking operations. Remember that researchers working in nature-inspired protocols always implicity trust the identity and routing information of *ant or bee agents*. This assumption is no more valid in real-world networks where compromised nodes can wreak havoc by launching malicious agents that can significantly alter the routing behavior of a protocol. The lack of any work in this important domain motivated us to undertake research to develop a simple, scalable, and efficient security framework for our *Bee-Hive* protocol.

We first provide a list of attacks that malicious nodes can launch on a network running the *BeeHive* protocol. We then introduce our *BeeHiveGuard* security framework, a signature-based security framework in which *bee agents* are protected by the use of the principles of Public Key Infrastructure (PKI) against tampering of their identity or routing information. An obvious disadvantage of this approach is that the size of *bee agents* increases manifold because the signatures are added to their payload. Moreover, complex decryption and encryption operations need to be performed at each intermediate node, which increases the processing complexity manifold. Our results indicate that the processing complexity of *bee agents* in *BeeHiveGuard* increased by more than 52,000% and the communication-related costs increased by more than 200% compared to *BeeHive*. Remember that *bee agents* are launched after every second; therefore, this overhead is definitely not acceptable. As a result, we have to look for other design paradigms that provide the same security level as *BeeHiveGuard* but with significantly smaller processing and communication costs.

After initial investigations, Artificial Immune Systems (AISs), inspired by immunology principles, provide a suitable framework for a simple, efficient, and scalable security framework. Our proposed framework, *BeeHiveAIS*, works in three phases: (1) In the learning phase it passively monitors the network traffic to learn the normal traffic patterns; (2) in the second phase it generates a set of detectors that are later used in the protection phase to classify agents that perform suspicious activities as malicious agents; and (3) during the protection phase it protects the system against malicious attacks. *BeeHiveAIS* has a simple anomaly detection algorithm that works without the need to transmit redundant information in the *bee agents*. Consequently, it has significantly smaller processing and communication costs compared to *BeeHiveGuard*.

We developed an empirical validation framework to verify that both frameworks are able to successfully counter a number of attacks launched by malicious nodes.

We tested both frameworks on topologies ranging from four to 150 nodes. The conclusion of the experiments is: BeeHiveAIS *provides the same security level as* BeeHiveGuard *does, but with significantly smaller processing and communication costs. Moreover, the relevant performance values of the* BeeHiveAIS *are within an acceptable range of those of* BeeHive *(without any attack).* The chapter contains extracts from our following published papers, reproduced with the kind permission of Springer Verlag.

1. H. F. Wedde, C. Timm, and M. Farooq. BeeHiveGuard: A step towards secure nature inspired routing algorithms. In Applications of Evolutionary Computing, volume 3907 of Lecture Notes in Computer Science, pages 243–254. Springer Verlag, April 2006.
2. H. F. Wedde, C. Timm, and M. Farooq. BeeHiveAIS: A simple, efficient, scalable and secure routing framework inspired by artificial immune systems. In Proceedings of the PPSN IX, volume 4193 of Lecture Notes in Computer Science, pages 623–632. Springer Verlag, September 2006.

Chapter 8: Bee-Inspired Routing Protocols for Mobile Ad Hoc and Sensor Networks

Towards the end of our book, we highlight the potential of nature-inspired routing protocols for emerging networks like Mobile Ad Hoc Networks (MANETs) and Wireless Sensor Networks (WSNs). Both types of networks are becoming popular because of their potential utility in war theaters, disaster management, security and tactical surveillance, and weather monitoring. The typical characteristic of these networks are that they are deployed in the real world without any requirement for an infrastructure. As a result, each node is delegated the task of routing as well. Due to mobility in MANETs and varying power levels in WSNs, the connectivity of nodes continuously keep on changing. This calls for energy-efficient power-aware self-organizing routing protocols.

We introduce our *BeeAdHoc* routing protocol for MANETs that delivers the same or better performance compared to existing state-of-the-art MANET routing protocols. But its energy consumption is significantly smaller than that of other protocols. Following our *Natural Engineering* approach we also implemented *BeeAdHoc* in Linux on mobile laptops. We developed a novel real-world testing strategy that gradually moves the testing environment from simulation to real MANETs. The results of our experiments indicate the same pattern: *BeeAdHoc delivered the same or better performance as compared to state-of-the-art algorithms but at significantly smaller communication costs.*

Finally, we describe our *BeeSensor* protocol for WSNs. *BeeSensor* tries to combine energy efficiency of *BeeAdHoc* with the scalability properties of *BeeHive*. We compared the *BeeSensor* with state-of-the-art nature-inspired routing protocols in a real target tracking application. The results indicate that *BeeSensor* delivers the same or better performance compared to existing algorithms but has significantly smaller processing and communications costs. The results reported in the chapter are intriguing enough to motivate researchers to develop self-organizing, simple, scalable, adaptive, and efficient routing protocols for emerging next-generation networks. The chapter contains extracts from our following published papers, reproduced with the kind permission of ACM, IEEE and Springer Verlag:

- H. F. Wedde, M. Farooq, T. Pannenbaecker, B. Vogel, C. Mueller, J. Meth, and R. Jeruschkat. BeeAdHoc: an energy efficient routing algorithm for mobile ad-hoc networks inspired by bee behavior. In Proceedings of GECCO, pages 153–161, Washington, June 2005.
- H. F. Wedde and M. Farooq. The wisdom of the hive applied to mobile ad-hoc networks. In Proceedings of the IEEE Swarm Intelligence Symposium (SIS), pages 341–348, Pasadena, June 2005.
- M. Saleem and M. Farooq. BeeSensor: A Bee-inspired power aware routing protocol for wireless sensor networks. In M. Giacobini et al. (Eds.), volume 4449 of Lecture Notes in Computer Science, pages 81–90. Springer Verlag, April 2007. 109.
- M. Saleem and M. Farooq. A framework for empirical evaluation of nature inspired routing protocols for wireless sensor networks. In Proceedings of Congress on Evolutionary Computing (CEC), pages 751–758. IEEE, Singapore, September 2007.

Chapter 9: Conclusion and Future Work

In this chapter, we summarize the contributions of our work. We stress the need for the *Natural Engineering* approach because this significantly helped us in successfully designing a dynamic, simple, efficient, robust, flexible, and scalable multi-path routing algorithm and then installing it in a real network of Linux routers. We believe that a similar approach can help in realizing other nature-inspired algorithms in their respective real environments.

We conclude the chapter with interesting future directions. The most important one is: design and development of a dedicated nature-inspired router in hardware which optimally runs nature-inspired routing algorithms. Before this step is taken, we have to reengineer *BeeHive* in such a fashion that it is capable of seamlessly replacing *OSPF* in the existing packet-switched IP networks.

2

A Comprehensive Survey of Nature-Inspired Routing Protocols

The major contribution of the chapter is a comprehensive survey of existing state-of-the-art nature-inspired routing protocols developed by researchers who are trained in novel and different design doctrines and practices. Nature-inspired routing protocols have become the focus of research because they achieve the complex task of routing through simple agents which traverse the network and collect the routing information in an asynchronous fashion. Each node in the network has limited information about the state of the network, and it routes data packets to their destination based on this local information. The agent-based routing algorithms provide adaptive and efficient utilization of network resources in response to changes in the network, catering to load balancing and fault management. The chapter describes the important features of stigmergic routing algorithms, evolutionary routing algorithms, and artificial intelligence routing algorithms for fixed telecommunication networks. We also provide a summary of the protocols developed by the networking community. We believe that the survey will be instrumental in bridging the gaps among different communities involved in research of telecommunication networks.

2.1 Introduction

The design and development of multi-path, adaptive, and dynamic routing algorithms has been approached by different communities of researchers, each having a strict traditional design philosophy, leaving little room for cross-fertilization of novel ideas between different research communities.[1] This provided us the grist for the mill for providing a comprehensive survey of routing protocols, designed and developed by different communities of researchers, for different types of telecommunication networks: circuit-switched and packet-switched. The major objectives of the survey are:

---

[1] Most of this chapter is reproduced by permission of the publisher, Elsevier, from our paper: *H. F. Wedde and M. Farooq. A comprehensive review of nature inspired routing algorithms for fixed telecommunication networks.* Journal of System Architecture, 52(8-9):461-484, 2006.

- To understand the basic design concepts and doctrines of the different communities, and then to contemplate the strengths and shortcomings of each approach.
- To create awareness among the researchers about state-of-the-art routing algorithms developed by other communities.
- To create a vision about future directions and challenges for routing protocols as they may be employed in totally different operating environments like sensor networks.
- To allow for cross-fertilization of ideas which will help in taking a comprehensive approach to counter the challenges of complex large-scale telecommunication networks.
- To create an intelligent and knowledge-aware network layer implicitly taking care of network management and traffic engineering by virtue of its intelligent routing algorithms.
- To lay the ground for a comprehensive performance evaluation framework for the comparative evaluation of routing protocols.

2.1.1 Organization of the Chapter

The rest of the chapter is organized as follows. Section 2.2 will provide major challenging requirements that a routing protocol should be able to meet, giving rise to a taxonomy of routing protocols in Section 2.2.2. We will first provide an overview of the Ant Colony Optimization (ACO) metaheuristic in Section 2.3, and then discuss in detail different routing algorithms inspired by ACO. Section 2.4 will outline important features of Evolutionary Algorithms (EAs) and then describe corresponding routing algorithms. Subsequently, we will conclude our survey of routing protocols for fixed networks in Section 2.5. We will briefly discuss the state-of-the-art routing protocols based on the traditional design paradigms of distance vector or link-state routing methods. Finally, we will conclude our survey by emphasizing the cross-fertilization of design principles of different approaches for the purpose of a comprehensive approach to solutions for the challenges of modern telecommunication networks.

2.2 Network Routing Algorithms

In this section, we briefly outline the challenges facing the telecommunication sector because of an ever-increasing demand for intelligent and integrated multimedia services from the user community. The solutions to such challenges lie in a multidimensional landscape of requirements for designing, developing, and implementing intelligent routing algorithms. These features are summarized in Section 2.2.1. In Section 2.2.2, we will outline a taxonomy of routing algorithms according to several criteria, reflecting different design doctrines, switching strategies, and network environments.

2.2.1 Features Landscape of a Modern Routing Algorithm

The design goals of a routing algorithm are summarized in the following:

- *Optimality* of a routing algorithm could be defined as the ability to select the best route [45]. The best route could be defined in terms of a quality metric, which in turn might depend on a number of parameters, i.e., hops, delay, or a combination of both. A routing algorithm can easily compute a best path in a static network but it becomes a daunting task in a dynamic network.
- *Simplicity* is a desirable feature of any routing algorithm. A routing algorithm should be able to accomplish its task with a minimum of software and resource utilization overhead. Simplicity plays an important role when a routing algorithm has to run on a computer with limited physical resources [45].
- *Robustness* of a routing algorithm could be described as its ability to perform correctly in the face of unusual or unforeseen situations like hardware failures, high load conditions, and incorrect implementations [45]. A router has to quickly react to the anomalies and reroute the packets on alternative paths. This property is also known as *faulttolerance*.
- *Convergence* is the process of agreement, by all routers, on optimal paths. In the face of router failures, a routing algorithm should be able to make all routers quickly agree, through transmitting update messages, on alternative optimal routes. Routing algorithms that converge slowly can cause loops or network outages [45].
- *Flexibility* is the ability of a routing algorithm to quickly and accurately adapt to a variety of network circumstances. They should be programmed to adapt to changes in the available network bandwidth, routers' queue size, and network delay, among other variables [45].
- *Scalability* is the ability of an algorithm to operate in large networks without an associated increase in demand for software or physical resources and in resource utilization overhead. The control packets should occupy a small bandwidth and have small processing overhead, and routing tables should occupy little memory.
- *Multi-path Routing* exploits the resources of the underlying physical network by providing multiple paths between source/destination pairs [145]. This requirement allows the protocols to achieve higher transfer rates than given by the bandwidth of a single link. The multi-path feature also helps in load balancing in the face of congestion, allowing for delivering more packets with smaller delays at the destination.
- *Reachability* is the ability of a routing algorithm to find at least one path between each source/destination pair [249].
- *Quality of Service (QoS)* is the ability of an algorithm to administer better service to selected real-time traffic like multimedia by providing dedicated bandwidth, controlled jitter, and latency [45].

2.2.2 Taxonomy of Routing Algorithms

Routing algorithms have been classified in [86] according to criteria reflecting fundamental design and implementation options such as:

- *Structure.* Are all nodes treated equally in the network?
- *State information.* Is network-scale topology information available at each node?
- *Scheduling.* Is routing information continually maintained at each node?
- *Learning model.* Do packets or nodes have an intelligent learning model?
- *Queue control.* Do nodes employ load balancing to manage growth of queues?

Such issues could be raised and discussed under all the following dimensions of networking, grouped below under the topics routing strategy/policy, design doctrine, and specific aspects of telecommunication networks.

Routing strategy/policy

Here we provide only a brief overview explaining the concepts of the taxonomy in [45].

- *Static versus Dynamic.* Static routing algorithms are simple table mappings established by network administrators before the routing begins. Such algorithms can react to changes only if the network administrator alters these mappings based on his experience with traffic patterns in the network. Dynamic algorithms update their routing tables according to changing network circumstances by analyzing incoming routing update messages and rerunning the algorithms to calculate new routes. This feature makes them suitable for today's large, constantly changing networks.
- *Single-path versus Multi-path.* Single-Path routing algorithms determine the best path to a destination while multi-path routing algorithms discover and maintain multiple paths to a given destination. This feature allows them to multiplex the traffic to the destination on multiple paths, as a result; both their throughput and reliability are higher than in the case of single-path routing algorithms.
- *Flat versus Hierarchical.* Flat routing algorithms consider all nodes in the network to be peers and they maintain an entry in their routing tables for all routers. This allows peers to discover a best route at the cost of transmitting more control packets and maintaining larger routing tables. Hierarchical routing algorithms form a logical group of routers and organize them into areas, domains, and autonomous systems. Such algorithms require two types of routers, *intra-domain routers*, which route traffic within a domain, and *backbone routers*, which route traffic between domains. The advantage of such an organization is that it mimics the traffic patterns of organizations in which most of communication occurs within small areas like factory locations in a big company. So each location could work with simple intra-domain routing algorithms. In this manner, such an organization requires significantly smaller routing tables which, in turn, require smaller memory storage, and result in little waste of bandwidth, for maintaining routes.

- *Intra-Domain versus Inter-Domain.* Intra-domain routing algorithms route data packets within the same domain only while inter-domain routing algorithms route data packets between domains. Within a domain or *Autonomous System* (AS), system administrators can select their own routing policy. Due to the different nature of such algorithms, an optimal intra-domain routing algorithm may not necessarily be an optimal inter-domain routing algorithm.

- *Link-State versus Distance Vector.* In *link-state algorithms*, each node floods the status of its links to all nodes of the network. Then each router constructs a graph of the complete topology and applies the Shortest Path First routing algorithm for obtaining the next hop on a shortest path to each destination and storing it in its routing table. In *distance vector* algorithms, routers send updates only to their neighbors. *Link-state algorithms* converge quickly and scale better, but require more CPU power and memory than distance vector algorithms; therefore, they are expensive to implement and support.

- *Host-Intelligent versus Router-Intelligent.* In *host-intelligent algorithms*, a host determines the entire route to a destination and appends it as a header to each packet, a process known as *source routing*. Other routers in the system simply forward the packets to the next hop contained in the header of the packet. In *next hop routing algorithms*, routers are intelligent and they discover and maintain paths while executing their algorithms; therefore, they are called *router-intelligent algorithms*.

- *Global versus Local.* In *global routing algorithms*, each node requires the information about all nodes, their inter-connectivity, and cost of links for constructing a graph and then applying path-finding algorithms on it. In contrast, *local algorithms* do not have access to information about the complete topology; rather, they work with a local traffic model, maintained at each router, for reaching a routing decision.

- *Deterministic versus Probabilistic.* *Deterministic algorithms* associate, with every destination in the routing table, an outgoing interface identifier and a cost associated with choosing that interface. *Probabilistic algorithms* associate probability values, depending on the costs of the links to the neighbors, with all neighbors of a node through which a packet could reach its destination. A neighbor with a higher probability value is supposed to be on a better path than a neighbor with a lower probability value. The probabilities of all neighbors are normalized such that their sum always remains one. *Probabilistic algorithms* multiplex the network traffic on different paths, depending on their probability value, and hence have better performance than *deterministic algorithms*, but they require more memory and CPU power [249].

- *Constructive versus Destructive.* *Constructive algorithms* begin with an empty set of routes and incrementally add routes till final routing tables have been constructed. In contrast, *destructive algorithms* start with a fully connected graph as an initial condition in which all routes are available, and gradually those paths that do not exist in the network are removed from the routing tables [254].

- *Best-Effort versus QoS.* *Best-effort algorithms* do not provide any guarantee that the demands of the applications will be met while *QoS algorithms* reserve the resources in the network to meet the demands of the applications. *QoS algorithms* provide guarantees to the applications through a policy of admission control.

Design doctrine

Routing algorithms could alternatively be classified on the basis of the design philosophy of their developers. The researchers in each community have been trained with a certain design and analysis doctrine which leaves little room for cross-fertilization of ideas from other communities. In this subsection, we briefly provide an overview of these communities that will help the reader in understanding the design principles of different types of routing algorithms. Given a mutual understanding of the various backgrounds of these communities there is a chance for developing state-of-the-art routing algorithms for the networks of the new millennium. We have categorized important routing algorithms according to their design doctrine in Figure 2.1. This figure can also be used as a road map for our survey of routing protocols for fixed telecommunication networks. The communities are discussed in the sequel:

(a) The *networking community* has pioneered the work in the field of packet-switched networks. The roots of this work go back to the development of ARPANET and a novel routing algorithm, which is based on an asynchronous Bellman-Ford algorithm [19, 168]. Since then many dynamic and multi-path routing algorithms have been developed by following the classic methodology for routing protocol development: non-intelligent link-state packets are used to collect information about the costs of neighbors to then propagate them in the whole network. Consequently, they all suffer from the same shortcomings: "wrong" or "out-of-order" local estimates have a global impact [58], and the algorithms require a global system model to execute Dijkstra's shortest-path algorithm [51]. The algorithms could be classified as global and deterministic routing algorithms.

(b) The *Artificial Intelligence routing community* works in two different areas: *Machine Learning* and *Agent-Based Learning*. The first community uses Reinforcement Learning (RL) [130] techniques, developed as a branch of *Machine Learning*, in order to propose routing algorithms for packet-switched networks. Examples are Q-routing [27] and PQ-routing [44]; both are based on Q-learning [265, 266]. Such algorithms are adaptive, decentralized, dynamic, local, and deterministic. Agent-based learning methods resulted in specific routing algorithms [250, 51, 107, 164, 58]. The major advantages of such algorithms are summarized as follows:

- The algorithms do not require an a priori global system model of the network; rather, they utilize a local system model as observed by the agents.
- The agents gather the network state in a decentralized fashion and leave the corresponding information on visited nodes. This enables them to make routing decisions in a decentralized fashion, without the need of a global controller.

- The algorithms have the ability to adapt autonomously to changes in the network, or in traffic patterns.
- The management of the network comes as a complimentary benefit of using such mobile agents.

The major emphasis of such routing algorithms is on designing intelligent agents for routing, management, and control of networks *in an autonomous manner*. The multi-agent systems provide a good infrastructure for design and development of such mobile agents [238, 280, 101, 250, 107, 141, 164]; however; the intelligence is achieved at the cost of complex design paradigms [115, 39, 292, 291, 182, 302, 303, 121, 124, 29, 71, 225].

(c) The *Natural Computing* research has two major directions: Evolutionary Computing [98, 116] and Swarm Intelligence [21]. Evolutionary Computing takes the evolution process in living cells as a basis for developing algorithms and systems. Consequently, evolutionary routing algorithms employ the evolutionary operators of *selection, crossover,* and *mutation* for online adaption to cope with changes in network environments. *DGA* (Distributed Genetic Algorithm) [147] is one such routing algorithm. The second emerging area, Swarm Intelligence, studies different self-organizing processes in nature and utilizes their principles as an inspirational metaphor to propose novel solutions to different daunting classical scientific problems. The novelty comes again from the fact that such systems lack one central complex controller, which normally coordinates and schedules different tasks in the system by virtue of its access to the global system state. In contrast, these population-based systems have simple entities that have only local knowledge, but *together* they form an intelligent system [21], [247]. *ABC* [213], *AntNet* [62], and *BeeHive* [273] belong to this class of routing algorithms. Nature-inspired routing algorithms are mostly adaptive, decentralized, local, dynamic, and probabilistic.

Specific aspects of telecommunication networks

We will restrict our survey of routing algorithms to only two types of telecommunication networks, namely, connection-oriented circuit-switched networks and connectionless packet-switched networks. We also focus on the *Natural Computing* algorithms for these two types of the networks; however, where appropriate, we will provide a brief summary of the algorithms developed by other communities.

Each type of network comes with a different set of requirements that a routing algorithm should meet. Topology changes are less frequent in fixed telecommunication networks but the traffic patterns are nondeterministic. Therefore, a routing algorithm should be able to control congestion. In connection-oriented networks, a circuit is reserved for each connection between a pair of source and destination; therefore, a routing algorithm should have good admission control through efficient resource utilization in order to reduce the call-blocking probability.

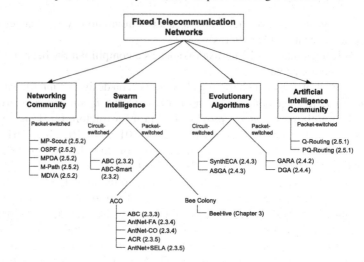

Fig. 2.1. A taxonomy of routing protocols for fixed telecommunication networks

2.3 Ant Colony Optimization (ACO) Routing Algorithms for Fixed Networks

We first briefly summarize important elements of the ACO metaheuristic in Section 2.3.1, and then provide a survey of two state-of-the-art routing algorithms designed on the basis of the ACO metaheuristic: *ABC* (Section 2.3.2), which is designed for circuit-switched telecommunication networks, and *AntNet* (Section 2.3.4), which is designed for packet-switched telecommunication networks.

2.3.1 Important Elements of ACO in Routing

The Ant Colony Optimization (ACO) metaheuristic has been inspired by operating principles of ants [22], which empower a colony of ants to perform complex tasks like nest building and foraging [76]. We summarize important elements of ACO, which have been utilized in routing algorithms, in the sequel.

Stigmergy

The ants are able to find the shortest path from their nest to a food source by sharing information through *stigmergy* [76, 57]. Stigmergy is a form of communication in which social insects like ants communicate indirectly through the environment [100, 57]. Ants lay pheromone while foraging. As a result, the concentration of pheromone on the shortest path is reinforced at a higher rate than on the other paths. Ants tend to prefer higher pheromone concentration paths, which results in a majority of ants using a shortest path for foraging in a steady state [57]. Stigmergy is the most important element of the ACO metaheuristic and has been instrumental in developing a society of mobile ant agents as they cooperate in solving discrete optimization and control problems [76, 75, 74, 79, 152, 77]. Here we limit our survey to applications of ACO to telecommunication networks.

Pheromone control

Bonabeau et al. have pointed out in [22] that the success of ants in collectively locating shortest paths is only statistical. If many ants initially happen to choose a nonoptimal shortest path, other ants will follow this path, which will result in pheromone reinforcement along this path. Consequently, ants will travel on a stagnating nonoptimal path in a steady state. However, if we assume that ants do find the shortest path in a steady state, then even this stagnation is not helpful because if all packets follow the shortest path, this will lead to congestion on it. Consequently, the path becomes nonoptimal, and other nonoptimal paths may become optimal due to changes in network conditions, or discovery of new paths after changes in the topology [221]. Therefore, it is extremely important to counter stagnation through intelligent pheromone control strategies. We outline some of these strategies here; however, the interested reader will find a detailed discussion in [221].

- *Evaporation.* In ACO algorithms, the value of pheromone t_{ij} in all links is decreased by a factor p such that $t_{ij} \leftarrow t_{ij}(1 - p)$ [75]. This helps in reducing the influence of past experience during decision making.
- *Aging.* The amount of pheromone that an ant lays on a path decreases with its age; an older ant lays less pheromone than a younger one [214]. Ants mostly assume symmetric links (in which the cost of links in both directions is the same). For asymmetric links, ants measure the costs during a forward trip and deposit pheromone on a backward trip. *Evaporation* and *Aging* favor present experiences which, result in discovery of new paths by avoiding stagnation.
- *Limiting and smoothing pheromone.* Some authors circumvent the problem of stagnation by setting an upper bound t_{max} of pheromone for every edge (i, j) [239]. This reduces the preference of ants for optimal paths over nonoptimal paths. However, one should be careful that *pheromone limiting*, if not used in conjunction with *evaporation*, will make all paths equal once the pheromone value reaches t_{max} for all links. *Pheromone smoothing* ensures that only a small amount of pheromone is permitted on paths where the current pheromone concentration is close to t_{max}. Consequently, a few dominant paths are not generated. But this feature might lead to a situation in which the number of ants that prefer to select a nonoptimal path keeps increasing because ants deposit more pheromone on these nonoptimal links, even though an optimal path might remain optimal in a steady or stable state.
- *Pheromone-heuristic control.* The authors of [77, 79] use the amount of pheromone in a link combined with a heuristic function to influence the decision of an ant. The heuristic function n_{jd} for telecommunication networks is determined by the queue length q_j (number of bits on network interface of neighbor j in a router). Finally, this heuristic, based on the current state of the network, is combined with

a long-term *learned goodness* p_{jd} of using neighbor j for reaching a destination d [62]. Such a hybrid approach enables a routing algorithm to be responsive to transient changes in the networks. However, setting the weight value in the formula in which the heuristic factor is combined with the long-term learned goodness of a link is very tricky. If a policy exceedingly emphasizes the weight of a heuristic component then it might cause oscillations, and inadequate emphasis would make the algorithm not react to transient changes in the networks.

- *Privileged pheromone laying.* The authors of *AntNet* [62, 65] enhance the ACO metaheuristic by a concept of *privileged pheromone laying.* In their algorithm, ants first evaluate the quality of their solution and then deposit an amount of pheromone based on the quality. They model the quality of a solution as a function of the trip time of a forward ant, the best known trip time, and a few other statistical parameters. The experiments reported in [62, 65] reveal that such a policy results in better convergence and performance. Later on the authors of [239] devised the *FDC fitness landscape* approach, which compares the fitness of a solution of each ant with an optimal solution and then deposits an amount of pheromone. The experiments reported in [239] confirm that *FDC* contributes to obtaining accurate and better results.

2.3.2 Ant-Based Control (ABC) for Circuit-Switched Networks

Schoonderwoerd et al. [214, 213, 212] were the first to apply the ACO metaheuristic to routing and loadbalancing problems in circuit-switched telecommunication networks. As a symmetric network, a circuit-switched network reserves a virtual circuit between a sender and a receiver by explicitly connecting them through crossbar switches. Consequently, the major challenge is to distribute the calls over multiple switches so that the system can support a maximum number of possible calls during peak hours. Such a network is not able to admit a call if all input ports of a crossbar switch are connected to its output ports. Consequently, congestion could be defined as a function of the number of used connections in a crossbar switch [91, 93, 205]. The performance of a switching algorithm is measured in terms of the number of calls which are blocked or failed due to congestion [9].

In the *ABC* algorithm, each node in the network stores the following attributes [214]:

- The capacity, for the number of simultaneous calls that a node (crossbar switch) can manage. The remaining free capacity of a switch is also stored.
- A pheromone-based routing table in which probability values, representing goodness of a node's neighbors for reaching each destination are stored. Each row i in the table represents a destination and each column j represents a neighbor. Each probability value p_{jd} represents the goodness of choosing j as the next hop for reaching destination d.
- A probability value of this node being the end point of a call.

In *ABC*, an ant, launched by node s and traveling towards destination d, will update the probability values for its source node at each intermediate node passed. We now

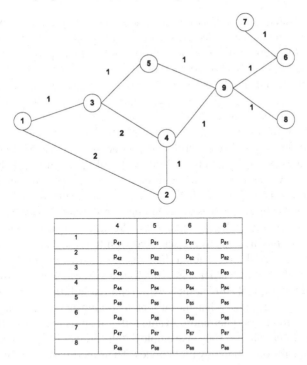

	4	5	6	8
1	p_{41}	p_{51}	p_{61}	p_{81}
2	p_{42}	p_{52}	p_{62}	p_{82}
3	p_{43}	p_{53}	p_{63}	p_{83}
4	p_{44}	p_{54}	p_{64}	p_{84}
5	p_{45}	p_{55}	p_{65}	p_{85}
6	p_{46}	p_{56}	p_{66}	p_{86}
7	p_{47}	p_{57}	p_{67}	p_{87}
8	p_{48}	p_{58}	p_{68}	p_{88}

Routing Table at Node 9

Fig. 2.2. Pheromone routing table in ABC

refer to Figure 2.2 to illustrate the relevant aspects of the *ABC* algorithm. Let us assume that an ant has been launched by Node 3, and its destination is Node 9. Once the ant reaches Node 5, it will update the p_{33} value in the routing table of Node 5, and once it reaches at Node 9, it will update the p_{53} in the routing table of Node 9. This will influence the ants traveling towards Node 3 and passing through Node 9. This approach, therefore, has been specifically designed for symmetric links [62, 221].

Schoonderwoered et al. used *aging*, *delaying*, and *noise* techniques to counter stagnation in the probability values in the routing tables. The amount of pheromone δp that an ant is allowed to deposit is given by the formula $\delta p = \frac{0.08}{age} + 0.005$. The purpose of *delaying* is to increase the transit time of certain ants proportional to the spare capacity of the node. The *delay* is defined as $delay = 80 \times e^{-0.075 \times r}$, where r is the remaining capacity of a node. Consequently, the rate at which ants are transmitted from congested nodes is reduced, and due to aging mechanisms the ants deposit less pheromone on the nodes that they subsequently visit. In this way the influence of the ants that visited a congested node is reduced on other ants. Finally, a certain ratio of ants do not follow the paths according to the pheromone values in the routing tables. Rather, they follow random paths where they may discover new and better routes in dynamic networks. Schoonderwoered et al. have experimentally verified that the *ABC* algorithm, on average, drops less calls compared to the algorithm of Appleby and Steward [9]. Moreover, it quickly reacts to changes in the topology. Bonabeau et al. extended *ABC* with the idea of *smart agents* [23], which utilize the concept of dynamic programming: the agents update the probability values at a given node for all visited nodes, rather than just for their source node. Consider an ant agent launched from Node 1 (see Figure 2.2) and traveling towards Node 9 via Nodes 3 and 5. At Node 3 it will update the probability value p_{11}; at Node 5 it will update the probability values p_{31} and p_{33}; and at Node 9 it will update the probability values p_{51}, p_{53}, and p_{55}. Consequently, *ABC* with *smart agents* reduced the number of calls which were dropped compared with *ABC*, and it was also able to react to changes in the topology. However, *smart agents* use a policy similar to that used by ants in *ABC* for updating the routing tables. Compared to *ABC* agents, *smart agents* have more complex behavior but the objective is achieved with fewer agents.

The authors of [209] have studied the behavior of *ABC* on a different network topology and confirmed the earlier results published by the authors of *ABC*. Recently, they enhanced the original *ABC* in their work reported in [208]: if the age of an ant that arrived at a node is greater than the current maximum age stored at the node, then it decreases the goodness (pheromone) value rather than increasing it. This concept is known as *anti-pheromone*. They also employed the probabilistic routing method as used for phone calls on a topology of 25 nodes. Their modified algorithm has shown a certain degree of improvement compared with the original *ABC*.

2.3.3 Ant-Based Control (ABC) for Packet-Switched Networks

Subramanian et al. [241] developed an algorithm for packet-switched networks on the basis of the ideas of *ABC*. They designed two types of ants: *regular* and *uniform*. Regular ants update the pheromone values in the routing tables based on the accumulated cost of traveling to a node. In Figure 2.2, an ant traveling from Node 3 to Node 9 via Node 5 will update the value p_{53} at Node 9 based on the accumulated cost of the path from Node 3 to Node 5 and then from Node 5 to Node 9. Uniform ants randomly select their next hop and they update the pheromone values in the routing tables based on the costs in the direction opposite to their travel. A uniform ant traveling from Node 3 to Node 9 will update the value p_{53} at Node 9 based on the cost of link from Node 9 to Node 5 and Node 5 to Node 3. The algorithm assumes that each node has determined the cost information of the link to its neighbors.

Heusse et al. [113], based on ideas of *ABC*, proposed an algorithm with *Cooperative Asymmetric Forwarding* (CAF) for routing in packet-switched networks. Their basic idea is: once a data packet is traveling from Node 5 to Node 9 then it carries with it the cost of link c_{59}, which is a sum of waiting and propagation delay from Node 5 to Node 9. At Node 9, it leaves this value in a reverse routing table. Once an ant traveling from Node 6 to Node 3 via Node 9 arrives at Node 9 and selects neighbor 5 as a next hop, it carries with it this value. At Node 5, it adds this cost to c_{96} to determine the accumulated cost c_{56}. It carries with it the estimates of reaching all nodes it has visited, and then updates all corresponding entries. At Node 5, it will update p_{99} and p_{96} depending upon c_{59} and c_{56} respectively. The algorithm will not work properly under low traffic load scenarios in which a small number of data packets are sent on the network. As a result, an ant will carry old values in the reverse routing tables, which might degrade the performance of the algorithm. Moreover, an additional reverse table is required to be maintained.

Van der Put and Rothkrantz [253, 252] designed *ABC-backward* based on the concept of forward- and backward-moving ant agents. The algorithm applies *AntNet* concepts (to be introduced shortly) to *ABC*. The algorithm can be used on cost-asymmetric networks. The authors have experimentally verified that *ABC-backward* has a better performance than *ABC* on both cost-symmetric and cost-asymmetric networks. *ABC-backward* solved a serious fax-distribution problem faced by KPN telecom (largest telephone company in Netherlands).

2.3.4 AntNet

AntNet was proposed by Di Caro and Dorigo in [61, 64, 60, 62]. It is inspired by the principles of the ACO metaheuristic but has additional network-specific enhancements as well. The algorithm is designed for asymmetric packet-switched networks, and the primary objective of the algorithm is to maximize the performance of a complete network. The algorithm implicitly achieves load balancing by probabilistically distributing packets on multiple paths.

In *AntNet*, the network state is monitored through two ant agents: *Forward_Ant agent*, and *Backward_Ant agent*. The agents are equipped with a stack on which the node's address and the trip time estimate to the visited nodes are pushed. A Forward_Ant agent is launched at regular intervals from a source to a certain destination depending upon the amount of traffic generated for the destination at the source. The probability p_{id} of launching a Forward_Ant agent to destination d at node i is $p_{id} = \frac{f_{id}}{\sum_{k=1}^{D} f_{ik}}$, where f_{id} is the number of bits flowing from node i towards node d, and D is total number of nodes in the network. Forward_Ant agent uses the normal queues to experience the true network conditions. If a Forward_Ant agent follows a cyclic path then the data about the nodes which lie on the cyclic path are removed from the stack. However, the agent is allowed to explore the network if the time it spent in the cycle is less than half the Forward_Ant agent lifetime. Once a Forward_Ant agent reaches its destination, it creates a Backward_Ant agent and transfers all information to it. The Backward_Ant agent visits the same nodes as the Forward_Ant agent but in a reverse order, and it modifies the entries (deposit of pheromone) in the routing tables in accordance with the trip time from the nodes to the destination. A Backward_Ant agent is only allowed to update entries in the routing tables of intermediate nodes if it discovers a good subpath from the intermediate node to the destination. The goodness is defined based on the trip time. The trip time values are calculated by taking the difference in entrance times of two subsequent nodes pushed onto the stack. The updating of routing tables only affects data packets and Forward_Ant agents traveling from node i to node d. The nodes in *AntNet* maintain the average trip times, the best trip times, and the variance of the trip times for each destination. In this way, information is statistically maintained at each node in the network for subsequent routing decisions. Backward_Ant agent uses the system priority queues to quickly disseminates the information to the nodes.

AntNet uses the heuristic function $l_j = 1 - \frac{q_j}{\sum_{k=1}^{N} q_k}$, where q_j is the number of bits in the queue of neighbor j and N is the total number of neighbors. The heuristic function favors neighbors with smaller queue lengths. P_{jd} is the goodness of neighbor j for reaching destination d. Backward_Ant agent enhances the goodness of neighbor j from where it arrived, using the formula $P_{jd} \leftarrow P_{jd} + r(1 - P_{jd})$, where r is a reinforcement factor, and it decreases the goodness value of other neighbors using the formula $P_{kd} \leftarrow P_{kd}(1 - r)$ $(k \neq j)$. P_{jd} is a long-term learned value which provides insight into the goodness of a neighbor for a particular destination. The reinforcement factor is defined as a function of the current trip time, the best trip time, and the statistical confidence intervals. Finally, this P_{jd} value is combined with the heuristic value l_j (just defined) to react to the current state of the network using the formula $P'_{jd} = \frac{P_{jd} + \alpha l_j}{1 + \alpha(N-1)}$, where α weighs the heuristic function with the probability values stored in the routing tables.

AntNet applies the concept of stochastic spreading of data packets along all paths according to the goodness of the paths. However, the goodness P_{jd} is further rescaled to reward better goodness solutions more than ones of lower quality. The rescaled values are stored in an another table, which is used during the switching of data packets. The concept of using two tables, one for ant agents and another for data packets, has been elaborated in [58].

Di Caro and Dorigo have conducted a number of experiments on different topologies like simpleNet, NSFNet, and NTTNet, which are reported in [61, 64, 60, 62]. They have chosen *throughput* and *90th percentile of packet delays as the performance* parameters. The experiments reported have shown that *AntNet* outperforms, with respect to throughput and delay, all other competitors, which consist of Q-routing, PQ-routing, Shortest Path First (SPF), and *OSPF*, except the Daemon algorithm. The improvement in performance is achieved at a cost of less than 1% of the bandwidth occupied by ant agents.

The authors proposed a variant of *AntNet*, known as *AntNet-FA* or *AntNet-CO*, in [65]. In *AntNet-FA*, Forward_Ant agents do not have to wait in the queue to measure the queuing delay. Rather, they use an estimation model to estimate the delay. This feature allows a Forward_Ant agent to use priority queues as well. The estimation model estimates the trip time t_{ij} from node i to j using the formula $t_{ij} = pd_{ij} + \frac{q_j}{b_{ij}}$, where pd_{ij} is the propagation delay of the link from node i to j, q_j is queue length in bits of neighbor j at node i, and b_{ij} is the bandwidth of the link from node i to j. Such a policy facilitates the quick spreading of the routing information, especially in large topologies. The authors have reported in [65] that the performance of *AntNet-FA* is significantly better than *AntNet* on a 150-node topology.

The *AntNet* algorithm utilizes the concept of privileged pheromone laying along with that of heuristic pheromone control to react to changes in the traffic patterns. Let us assume that a Forward_Ant wants to find a path from Node 9 (see Figure 2.2) to Node 2. Interestingly, *AntNet* will maintain the routing table entries p_{62} and p_{82}

for reaching Node 2, although it is impossible to reach Node 2 via Node 6 and Node 8. Consequently, p_{62} and p_{82} will be 0. However, it might be possible that q_6 and q_8 are significantly smaller than q_5 and q_4. Therefore, it is possible that once p_{62} and p_{82} are combined with a heuristic value, new values for p_{62} and p_{82} will be non-zero. This will lead to sending data packets with destination 2 to Node 6 and Node 8, which of course appears to be a simple oversight by the authors. After subsequent loops, a Forward_Ant or a data packet may ultimately reach Node 2 through Nodes 5 or 4. The results of our scalability experiments in Chapter 4, however, demonstrate that the impact of this and other relevant oversights becomes more profound in larger topologies.

The ant agents, by utilizing the stack and making forward and backward trips, may occupy a significantly large portion of bandwidth, in comparison to ant agents of *ABC* in large topologies. The agents perform complex computations once they arrive at a node. As a result, the processing complexity of ant agents here will be significantly higher than that of the ant agents of *ABC*. As mentioned before, we report in Chapter 4 our scalability study of *AntNet* in large topologies.

2.3.5 Ant Colony Routing (ACR) and AntNet+SELA QoS-Aware Routing

Di Caro has discussed ACR, which is a general framework for designing fully distributed and adaptive systems for network control and management, in [58]. ACR can be viewed as a distributed society of static agents, which are known as node managers, and mobile agents, which are proactively or reactively launched in the network. Node managers autonomously manage node activities by learning and then following stochastic management policies based on local pheromone values, which represent goodness of different control actions. However, they expand their "sensory field" to acquire information about their environment with an adaptive generation of mobile agents. Mobile agents take an active *preceptor* role on behalf of the node managers that launch them. These agents collect the important parameters that act as input parameters to learning strategies of node managers. A node manager, based on the feedback provided by preceptor agents, might alter its control actions.

Stochastic decision policies are well suited for non-stationary and distributed environments because they help in spreading data over multiple paths, thus implicitly providing loadbalancing as well. A preceptor agent may follow either a point-to-point mode by following pheromone tables for destinations already discovered, or a broadcast strategy for a destination not known at a node. The broadcast strategy helps in replicating the active perceptions to discover as many good options as possible. Active preceptors are situated at a lower level in the hierarchy than node managers, and hence must not be allowed to directly modify the internal state of the node managers. Preceptors simply communicate the collected information to the node managers, which may accept or reject the information. This contributes to designing secure systems, in which malicious agents cannot directly modify the internal state of node managers, and to practicing object oriented design (information hiding).

Effectors are mobile agents which have a deterministic precompiled behavior. They are used to carry out highly specialized tasks like allocation and deallocation of resources. In QoS routing, they will help in finding the paths to a destination in which enough network resources can be reserved to meet the QoS guarantees for the application. As a multi-path routing system, ACR will obviate the need to discover new paths in the case of router failures since "backup" paths are already available. In this way, multiple ant colonies can coexist together and manage the activities of a node in a social agreement with all other nodes.

AntNet+SELA [67] was designed to provide QoS guarantees to variable bit rate (VBR) traffic in Asynchronous Transfer Model (ATM) networks. ATM networks provide statistical guarantees by reserving virtual circuits either on a per-flow basis or on a per-destination basis. The node managers which do both admission control and routing are designed as *stochastic estimator learning automata* (SELA) [183] to be applied to a distributed routing system for ATM networks [10]. The node managers utilize active preceptors to proactively update a link-state database in which the goodness values of different paths leading to a destination are stored. Active preceptors utilize different routing tables than do data packets. When a new application arrives at a node, two groups of active preceptors are launched. The first group consists of path-probing setup ants which probe k different paths leading toward the destination, and the second group consists of path-discovering setup ants which make use of existing pheromone tables to discover new QoS-feasible paths for the applications. Both agent groups reserve resources temporarily as they traverse the paths. Path-discovering setup ants bias their decision on the current status of queue lengths to calculate the goodness of a neighbor (the value of α is set much higher; refer to Section 2.3.4). Moreover, they always choose the link that has the highest probability value. If the probability values of two best links differ minimally, then the ants are allowed to replicate themselves on both links; however, replication is allowed only once for better control of the number of ants. If a traversed path does not meet the QoS requests, an effector ant is generated that follows the same path but in reverse order to free the allocated resources. However, if an ant is able to locate a QoS-feasible path to a destination, then it comes back to the source node and provides information about the path to the node manager. If multiple ants have come back, a node manager may decide whether to split the traffic load from the application into multiple paths or not. Then the application can start sending the packets. Effector agents (monitor ants) are periodically launched, once the application is running, to provide feedback about the state of the links to the node managers. The feedback is helpful for load balancing if the current network load is not balanced. Once the application is finished, effector agents are launched over the paths used by the application to free allocated resources. The management scheme can handle QoS and best-effort traffic at the same time (by utilizing routing tables built by proactive monitor ants).

2.3.6 A Brief History of Research in AntNet

Oida and Kataoka [178] decided to improve an earlier version of *AntNet* in which the status of data link queues was not used in the goodness formula (yielding no pheromone-heuristic control). Without this queue dependency feature, *AntNet* will suffer from stagnation once the goodness of any link of a neighbor reaches 1. The authors of [178] modified the routing table updating rules to avoid the "locking" of routing tables. Their algorithms *DCY-AntNet* and *NFB-Ants*, upon comparison with an earlier version of *AntNet* [59], performed much better under challenging situations. Doi and Yamamura [72, 73] also proposed a few additional heuristics to avoid the above-mentioned locking problem in *AntNet-FA*. In fact *AntNet-FA* does not suffer from the locking problem [58]. Their algorithm, *BNetL*, showed similar performance compared to *AntNet-FA*. The authors of [89, 90] have developed a multi-agent system, which consists of static and mobile agents (ACR concept), for multiple-criteria load balancing on a network of processors.

Oida and Sekido [179, 180] proposed the *Agent-based Routing System* (ARS) as an enhancement of *AntNet* for QoS routing. Each service class (in terms of bandwidth) has its own colony of ants. The ants move in a "virtually constrained network" and make their decisions based on the values in the routing tables and the amount of bandwidth already reserved by the ants of the colony. The ants of a colony use only those links whose available bandwidth is greater than the bandwidth constraint assigned to the colony. If the available bandwidth of a link is very small then the probability of selecting it is already low. If an ant takes too many hops or if none of the out-going links has enough remaining bandwidth, then the journey of the forward ant is terminated.

Baran and Sosa [17] made the following improvements to *AntNet-FA*:

- Intelligent initialization of routing tables is done in which the entries in the routing tables are not uniformly initialized. Rather, the probability values for these destinations, which happen to be neighbors of a node, are initially given a higher value.
- The algorithm explicitly sets the pheromone value of a neighbor for reaching a destination to 0 if the link to the neighbor or the neighbor crashes. This pheromone value is evenly distributed among the remaining neighbors through which a destination is still reachable. This feature makes *AntNet-FA* fault-tolerant.
- *Uniform ants*, like those proposed in [241], are introduced to counter the stagnation of the entries in the routing tables of the nodes. However, we must again emphasize that *AntNet-FA* does not suffer from stagnation because of its heuristic pheromone control and privileged pheromone-laying features.
- The ant agents make greedy deterministic decisions instead of random proportional ones. The policy might make ant agents infinitely loop if its greedy deterministic decisions force it to follow a cyclic path.
- The number of ants living in the network has been arbitrarily limited to four times the number of links. The authors did not provide a reason for this value. This approach might help in reducing the bandwidth costs of ant agents when the network is experiencing congestion, but it will also impair the responsiveness of the algorithm to dynamic network traffic situations.

The authors of [161] replaced the stochastic decision policy of *AntNet* with a deterministic greedy policy, which does not use a queue heuristic. The authors compared this version of *AntNet* with *OSPF* on small tree, ring and star topologies by simulating FTP traffic using TCP Tahoe. In a steady state, both algorithms showed a similar performance. Also, a hybrid QoS-aware routing algorithm was developed [161, 162] by combining useful features of *AntNet* and *ABC*. This algorithm provides soft guarantees on two parameters: end-to-end delay and throughput. The algorithm utilizes two types of ants, one for each constraint. The *delay ants* are similar to the ant agents utilized by *AntNet*. The *throughput ants* inherit behavior of ants in *ABC*: they are artificially delayed at each node for time proportional to the occupied bandwidth, which is measured as a local exponential average of the link utilization. Their virtual delay is a measure of the available bandwidth in the network. The experiments conducted used the same traffic patterns and topologies as mentioned before. The results revealed that the performance of the AntNet-like algorithm is similar to *OSPF*; however, it scales better than *OSPF* under an increase in traffic load.

The authors of [132] have provided a brief overview of so-called Swarm Intelligence (in practice, ACO) algorithms for routing in networks. They propose an *Adaptive-SDR* algorithm in [131] that organizes the network into clusters by using a centralized k-mean algorithm. Once the partition process is complete, the algorithm maintains inter-clustering and intra-clustering routing tables at each node. Multiple colonies of ants are used to discover and maintain these different routing tables. In this manner the number of ants that need to be sent in the network is reduced because a node only maintains routes to the nodes within the cluster and not to all nodes in the network. However, the implementation of *AntNet* in [131] is not correct because it was asserted that *AntNet* deterministically routes data packets on highest-probability paths, which is in contradiction to the stochastic spreading feature of *AntNet* [58]. The comparison of Adaptive-SDR with *AntNet* (in the erroneous version explained above), *OSPF* and RIP showed that Adaptive-SDR achieves the best results with reference to the throughput and average delay. The experiments were conducted on 16- and 48-nodes network topologies in NS-2 simulator.

Jain [123] implemented a version of *AntNet* quite similar to that reported in [161] on the NS-2 simulator. In her algorithm, data packets are deterministically forwarded in a greedy fashion to the highest-probability neighbor. The experiments were run on grid networks of different sizes. The two *AntNet* algorithms exhibited the same behavior under low load scenarios, but this single-path *AntNet* algorithm [123] quickly adapted to new situations. Sim and Sun [221] proposed the *Multi-Ant Colony Optimization* (MACO) approach, which utilizes multiple colonies each laying its own type of pheromone, for loadbalancing in connection-oriented networks. An ant is expected to choose a path that has a higher pheromone concentration of its own type, and due to the concept of pheromone repulsion, an ant is less likely to prefer a path with a higher concentration of pheromone laid by ants of other colonies in order to find good (disjoint) paths. The advantages of using MACO in circuit-switched routing is that it spreads data packets over multiple paths without significantly increasing the routing overhead.

Tadrus and Bai [244] developed the *QColony algorithm* for QoS routing based on principles similar to that of the ACR framework. QColony nodes maintain different types of pheromone tables for different types of constraints. The algorithm utilizes different classes of agents, each of which has a different priority and serves different tasks, e.g., searching for a best-effort or QoS path. Each ant is routed with the pheromone table that corresponds to its type, and it updates the routing table with a weight dependent on its age and priority. The ants build tables in a proactive and on-demand manner that are used to find feasible paths at session startup time. An application session can specify its acceptable range of bandwidth and maximum number of hops. Once the session has started, QColony also sends *soldier ants*, who provide multiple paths for loadbalancing and fault-tolerance and favor paths having smaller hop count values. The experiments described in [244] show that the performance of QColony is comparable to ARS and to the selective flooding algorithm [43] for small topologies and under low network traffic load; but its performance is significantly better for large networks and heavy traffic loads. The authors of [240] and [37] have proposed simpler algorithms for QoS routing as compared with QColony.

Carrillo et al. [38] have done a preliminary study on the scalability of *AntNet* that is based on a simple theoretical formulation. It is not verified through extensive experiments in large-scale topologies. They have argued through their theoretical model that *AntNet* is scalable. However, the correctness of the findings, without experimental verification, is a serious shortcoming of their study.

Zhong and Evans [301] did a preliminary study in which they outlined important attacks that ant agents, launched by malicious nodes, could make. They did point out that using certificates is not a feasible option for the *AntNet* algorithm because of the processing complexity of the approach. However, they did not provide a technical solution to the problem that could be implemented in the system. Their idea is to send a verification ant that follows the best path toward a destination when the goodness of a neighbor increases above a threshold. The trip time could be calculated either by dividing the round-trip time by two, or by measuring the difference between the entrance time at destination and launching time at source, assuming that the clocks are perfectly synchronized using the Global Positioning System (GPS) service. Similarly, Yang et al. [294] are the first ones who implemented *AntNet* in the application layer of the network stack and then did experiments on a small topology of five nodes. The results of their experiments show the advantages of dynamic reinforcement, which is based on the trip time of ants, over constant reinforcement.

We have summarized the history of ant algorithms for best-effort routing in Table 2.1, and of ant algorithms for QoS routing in Table 2.2.

2.4 Evolutionary Routing Algorithms for Fixed Networks

In this section, we provide a brief survey of routing algorithms, which have been developed on the background of *Evolutionary Algorithms* (EAs), which in turn are inspired by the evolution process in living cells. A description of the principles of evolutionary algorithms, and their application to different optimization problems, is discussed in [98, 116], and a comprehensive survey about evolutionary strategies is provided in [20].

Authors	Algorithm name	Year	References
Di Caro and Dorigo	AntNet,AntNet-FA	1997	[59, 62, 61, 65, 64, 60, 63]
Subramanian, Druschel, and Chen	ABC Uniform ants	1997	[241]
Heusse, Snyers, Guerin, and Kuntz	CAF	1998	[113]
van der Put and Rothkrantz	ABC-backward	1998	[253, 252]
Oida and Kataoka	DCY-AntNet,NFB-Ants	1999	[178]
Gallego-Schmid	AntNet NetMngmt	1999	[96]
Doi and Yamamura	BntNetL	2000	[72, 73]
Baran and Sosa	Improved AntNet	2000	[17]
Jain	AntNet Single-path	2002	[123]
Zhong and Evans	AntNet security	2002	[301]
Kassabalidis et al.	Adaptive-SDR	2002	[132, 131]
Yang et al.	AntNet on LAN	2002	[294]

Table 2.1. Wired Best-effort networks. The table is reproduced from the thesis of Di Caro [58] with his kind permission

Authors	Algorithm name	Year	References
Schoonderwoerd et al.	ABC	1996	[214, 213]
White, Pagurek, and Oppacher	ASGA	1998	[287, 288]
Di Caro and Dorigo	AntNet-FS	1998	[63]
Bonabeau et al.	ABC Smart ants	1998	[23]
Oida and Sekido	ARS	1999	[180, 179]
Di Caro and Vasilakos	AntNet+SELA	2000	[67]
Michalareas and Sacks	Multi-swarm	2001	[161, 162]
Sandalidis, Mavromoustakis, and Stavroulakis	Ant-based routing	2001	[208, 209]
Subing and Zemin	Ant-QoS	2001	[240]
Tadrus and Bai	QColony	2003	[244]
Sim and Sun	MACO	2003	[221]
Carrillo, Marzo, Fabrega ,Vila and Guadall	AntNet-QoS	2004	[37]

Table 2.2. Wired QoS networks. The table is reproduced from the thesis of Di Caro [58] with his kind permission

2.4.1 Important Elements of EA in Routing

We first summarize the important features of evolutionary algorithms which have been successfully employed in routing algorithms.

Chromosomes

The algorithms model the solutions of a problem by encoding it as a gene, a chromosome, or an individual. The algorithm randomly generates a population of the individuals by randomly altering different genes (options) in the individuals. In a routing algorithm an individual is a string that consists of a sequence of nodes.

Fitness

The agents are launched by the nodes; they traverse the sequence of nodes encoded in the chromosomes. Once an agent returns to its source node, the fitness of its corresponding chromosome is evaluated on the basis of the routing information collected by the agent. The fitness can be defined as a function of trip time or hop count. Its definition plays an important role in the performance of an algorithm.

Evolutionary operators

The algorithms apply selection, crossover, and mutation operators on individuals that are based on the Darwinian notion in biology. After the fitness evaluation of all individuals of the first generation, the n fittest individuals are selected for replication in the new generation. Some of them are taken as parents for a crossover operation in which a partial solution of one individual is combined with the partial solution of another individual. Finally, a part of a solution in an individual is replaced by some other random value, and this is termed mutation. Mutation and crossover operators provide diversity within a population. The selection operator ensures that a portion of a population consists of the fittest individuals from the previous generation. In this way an algorithm is not only able to strive for the optimal solution but also avoids stagnation. In routing algorithms, it corresponds to keeping the best routes discovered so far and to discovering/evaluating new routes through mutation and crossover operators. Extremely poor routes are to be extinguished through continuous application of genetic operators. Evolutionary algorithms thus provide adaptation in dynamically changing environments. The approach Munetomo took in [171, 172] is promising because evolution is a distributed process in which each individual independently adapts to its environment without the need for its having explicit communication with other individuals. An evolution process is also robust against changes in the environment. These features make evolutionary algorithms appealing for telecommunication networks. A detailed survey has been provided in [224]. Here we will just provide a brief summary of three state-of-the-art routing algorithms, namely *GARA* [173], *ASGA* [282], and *DGA* [147].

2.4.2 GARA

Munetomo [173, 171] developed the *Genetic Adaptive Routing Algorithm* (GARA), which utilizes *path genetic operators* to identify a subset of routes that should be monitored. The fitness of a route is calculated by normalizing observed communication latencies among alternative routes to the same destination. The algorithm periodically applies path genetic operators, which remove the entries for the routes whose destinations do not frequently receive data packets and for the route with a worst fitness value. As a result, the routing table only contains routes to the destination where packets are frequently sent. In *GARA*, path mutation and path crossover operators are based on the topology. Through a mutation operator, a mutation node is randomly selected from a route leading to a particular destination. In the next step, a neighbor of a mutation node is randomly selected and then the source node, the selected neighbor, and the destination node are connected by Dijkstra's shortest-path algorithm. In the crossover operator, a router, which exists in two routes leading to the same destination, is selected as a crossover point. The sub-routes after the crossover point are then exchanged. This limits the crossover to those routes which have at least one common node.

We will illustrate both operations in the topology in Figure 2.3. Let us assume that we want to apply the mutation operator to route 9-4-2 and we select Node 4 to be the mutation node. We select neighbor Node 3 of Node 4 as a replacement. Now once we try to join Node 9 with Node 3 and Node 3 with Node 2 we get the new route 9-4-3-1-2. Let us assume that we want to apply the crossover operator to routes 8-9-4-3 and 8-4-2-1-3. We again select Node 4 as a crossover point and achieve two new routes: 8-4-3 and 8-9-4-2-1-3. The algorithm "learns" about new routes by utilizing the above-mentioned strategies. *GARA* is a *host-intelligent* routing algorithm. It requires that the sender node put the complete route in the header of each data packet. As a source routing approach, it does not scale well for large networks where the overhead of putting the complete route into each packet will considerably increase the size of the packet, and hence cause congestion and waste of bandwidth. Therefore, the authors decided to switch to first partial source routing, in which only a few initial hops are put in the source header, and thereafter to next hop routing [172].

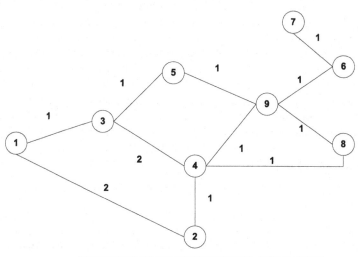

Agent ID	Agent Fitness	node ID and Trip time (msec)
85	0.32	(4,10),(3,30),(1,35),(2,55),(4,65), (8,65)
234	0.45	(6,10),(7,20)
31	0.66	(4,10),(8,15),(9,20),(6,25),(7,30)
...	...	...
25	0.81	(5,10),(3,15),(4,25),(2,30),(1,35)

Routing Table at Node 9

Fig. 2.3. Routing table in DGA

2.4.3 ASGA and SynthECA

White et al. combined important concepts of the Ant System [78] with the ideas of genetic algorithms into the routing algorithm, *ASGA* (Ant System with Genetic Algorithm), for circuit-switched networks [282, 287, 288]. The algorithm can be utilized for point-to-point, point-to-multipoint, and multi-path routing in circuit-switched networks. The algorithm follows a standard genetic algorithm in which an initial population of ant agents is generated; the ants are assigned random values for pheromone (α_d) and cost sensitivity (β_d) parameters for reaching destination d. These *explorer agents* are launched in the network and they follow their journey according to the pheromone values in the routing tables. During exploration they maintain an internal cost path variable C_d, instead of the trip time t_d. However, the authors do not clearly define C_d and l_{ij}, the cost function of a link between node i and node j, for circuit-switched networks. After crossing a link from node i to node j, the ants update their cost variable using the formula $C_d = C_d + l_{ij}$. Once the explorers reach their destination they start their return trip and update the pheromone tables by using modified Ant System equations. After they arrive at their source node, the paths found are written into a buffer, their fitness is evaluated, and they are associated with (α_d, β_d) parameters in an another table. Finally, for the second iteration, the genetic algorithm applies selection, crossover, and mutation operators on the current generation to create a population for the second iteration. The new generation of explorer agents is again assigned (α_d, β_d) values. The genetic algorithm empowers the explorers to keep on exploring alternate paths; this feature, coupled with evaporation, privileged pheromone-laying and pheromone-heuristic control, enables the algorithm to avoid stagnation. In the *ASGA* algorithm, a source node decides, depending on the percentage of ants that followed the same path, whether the network or router resources should be reserved for a call. This objective is achieved by launching *allocator agents* [288], which allocate resources along the paths selected. Similarly, when a path is no longer needed, *deallocator agents* are launched. They deallocate the network resources used in the nodes and links. The system also utilizes *evaporator agents* which circulate in the network and evaporate the pheromone concentrations laid on a path to promote exploration. The authors have evaluated and described the merits of *ASGA* in [111, 286]. They conducted preliminary experiments on smaller, simple topologies, and the results show that the algorithm is dynamically able to compute shortest paths, and that the genetic adaptation of (α_d, β_d) considerably contributes to improving the performance of the algorithm. However, the considerable overhead in terms of bandwidth and processing was not evaluated. Moreover, the performance of *ASGA* was not compared to other state-of-the-art nature-inspired routing algorithms.

Based on *ASGA* a general framework *SynthECA* (Synthetic Ecology of Chemical Agents) [283, 111, 286, 281] has been developed. A detailed review is made in [221, 283, 285, 284], and a detailed description is provided in [281]. *SynthECA* manages point-to-point, point-to-multipoint, and multi-path routing like *ASGA*, with an additional feature for fault location detection and management [283, 285].

The agents in *SynthECA* are described by a tuple, which consists of *emitters, receptors, chemistry, Migration Decision Function* (MDF) and *memory*. The emitters associated with the agents generate chemicals, their production rate is controlled by an *Emitter Decision Function* (EDF), and they are deposited in the ambient environment of an agent. Receptors sense chemicals in the agents' environment according to their *Receptor Decision Function* (RDF) and then take appropriate actions. The emitters and receptors are digitally encoded with 0,1, or #, where # is a wild card. For example, a receptor with encoding 10## can detect chemicals 1000, 1001, 1010, and 1011 [283]. The chemistry associated with an agent is a set of rules that determine how different types of chemicals can react together to produce different chemicals, thereby changing the local environment of an agent. The agents in *SynthECA* utilize five types of chemical reaction rules among pheromones:

1. $X \rightarrow$ 'nothing' (evaporation property)
2. $X + Y \rightarrow Y$ (survival of the fittest)
3. $X + Y \rightarrow Z$ (reproduction/stigmergy)
4. $X + Y \rightarrow X + Z$ (survival of the fittest)
5. $X + Y \rightarrow W + Z$ (reproduction/stigmergy)

Rule 1 ensures dynamism, and all other rules allow the receptor ants to communicate important network state information to other ants. Rules 2 and 4 can allow high priority traffic to use the resources in the network while low-priority traffic could be either discarded or diverted to another route. Rules 3 and 5 can be used to communicate inhibitory or excitatory messages to other ants, which will cause them to detour from faulty portions in the network in the first case, or be attracted to good paths in the second case. The memory within an agent stores special chemicals, for which no emitter and receptor is defined, to determine next hop for the agent: MDF (the migration detection function mentioned above) determines the next hop of the agent as a function of chemical values and link costs.

The system consists of three classes of agents: *Route Finding Agents* (RFAs), *Connection Creation Monitoring Agents* (CCMA), and *Fault Detection Agents* (FDA). The route finding agents have already been described in the beginning of the section, and they are explorers, allocators, deallocators, and evaporators. The purpose of CCMA agents is to monitor the quality of connection parameters using special q-chemicals [283]. Finally, fault detection agents observe the quality of different links by accessing q-chemicals laid by CCM agents. Their job is to look for high q-chemical values above a threshold, and take remedial actions if needed.

SynthECA consists of a colony of different types of ant-like agents that utilize chemical features along with ACO principles (Section 2.3) to solve a problem. Moreover, the agents undergo continuous evolution through an evolutionary algorithm at each node. The system is quite complex compared with ACR (Section 2.3.5). We believe a thorough analysis about the complexity of such a system is necessary, both in terms of processing power and network resources, before a clear judgment about its benefits can be made in comparison to all previously discussed approaches.

2.4.4 DGA

Liang et al. have studied the impact of the size of the routing table on the performance of AntNet [146]. They reduced the number of entries in the routing table of AntNet, thus decreasing routing table information, hence the overhead for making routing decisions; and termed the new algorithm AntNet-local. The experiments conducted in NTTNet [146] reveal that AntNet-local has significantly poor performance as compared with AntNet-global both in terms of throughput and delay. However, the authors did not discuss the trade-off between reduced overhead and quality of paths selected.

They then developed the *Distributed Genetic Algorithm* (DGA) [147] based on the concepts of *GARA* [147]. The following features of *DGA* allow for easy application of genetic operators:

- Each node initializes a population of agents with size $size = links^2 \times c_1$, where c_1 is a constant. The formula is motivated by the fact that higher-degree routers have to explore more links and connections between and across neighbors. Initially, only half the agents are launched in the network.
- Each agent in *DGA* is modeled as a chromosome of integers, which represent next hop offsets. Once an agent enters a node i, it picks up the offset number m_i from the chromosome, and then applies the formula $index_i = m_i \% N_i$, where N_i is number of neighbors of node i. The agent then identifies a link with offset $index_i$, counting clockwise from the entering port. The link is selected as the next hop. The number of entries in the chromosomes are restricted to m (currently 6). The size of a chromosome determines how many hops, starting from the source node, an agent is allowed to take. This representation is then independent of the network connectivity, and hence simplifies the design of genetic operators. Each agent is equipped with a stack which carries the address of the visited nodes and the trip time value to the visited nodes from the source node.
- An agent terminates its journey if it has visited the last entry in the chromosome. However, if $index_i$ ends up being at the same link from where the agent arrived, then the next hop is selected from the remaining neighbors in a random fashion and accordingly the entry in the chromosome is altered. If no next hop can be selected because the agent has already visited all the neighbors, then the chromosome is truncated and the forward agent is converted into a backward agent.
- The backward agent only modifies the routing tables at its source node. Its fitness function is defined as $f = \frac{\sum_{k=1}^{D_i} \alpha_k^i \times t_k}{\sum_{k=1}^{D_i} t_k}$, where D_i is number of explored destinations at node i, and t_k is the trip time value for destination k. α_k^i is defined as $\alpha_k^i = \frac{m_k^i}{T_i}$, where m_k^i is the total number of packets generated for node k at node i while T_i is the total number of packets generated for all discovered destinations at node i. The definition assigns a high fitness value to an agent which has discovered a low-latency path to a destination (where more data packets are sent). The ID of an agent, its fitness, the nodes it visited, and the trip time to the nodes are stored in a routing table (see Figure 2.3).

- The authors of *DGA* have introduced the concept of aging by periodically decreasing the fitness values (f) of the agents and at the same time increasing the trip time values (t_k) through the formulas $f = f \times c_2$ and $t_k = \frac{t_k}{c_2}$. This will avoid stagnation in the routes. c_2 is set between 0.8 and 0.9.
- Once four agents return to a node, selection, crossover, and mutation operators are applied to the two best agents. Since chromosome representation is not dependent on the topology, one can simply use the traditional genetic operators. The two new agents are inserted into the node population after deleting the two worst agents from the population.
- Periodically, every 500 or 700 msec, each node passes its best 3–5% individuals to its neighbors.
- Once a node wants to forward a data packet whose destination has been discovered, it is forwarded through the agent which has the shortest trip time value to the destination. However, if the destination is still not discovered, the packet is routed through the agent which has the highest fitness value.

The authors compared their algorithm with AntNet-local on NTTNet, under a low traffic load of about 35 packets/sec. The results demonstrate that *DGA* is able to deliver more packets compared with AntNet-local, but with a higher delay. The authors dropped data packets which followed a cyclic path, but did not provide a proper justification for it. We provided a detailed critical review on *DGA* in [268]. *DGA* is a complex and sophisticated algorithm which launches half of the population at start-up. Consequently, agents occupy approximately 50% of the bandwidth, which is really not acceptable. The authors did not provide results for *OSPF* and AntNet-global. Our study [268] shows that at 35 packets/sec the performance of *DGA* is significantly inferior compared to both *OSPF* and AntNet-global. Another important drawback of *DGA* is that its communication costs increase with a decrease in traffic load. Ant agents use the same buffers as data packets, and a next generation of agents is launched from a node once it receives four agents from the previous generation. Consequently, more agents will be traveling on the network if their trip time is small, and this happens on a small topology or under low traffic loads, or in both scenarios. This behavior is in contrast to the expectation: more exploration under high traffic load and low exploration under low traffic load. Last but not least, the authors themselves were not sure about the complexity of searching and storing the population-based routing tables employed by *DGA*. An exemplary routing table is shown in Figure 2.3.

2.5 Related Work on Routing Algorithms for Fixed Networks

We now provide a very brief review of the algorithms developed by the Artificial Intelligence (AI) community and the networking community. The motivation of doing this is to introduce state-of-the-art routing algorithms developed by these communities, which will provide the basis for cross-fertilization of ideas among all communities.

2.5.1 Artificial Intelligence Community

The artificial intelligence community has applied Reinforcement Learning (RL) [130], developed as a branch of machine learning, to propose routing algorithms for packet-switched networks. The two well-known algorithms are Q-routing [27] and PQ-routing [44], which are based on the Q-learning [265, 266] algorithm.

Q-routing employs an online asynchronous decision policy based on local information. Every router maintains Q-values, which represent the goodness of a neighbor for reaching a particular destination. The value $Q_i(j, d)$ is an estimate of the time at node i that a packet will take for reaching destination d via neighbor j. Once the neighbor j receives a packet it will immediately send a feedback packet to node i with a new estimate $Q'_j(d) = \min_{z \in N(j)} Q_j(z, d)$, which is the best trip time estimate held at node j for destination d. If the feedback packet took t_{ji} time, which is the sum of the propagation delay of the link and the queuing delay at node i, then node i could revise its estimate according to the formula $\delta Q_i(j, d) = \eta(Q'_j(d) + t_{ji} - Q_i(j, d))$, where η is the learning rate, a standard feature of iterative algorithms, generally set to a value which satisfies the stochastic approximation convergence [254]. The authors used a value of 0.5 in their experiments. In this way the time-to-go estimates are updated using exponential averaging. Finally, the data packets are deterministically routed through the neighbor which has the lowest associated Q-value (highest goodness). The deterministic routing policy will keep on sending the data packets through the neighbor with the lowest Q-value until the Q-values of the other neighbors drop below the Q-value of the selected neighbor. If a neighbor recovered from a transient overload, it would never be selected as a next hop until the Q-value of all other neighbors become worse than its own. This feature provides no room for load balancing. The authors conducted their experiments on an irregular 6×6 grid. The results show that Q-routing performs similarly to the Shortest Path First routing algorithm under low network loads, and performs significantly better under higher network loads. Moreover, the control overhead is directly proportional to the number of data packets switched by a node, which under a high network load could be unacceptable.

Choi and Yeung [44] proposed the Predictive Q-routing algorithm, known as PQ-routing, which overcomes the above-mentioned problem. Moreover, they take into account in [44] that Q-routing does not always converge to the shortest path under low loads. In PQ-routing, they do *controlled exploration* of congested paths by occasionally sending probe packets along the paths. The probing frequency depends on the network traffic and recovery rate of a path. Q-values are updated in a similar way as that in Q-routing. However, a more sophisticated routing policy is employed. The recovery rate of each neighbor is determined based on the difference in two subsequent δQ values for each neighbor. Then, based on the recovery rate of all neighbors, existing Q-values, and best estimated Q-values, the next hop leading to a particular destination is selected. The authors conducted a number of experiments on a 13-node topology and a 6×6 irregular grid. The results demonstrate that the adaptation time of PQ-routing is significantly smaller once traffic patterns or topologies change because PQ-routing utilizes the concept of recovery rates. PQ-routing is generally better than Q-routing under both low load and varying network conditions, but its performance becomes comparable with Q-routing under high load conditions.

2.5.2 Networking Community

The most important work in the field has been contributed by the networking community, which also considers itself responsible for pioneering packet-switched networks. The roots of its work go back to the development of ARPANET and a routing algorithm for it based on the asynchronous distributed Bellman-Ford algorithm [19, 168]. The basic idea of the algorithm is that each router maintains only the best known cost paths to each destination. Each router forwards its current routing table to all its neighbors as a vector of distances to all nodes in the network. Once the neighbors receive the estimates, they compare them with their own estimates. If a neighbor's cost estimate to a destination is less than the current estimate of a node, then the node accordingly updates its routing table with the new estimate. The algorithm iteratively progresses toward stability. After the first iteration, the routers know the current best path costs to all routers within a one hop diameter. The diameters keep expanding by 1 with each iteration until each router has routing information for all nodes in the network. The algorithm, however, suffers from count-to-infinity and looping problems [19, 168, 189].

As ARPANET grew bigger, many researchers proposed novel adaptive routing algorithms in [36, 95, 94, 135]. In the 1980s, ARPANET was transformed into NSFNET, which became the T1 US backbone. *OSPF*, which is a link-state routing protocol, was developed for NSFNET. The link-state routing algorithms avoid looping and count-to-infinity problems. *OSPF* is currently the state-of-the-art routing algorithm employed as an Interior Gateway Protocol (IGP). The Routing Information Protocol (RIP) [108, 151] algorithm based on the asynchronous distributed Bellman-Ford algorithm is used for routing within an AS (autonomous system) in the Internet [169]. *OSPF* stores the entire topology of a network in a weighted directed graph, in which each edge corresponds to a link and each node corresponds to a router. The cost of a link is a function of the propagation delay of the link. However, the network administrators are also allowed to change these costs based on their on-field knowledge about network traffic loads. Each node in the *OSPF* algorithm estimates the costs of the links to its neighbors. It then encapsulates the addresses of its neighbors and the costs of the links to the neighbors in a link-state packet and broadcasts it to all of them. The neighbors, in turn, send the link-state packet to their neighbors, and so on until all nodes in the network get the packet. Each node builds the network topology in a distributed fashion. Finally, each node builds a shortest-path tree to all destinations by considering itself as a root. The shortest paths from the root to all nodes are calculated using the deterministic Dijkstra Algorithm [69]. The next hop on the shortest path to each destination is stored in a routing table. The interested reader can find a detailed description of this algorithm in [169] and its complete implementation in [170]. However, none of the algorithms try to maintain multiple paths to a destination at a given node. This shortcoming results in an underutilization of network resources which consequently results in their poor performance.

In the 1990s research started on multi-path routing algorithms. Chen, Durschel, and Subramanian developed an algorithm *MP-Scout* [41, 43, 42] in which multiple paths are maintained at a given node for each destination node. *MP-Scout* is based on the concept of backward learning for determining loop-free multipaths. A simple version of *MP-Scout* is known as *SP-Scout*, in which a destination node periodically floods scout messages. Each scout message is uniquely identified with a tuple $< d, C_d, x >$, where C_d is the cost to reach d and x is an increasing sequence number. C_d is initially set to 0. The time interval between two scout floodings is known as a broadcast interval (BI). When a node i receives a scout message from its neighbor j for destination d, then the node first updates the cost parameter of the scout to C'_d, $C'_d = C_d + C_{ij}$, where C_{ij} is the cost of the link from node i to neighbor j. During the first BI, the node immediately forwards the first received scout to all neighbors of the node except the one from which it arrived. The node might receive more scouts from other neighbors within the same BI, but they are not forwarded once a scout has already been forwarded. Consequently, it will just remember the best scout, and the neighbor from which it has been received is termed as the *designated neighbor* (next hop) for its source node. In subsequent intervals, only the scouts from the designated neighbors are forwarded. If i did not receive any scout from the designated node in the last BI, then it will again flood the first scout and update its *designated neighbor* for the source node of the scout message. In *MP-Scout*, the scout message is identified with a tuple $< d, C_d, sID, pID >$, where pID corresponds to a path ID and sID to the scout ID, d is its launching node, and C_d is the cost to reach d. The launching node modifies the pID for scout messages launched on different links. Based on the pID it can identify whether the paths through different scouts can lead to a loop. The algorithm also applies two types of thresholds: scout and data. A scouting threshold sets a limit K for the best known routes to be maintained at an intermediate node. However, this may result in propagating the scouts which have traversed inferior paths. The algorithm applies the concept of data threshold in which a scout that advertises a path with a greater cost (by a given percentage) than the minimum cost path is discarded. Each data packet carries an additional path ID field in its header to enable the routers to forward it to a next hop node which lies on this path ID. *Equal-cost multi-path* (ECMP)[169] is a routing technique in which multiple packets are distributed, typically in a simple round-robin fashion, on multiple paths by assigning equal costs to them. Optimal splitting by using ECMP is not possible as the costs are not updated and do not model traffic loads. A better optimal splitting algorithm, *OSPF-OMP*, has been proposed in [257]. *OSPF-OMP* samples the current network traffic load via opaque Link-State Advertisements (LSAs) within an *OSPF* area. The information carried can be link loading, link capacity, and a measure of packets dropped. The information contained in a LSA is used to change load-splitting decisions.

Recently, Vuturkey has proposed three multi-path routing algorithms: The *Multipath Partial Dissemination Algorithm* (MPDA) [263], *MPATH* [262, 261], and the *Multi-path Distance Vector Algorithm* (MDVA) [260]. All of these algorithms make use of *loop-free invariants* (LFI) discussed in [263], to ensure loop freedom in every instance. The condition is: for each destination d, a node i can choose a successor n

whose distance to d, as known to i, is less than the distance of node i to node d. The interested reader will find a complete review of these algorithms in [264].

MDVA [260], as the name suggests, is a multi-path distance vector algorithm which tries to overcome count-to-infinity and the looping problems in the distance vector algorithms as discussed in the beginning of the subsection. The algorithm avoids looping by sending the cost estimates along a *Directed Acyclic Graph* (DAG) rooted at a destination. Each node in DAG computes its cost to the destination by using the costs reported by "downstream" nodes, and it reports its costs to the "upstream" nodes. This method is called *diffusing computation* and was suggested in [70] to ensure that a distributed computation will always terminate if the computations are performed in an acyclic fashion. MDVA uses a RIP-like algorithm to compute the cost D_d^i of reaching destination d from node i, and SG_d (DAG), which is defined by a link set consisting of successor nodes leading to destination d. The paths in SG_d should have the *loop-freedom* and *connectivity* characteristics for efficient routing.

Each node in the network maintains D_d^i, the successor set S_d^i, the feasible distance FD_d^i, the reported distance RD_d^i, and SD_d^i, which is the shortest possible distance through the successor set. The table also stores a set of waiting neighbors in a diffusing computation. Each node also maintains a neighbor table in which D_{jd}^i, the values for the distance of neighbor j to d as communicated by j, are stored. The link table stores the cost l_j^i for the adjacent link to each neighbor. At startup time, a node initializes all distances in its tables to infinity, and all successor sets to null. If a link is down or a node is unreachable, then its cost is considered infinity. Nodes executing MDVA exchange messages of the form $(type, d, c)$ where $type$ could be QUERY, UPDATE, or REPLY, d is destination node, and c is the distance to the destination node. Upon arrival of a message, a node updates its routing tables. This step is repeated if the cost or status (up or down) of an adjacent link changes. When an adjacent link becomes available, the node sends an update message of type $(UPDATE, d, RD_d^i)$ for each destination d over the link. When an adjacent link to a neighbor j fails, the neighbor table associated with neighbor j is deleted and the cost of the link is set to infinity. Similarly, when the cost of the link to j (l_j^i) changes, l_j^i is set to the new cost, and the vector tables for each destination are updated. A node can be either in ACTIVE or PASSIVE state with respect to a destination. Initially, all nodes are in the PASSIVE state, and as long as the link cost decreases, MDVA works like RIP and nodes will remain in the PASSIVE state. However, if the distance to a destination increases, either because an adjacent link cost changed or a message is received from a neighbor, then the diffusing computation (ACTIVE state), as described before, is started by sending QUERY messages to all neighbors with the shortest distance SD_d^i through S_d^i, and then waiting for the neighbors to reply. If the increase in distance is due to a query from a successor, the neighbor is added to the list of neighbors waiting for a reply. When all replies have been received via the REPLY message, the node can be sure that the neighbors have incorporated the distances that the node reported, and it is now safe to migrate to the PASSIVE state. If a node in the ACTIVE state receives a QUERY message from a neighbor which is not in S_d^i, then a REPLY is immediately sent. However, if it is in S_d^i, a test is made to verify if SD_d^i increased beyond the previously reported distance. If it did not, a reply is immediately sent. However, if it did increase, the QUERY is blocked and the neighbor is added to the neighbors' list. The replies to such neighbors are deferred until the node is ready to go to the PASSIVE state. When all replies have been received and the distance D_j^i increased again, the ACTIVE phase is extended by sending a new set of queries; otherwise, the active phase ends. If the ACTIVE phase continues, no REPLY messages are sent to the pending queries; otherwise, all replies are sent and the node enters the PASSIVE state.

MPDA [263, 264] is obtained by applying loop-free invariant conditions to the *PDA algorithm*, a link-state routing algorithm in which "enough" routing information in the network is propagated so that each router has *sufficient* link-state information to compute the shortest paths to all destinations. In PDA, a router communicates information to its neighbors about only those links that are part of a minimum-cost routing tree, and the router validates the link information based on the distances to the head of the links, and not on sequence numbers. In PDA, a router maintains the following information:

1. The *main topology table*, T_i, stores the characteristics of each link known to router i. Each entry in T_i is a triplet (h, t, c) where h is the head, t is the tail and c is the cost of the link $h \rightarrow t$.
2. The *neighbor topology table*, T_j^i, is associated with each neighbor j. The table stores the link-state information communicated by neighbor j.
3. The *distance table* stores the distances from router i to each destination based on the values in T_i, and the distances from each neighbor j to each destination based on the values in T_j^i for each j. The distance of node i to node d in T_i is denoted by D_d^i; the distance from j to d in T_j^i is denoted by D_{jd}^i.
4. The *routing table* stores, for each destination d, the successor set S_d^i and the feasible distance FD_j^i, which is used by *MPDA* to enforce LFI conditions.
5. The *link table* stores, for each neighbor j, the cost l_j^i of the adjacent link to the neighbor.

The routers exchange *link-state update* (LSU) messages which contain one or more entries specifying addition, deletion, or change in cost of a link in the router's main topology table T_i. Initially, a router initializes all distance type variables with infinity and node type variables with null. Once a router receives an LSU from its neighbors, it updates its routing tables and, based on the information, constructs a shortest path tree. A router only communicates the differences in the tree to its neighbors. When two or more neighbors report different costs of the same link, the conflict is resolved in favor of the neighbor reporting the shortest distance. In case both report the same cost, the tie is broken in favor of the lowest address. In *MPDA*, the routers operate in two states: ACTIVE and PASSIVE. If a router is in the PASSIVE state and there is no change in its topology T_i, then the router has nothing to report and it remains in this state. However, if a router in the PASSIVE state receives an event that changes its topology, the router sends those changes to its neighbors and goes into the ACTIVE state and waits for acknowledgments (*ACKS*). Remember that LSUs are acknowledged in MDPA and inter-neighbor synchronization spans only a single hop, unlike the synchronization in [70], which potentially spans the whole network. In the ACTIVE state the router updates only T_j^i and l_j^i. The topology table T_i is updated at the end of the ACTIVE phase, after acknowledgments from all neighbors have been received. In this way router i incorporates the latest changes that occurred during the ACTIVE phase in T_i. If no changes occur in T_i then the router does not have to report anything and goes back into the PASSIVE state; otherwise, a new LSU is sent to the neighbors and the router immediately goes into the ACTIVE state again. When a router detects that an adjacent link failed, any pending ACKS from this neighbor are considered to have been received. MPDA cannot suffer from deadlocks because all LSUs are acknowledged in a finite time.

MPATH [262, 261] belongs to the class of routing protocols in which distance-vectors are combined with the identity of predecessor nodes, which are just before the destination node on a shortest path. The following information is stored at each node.

1. The *main distance table* contains D_d^i and p_d^i, where D_d^i is the distance of node i to destination d and p_d^i is the predecessor to destination d on a shortest path from node i to node d. The table also stores, for each destination d, the successor set S_d^i, the feasible distance FD_j^i, the reported distance RD_j^i, and two flags, *changed* and *report-it*.
2. The *main link table* T_i is the node's view of the network and contains links represented by (m, n, c) where (m, n) is a link with cost c.
3. The *neighbor distance table* for neighbor j contains D_{jd}^i and p_{jd}^i, where D_{jd}^i is the distance of neighbor j to d as communicated by j, and p_{jd}^i is the predecessor of d on a shortest path from j to d as notified by j.
4. The *neighbor link table* T_j^i is the neighbor j's view of the network as known to i and contains link information.
5. The *adjacency link table* stores the cost l_j^i of the adjacent link to each neighbor j. If a link is down, its cost is infinity.

Each message contains an update entry like (d, c, p), where c is the cost of the node sending the message to destination d, and p is the predecessor node on the path to d. Each message carries two flags: *query* and *reply*. MAPTH is based on PATH, which essentially uses the same update procedures as PDA but differs only in the types of messages exchanged. We skip the details for the sake of brevity, but an interested reader will find the details in [262, 261]. MPATH maintains a routing graph in the form of a link set SG_d for destination d that remains a DAG at every instant. Consequently, the shortest-path trees are also shortest-path multipaths. The routers operate in two states: ACTIVE and PASSIVE. The router only switches from the PASSIVE state, which is a steady state, to the ACTIVE state once it receives a message indicating that its cost to a destination has increased above the previously reported RD_d^i. The node then sends an update message by setting a *query* flag in it. While in the ACTIVE state, a node is allowed to update S_d^i after receiving replies form all its neighbors. The router goes back to the PASSIVE state if none of the distances increased beyond RD_d^i. Otherwise, the router will remain in the ACTIVE state and start the above-mentioned cycle again. When a router detects that an adjacent link failed, an implicit reply with infinite distance to the destination is assumed. This mechanism ensures freedom from deadlocks and that all routers will switch to the PASSIVE state with correct distances to destinations. All of these algorithms follow the classic model of a network routing protocol development: they use non-intelligent link-state packets to collect the information about the costs to neighbors and then propagate them in the complete network. Consequently, they suffer from the same problems: "wrong" or "out-of-order" estimates have a global impact, and the link-state algorithm requires a global system model to execute Dijkstra's shortest-path algorithm. The algorithms are complex and they react slowly to changes in the topologies.

The algorithms discussed in this chapter can be easily classified along two dimensions: route discovery and packet switching. Some algorithms discover routes in a probabilistic manner and some in a deterministic manner, and this holds for packet switching as well. Figure 2.4 classifies the representative algorithms along these lines. The classification of the routing algorithms with respect to the other design axis, introduced in Section 2.2.2, is provided in Table 2.3.

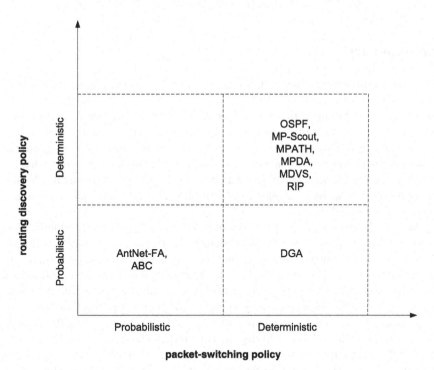

Fig. 2.4. Routing classification

Feature	Routing Algorithms								
	AntNet-FA	ABC	DGA	BeeHive	Q-Routing	MDVA	MPDA	MPATH	OSPF
Static (S) vs. Dynamic (D)	D	D	D	D	D	D	D	D	D
Host-Intelligent (HI) vs. Router-Intelligent (RI)	RI	RI	RI	RI	RI	RI	RI	RI	RI
Single-path (SP) vs. Multi-path (M)	M	S	S	M	M	M	M	M	S
Constructive (Co) vs. Destructive (De)	De	De	De	Co	De	De	Co	Co	Co
Fault-Tolerant	No	No	Yes	Yes	No	Yes	Yes	Yes	Yes
Global (G) vs. Local (L)	L	L	L	L	L	G	G	G	G
Flat (F) vs. Hierarchical (H)	F	F	F	Hybrid	F	F	F	F	F
Loop Freedom	No	No	No	No	No	Yes	Yes	Yes	Yes
Load Balancing	Yes	Yes	No	Yes	No	Yes	Yes	Yes	No
Stigmergy (St) vs. Direct Communication (Dc)	St	St	-	Dc	-	-	-	-	-
Best Effort (B) vs. QoS (Q)	B	B	B	B	B	B	B	B	B

Table 2.3. Classification of routing algorithms for fixed networks

2.6 Summary

The efficient utilization of network resources is becoming an important issue in traffic engineering. One solution to such challenges is to design efficient, decentralized, fault-tolerant, and multi-path routing algorithms which accomplish the task of routing with no access to global topological information. In this chapter we have provided a comprehensive survey of state-of-the-art routing algorithms designed and developed by communities that have different design doctrines. We believe that the survey will be instrumental in initiating an integrated approach to routing in telecommunication networks by allowing cross-fertilization of design principles from different design philosophies.

We have briefly introduced two types of nature-inspired routing algorithms: ACO and evolutionary. The agents in ACO routing algorithms communicate indirectly through the environment (stigmergy) and the agents provide positive feedback to a solution by laying pheromone on the links. Moreover, they have negative feedback through evaporation and aging mechanisms which avoids stagnation. The evolutionary algorithms achieve adaptivity by applying the genetic operators of *crossover, mutation*, and *selection* to their population of agents. We believe that the detailed survey was instrumental in identifying the benefits and shortcomings of the existing state-of-the-art nature-inspired routing algorithms. Consequently, we were able to perform comprehensive requirements engineering for our *BeeHive* algorithm.

3

From *The Wisdom of the Hive* to Routing in Telecommunication Networks

The major contribution of the chapter is a dynamic, simple, efficient, robust, flexible, and scalable multi-path routing algorithm, BeeHive, inspired by the foraging principles of honeybees. The communicative model of bees was instrumental in designing intelligent bee agents, which are suited for large and complex topologies. The results obtained from extensive simulation experiments conclude that bee agents *occupy smaller bandwidth and require significantly less processor time compared to the agents of existing state-of-the-art algorithms. However, even with such simple agents,* BeeHive *achieves similar or better performance compared to state-of-the-art routing algorithms like* AntNet.

3.1 Introduction

The major contribution of the work presented in this chapter is a dynamic, simple, efficient, robust, flexible and scalable multi-path routing algorithm, *BeeHive*, inspired by the wisdom of the hive. We turned this wisdom into an engineering approach, thus allowing ourselves to realize the resulting algorithm in a network stack of Linux.[1] The engineering approach is a result of discussions with network engineers in our system management group who have extensive experience working with Cisco routers. Consequently, we adopted only those features in the algorithm design phase that are easily realizable in a real Linux router.

We set the following challenging requirements for *BeeHive* in order to simplify its implementation in the network stack of a Linux system:

1. *BeeHive* must utilize only forward-moving agents to accomplish the routing task.
2. *Bee agents* must not contain any stack for carrying out their duties.
3. The portion of bandwidth occupied by *bee agents* should be less than 1% of the bandwidth.

---

[1] Linux was chosen because it is an open source free operating system. The availability of source code significantly helped in replacing *OSPF* with *BeeHive* in the routing framework.

4. The algorithm must not maintain any statistical variables to calculate the quality of a link.
5. The time needed to process *bee agents* at a node should be kept to a minimum.
6. The size of a routing table in *BeeHive* must be of the same order as that of *OSPF*.
7. *BeeHive* must be able to scale to large network topologies.
8. *BeeHive* must be able to handle router and link failures in the networks.
9. The performance of *BeeHive* must be at least as good as that of existing state-of-the-art routing algorithms like *AntNet* and *DGA* under high traffic loads, and better than that of *OSPF* under low or static traffic loads.
10. The implementation of *BeeHive* in a network simulator must not use any simulator-specific features that are not available inside the network stack of the Linux kernel.

The motivation behind challenges 1 and 2 is twofold: firstly, the size of an agent is not dependent on the number of hops it makes, and this results in significantly less waste of bandwidth of the network (challenge 3). Secondly, the time to process a stack-less agent at a node turns out to be significantly less than in the presence of a stack. The result of meeting challenges 2 and 4 is of direct impact on the meeting of challenge 5. The smaller routing tables (challenge 6) provide two benefits: they occupy less memory and they can be easily loaded into the cache of a router for efficient packet switching. If *BeeHive* is able to meet challenges 1 to 6, then it is expected to meet challenge 7 automatically.

Our results from extensive simulations clearly demonstrate that the bandwidth *bee agents* occupy is significantly smaller than the one used by *ant agents* in *AntNet*. The time to process agents in *BeeHive* is also significantly less than that of *ant agents* in *AntNet*. Also, the size of the routing tables is significantly smaller than in *AntNet*. However, even with a simpler agent and learning model, *BeeHive* achieves similar or better performance compared with *AntNet*. The advantages of *BeeHive* over *AntNet* become more obvious in larger topologies.

3.1.1 Organization of the Chapter

We will provide a short review of working principles of a honeybee colony in Section 3.2 that will help the reader in comprehending bee behavior and how it differs from ant behavior. It will also assist the reader in understanding the mapping of concepts from nature to networks as described in Section 3.3. We will define our agent model in Section 3.4, based on which we will introduce our algorithm in Section 3.5. In Section 3.6 we will introduce our performance evaluation framework. This constitutes another important contribution of the work presented in this chapter. It is meant as a basis for an unbiased evaluation of the routing algorithms. We will provide a brief description of existing state-of-the-art routing algorithms in Section 3.7, with which *BeeHive* is compared, emphasizing the novel direction of our work. Section 3.8 will describe our simulation environment for the extensive experiments reported in Section 3.9. Finally, we will provide a short summary that concludes the chapter.

3.2 An Agent-Based Investigation of a Honeybee Colony

In this section we briefly outline the organizational principles of a honeybee colony that will enable computer scientists to develop agent-based algorithms for different optimization and real-world problems. A honeybee colony pragmatically solves the most interesting multi-objective optimization problem: how to allocate resources to different tasks under a constantly changing operating environment so that the colony maximizes its profit. The interested reader can find details in [216, 259]. This is the same problem that many researchers try to solve in the field of multi-objective optimization [233, 56, 32] under dynamic and time-varying environments.

3.2.1 Labor Management

A honeybee colony is organized with morphologically uniform individuals but with different temporary specializations. The benefit of this approach is that it enhances a colony's flexibility to adapt its response to an ever-changing environment while at the same time doing the tasks with an acceptable level of efficiency. For example, a nectar forager is able to extensively forage at a discovered flower site as she does not waste time in storing the nectar inside the hive. Moreover, a bee colony is able to adapt the activity level of its specialists according to the group's needs. For example, nectar foragers may become pollen foragers if the amount of protein that they receive from nurse bees falls below a threshold level, or nurse bees might take the role of food-storer bees if the rate of processing nectar is slower than the rate of collecting nectar (foragers indicate this through a tremble dance).

3.2.2 The Communication Network of a Honeybee Colony

A honeybee colony utilizes a hybrid communication network that consists of *signals* and *cues* for information exchange among its members. *Signals are information-bearing actions or structures that have been shaped by natural selection to convey specific information in a unique, wise manner. Cues are variables that likewise convey information about the state of the colony, but they have not been modeled by natural selection to convey that information* [223]. Signals enable direct communication among the members of a honeybee colony via waggle dances, tremble dances, and shaking signals, while cues enable indirect communication among the members through the environment shared by them. Both provide an efficient information exchange mechanism that empowers the members to mostly communicate indirectly (group-to-one paradigm), and when required, directly using the one-to-one paradigm.

A good example of a cue is the search time for finding a food-storer bee that a forager experiences once she wants to unload her nectar. A nectar forager uses this cue to get an estimate of both nectar-collecting and nectar-processing rates of the colony; higher search time to find a food-storer bee is an indicator to the forager that the rate of processing nectar is slower than the rate of collecting nectar. Consequently, she will decide whether to perform a tremble dance instead of a waggle dance. The forager, by doing a tremble dance, will achieve two objectives: one, she will recruit more food-storer bees to increase the rate of processing nectar, and two, she will request other foragers, on the dance floor, to stop performing waggle dances; as a result, the rate of collecting nectar will be decreased.

3.2.3 Reinforcement Learning

A colony experiences a strong fluctuation in the external supply or internal demand (or both) of its commodities: nectar, pollen and water. The feedback signals, both negative and positive, are important to regulate their amount so that the colony has sufficient stockpile of each of these commodities. This is achieved by recruiting unemployed foragers for finding good supply sites through waggle dances. A forager decides to dance only if the quality of the food site visited by her is above a certain threshold if she experiences a very small search time to locate a food-storer bee for her commodity (a cue that the colony needs the commodity). By keeping the search time within certain thresholds, the honeybee colony reinforces the foraging labor at a site in times of need and vice versa. A stochastic model for the foraging behavior has been presented in [217]. Sumpter used this basic model to come up with an agent-based model in [242]. Sumpter's model provides a solid foundation for developing an agent-based reinforcement learning algorithm [130].

3.2.4 Distributed Coordination and Planning

A honeybee colony achieves coordination among its thousands of members without any central authority. The colony does not have a hierarchy where some individuals require information and then allocate tasks to different members and monitor them. Each individual decides to do a job depending upon the need of the colony that it estimates using the above-mentioned communication facilities and measures.

3.2.5 Energy-Efficient Foraging

The foragers tend to optimize the *energetic efficiency* of foraging at a flower site. For example, if during an average foraging trip a forager collects G units of energy, expends C units of energy, and spends time T, then *energetic efficiency* could be defined as $(G - C)/C$. Consequently, the sites that are in the neighborhood of the hive get preference to the sites far away from it [216]. That is why von Frisch believed that (short distance) foragers which return from nearby sites perform round dances while other (long distance) foragers perform waggle dances [259]. This principle enables a colony to collect the commodities at an optimum expenditure of energy.

3.2.6 Stochastic Selection of Flower Sites

The unemployed foragers on the dance floor observe, at the maximum, two or three dancers and then choose one from among them at random. They do not broadly survey the dance floor to identify the best flower site. This concept is contradictory to that of human society where well-informed customers are crucial to proper functioning of competitive markets. According to Seeley, this stems from the fact that whereas the individual human tries to maximize her or his own profit, the individual forager seeks to maximize her colony's profits [216]. This "sacrifice for group principle" enables a colony to distribute its forager force over different flower sites rather than allocating it to the best site only. Such a policy results in quick reallocation of foragers foraging at the best site to other discovered sites once the best flower site is about to fade away.

3.2.7 Group Organization

The employed foragers that collect nectar from the same type of flowers recognize one another in the hive through the flower fragrance that clings to their body. Only group companions respond to the dances, and they show no interest in the dances performed by the foragers of other groups, which have been foraging at other types of flowers. However, the employed foragers might switch to another group if the quality of their flower site degrades to an extent where it is no more profitable to continue foraging at this site [259].

In the next section we will elaborate the most important step of our engineering approach: the mapping of concepts in a honeybee colony, discussed in the current section, to an operating environment of real packet-switching networks. This step will help the reader in understanding our algorithm described in Section 3.4.

3.3 BeeHive: The Mapping of Concepts from Nature to Networks

In this section we briefly illustrate the mapping of concepts from nature to networks, one of the most important steps in *Natural Engineering* that will simplify for the reader tracing back the origin of important features of our *BeeHive* algorithm to the principles of a honeybee colony.

1. We could consider each node in the network as a hive that consists of *bee agents*. Each node periodically launches its *bee agents* to explore the network and collect the routing information that provides the nodes visited with partial information on the state of the network. These *bee agents* can be considered as scouts that explore and evaluate the quality of multiple paths between their launching node and the nodes they visit.
2. *Bee agents* provide the nodes they visit with information on the propagation delay and queuing delay of the paths they explored. These lead to their launching node from the visited nodes. One could consider the propagation delay as distance information, and the queuing delay as direction information (remember that bee scouts also provide these parameters in their dances): this reasoning is justified because a data packet is only diverted from the shortest path to other alternate paths when large queuing delays exist on the shortest path.

3. A *bee agent* decides to provide its path information only if the quality of the path traversed is above a threshold. The threshold is dependent on the number of hops that a *bee agent* is allowed to take. Moreover, the agents model the quality of a path as a function of the propagation delay and the queuing delay of the path; lower values of the parameters result in higher values for the quality parameter.

4. The majority of the *bee agents* in the *BeeHive* algorithm explore the network in the vicinity of their launching node and very few explore distant part of the network. The idea is borrowed from the honeybee colony, resulting not only in reducing the overhead of collecting the routing information, but also in helping maintain smaller and local routing tables.

5. We consider a routing table as a dance floor where the *bee agents* provide the information about the quality of the paths they traversed. The routing table is used for information exchange among *bee agents*, launched from the same node but arriving at an intermediate node via different neighbors. This information exchange helps in evaluating the overall quality of a node (as it has multiple pathways to a destination) for reaching a certain destination.

6. A nectar forager exploits the flower sites according to their quality while the distance and direction to the sites are communicated to it through waggle dances performed by fellow foragers on the dance floor. In our algorithm, we map the quality of paths onto the quality of nodes for utilizing the bee principle. Consequently, we formulate the quality of a node for reaching a destination as a function of proportional quality of only those neighbors that possibly lie in the path toward the destination.

7. We interpret data packets as foragers. Once they arrive at a node, they access the information in the routing tables, stored by *bee agents*, about the quality of different neighbors of the node for reaching their destinations. They select the next neighbor toward the destination in a stochastic manner depending upon its goodness. As a result, not all packets follow the best paths. This will help in *maximizing the system performance although a data packet may not follow the best path*, a concept directly borrowed from a principle of bee behavior: *a bee can only maximize her colony's profit if she refrains from broadly monitoring the dance floor to identify the single most desirable food source* [216] (see Section 3.2).

Now we are in a position to introduce our *bee agent model* and *BeeHive* algorithm in the following sections.

3.4 The *Bee Agent* Model

Our *bee agent* model consists of two types of agents: *short-distance bee agents* and *long-distance bee agents*. Both agents have the same responsibility: exploring the network and evaluating the quality of the paths that they traverse. They only differ in the distance (hops) they are allowed to cover starting from their launching node. *short-distance bee agents* collect and disseminate routing information in the neighborhood of their source node (up to a specific number of hops) while *long-distance bee agents* collect and disseminate routing information typically to all nodes of a network. This helps in collecting the routing information as quickly as possible with a minimum processing and bandwidth overhead.

The *bee agents* that are launched from the same node form an *affinity group* in which they show interest in each other's information. Once the *bee agents* of the same group arrive at the same node, but via different neighbors of the node, they access the routing information, collected by their fellow *bee agents* in the group, in the routing table. They will decide to discontinue their exploration of the network after storing their information in the routing table if one of their members has already arrived at the node. The communication model among *bee agents* is realized as a *blackboard system* [177].

Definition 2 (Foraging region) *The network is organized into fixed partitions called* foraging regions. *A partition results from particularities of the network topology and the number of hops that a* short-distance bee agent *is allowed to traverse. Each* foraging region *has one representative node. Currently the lowest IP address node in a* foraging region *is elected as the representative node. If this node crashes then the next higher IP address node takes over the job.*

Definition 3 (Foraging zone) *A* foraging zone FZ_i *of a node* i *consists of all nodes from which* short-distance bee agents *can reach this node. A* foraging zone *may span over multiple* foraging regions.

The basic motivation behind the two definitions above is to combine the benefits of a flat routing scheme, in which all routers are equivalent, with a hierarchical (cluster) routing scheme, in which cluster heads (or *representative nodes*) have more functions than ordinary routers in the cluster. In our hybrid scheme, each node maintains current routing information for reaching nodes within its *foraging zone* and for reaching the *representative nodes* of *foraging regions*. This mechanism enables a node to route a data packet whose destination is beyond the *foraging zone* of the given node along a path toward the *representative node* of the *foraging region* containing the destination node. Consequently, our algorithm requires routing tables whose size is of the same order as that of *OSPF*; but a *representative node* has no special management functions except launching *long-distance bee agents*. Remember that in a hierarchical (cluster) routing scheme, a packet whose destination is outside the current cluster is sent to the *cluster head*. The cluster head forwards it to the cluster head of the cluster which contains the destination node. Finally the cluster head of the cluster containing the destination node forwards it to the destination node. The concepts of *foraging region* and *foraging zone* provide the benefit of smaller routing tables but without the overhead of routing through cluster heads.

Researchers have proposed a number of algorithms for partitioning a network into clusters [160]. The basic feature of the algorithms is that a cluster should be formed in such a manner that the cluster head (or representative node) should be at the centroid of the cluster. We deliberately did not use the concept of centroid because we wanted to investigate the performance of *BeeHive* without any optimization of the clustering algorithm. Nevertheless, we would like to mention that choosing a representative node based on the centroid concept did not significantly improve the performance of *BeeHive*. The minor improvement in the performance came at a greater cluster management overhead. Therefore, we did not incorporate any optimizations in the *foraging region* formation algorithm. Our results from the extensive experiments clearly demonstrate that *BeeHive* is able to provide similar or better performance compared with existing state-of-the-art routing algorithms without any such optimizations, which clearly is a proof of its robustness. Informally, the *BeeHive* algorithm and its main characteristics can be summarized as follows:

1. All nodes start the *foraging region* formation process during the start-up phase. They try to form a *foraging region* with the same address as their own and consider themselves to be the *representative node* of the *foraging region*. Then they launch the first generation of *short-distance bee agents* to propagate their nomination in their neighborhood.

2. If a node receives a *short-distance bee agent* from a node whose *representative node's* address is smaller than that of the receiving node, then it discontinues its efforts to be a *representative node* and instead joins the *foraging region* of the *representative node* with the smaller address.

3. If a node later on learns that its *representative node* has joined another *foraging region*, as indicated by the *short-distance bee agents* of the *representative node*, then the node starts the same election process explained in Step 1.

4. The nodes keep launching the next generations of *short-distance bee agents* by following Steps 1, 2 and 3 until the network is divided into *foraging regions* and *foraging zones*. Finally, each node informs all other nodes in the network to which *foraging region* it belongs. This step is repeated only if *foraging regions* are reshaped because of link or node failures in the network.

5. At the end of Step 4 the algorithm enters into a "normal" phase (it lasts 30 seconds) in which each nonrepresentative node periodically sends a *short-distance bee agent* by broadcasting replicas of it to each neighbor site.

6. When a replica of a particular *bee agent* arrives at a site it updates the routing information there, and the replica will be flooded again; however, it will not be sent to the neighbor from where it arrived. This process continues until the life-time of the agent has expired, or if a replica of this *bee agent* had been received already at a site, in which case the new replica will be killed.

7. Representative nodes launch only *long-distance bee agents* that would be received by the neighbors and propagated as in 6. However, their life-time (number of hops) is limited by the *long-distance limit*.

8. The idea is that each agent, while traveling, collects and carries path information, and that it leaves, at each node visited, the trip time estimate for reaching its source node from this node over the incoming link. *Bee agents* use priority queues for quick dissemination of routing information.

9. Thus each node maintains current routing information for reaching nodes within its *foraging zone* and for reaching the *representative nodes* of *foraging regions*. This mechanism enables a node, as explained before, to route a data packet (whose destination is beyond the *foraging zone* of the given node) along a path

toward the *representative node* of the *foraging region* containing the destination node.

10. The next hop for a data packet is selected in a stochastic fashion according to the quality measure of the neighbors. The motivation for this routing policy has already been explained in Section 3.3. Note that the routing algorithms currently employed in the Internet always choose a next hop on the shortest path [189].

Figure 3.1 provides an exemplary working of the flooding algorithm. *short-distance bee agents* can travel up to three hops in this example. Each replica of the shown *bee agent* (launched by Node 10) is specified with a different trail to identify its path unambiguously. The numbers on the paths show their costs. The flooding algorithm is a variant of breadth-first search algorithm. The network is partitioned into two *foraging regions* 0 and 8 by following the above-mentioned Steps 1, 2, 3, and 4. The *foraging zone* of Node 10, which spans over both *foraging regions*, consists of Nodes 2, 3, 4, 5, 6, 7, 8, 9, and 11.

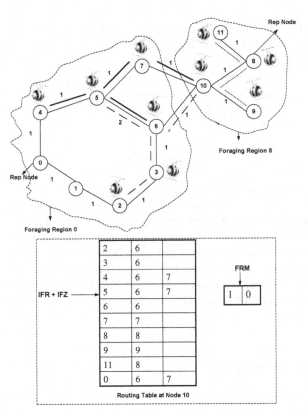

Fig. 3.1. Bee agents' flooding algorithm

3.4.1 Estimation Model of Agents

We now briefly explain the estimation model used by *bee agents* to estimate the trip time required by a data packet to reach its launching node from the current node. In *BeeHive*, the delay t_{in} that a data packet will experience in reaching a neighbor n from node i is modeled as follows:

$$t_{in} = q_{in} + tx_{in} + pd_{in} + pr_i + pr_n, \tag{3.1}$$

where q_{in} is the queuing delay for neighbor n at node i, tx_{in} and pd_{in} are the transmission delay and the propagation delay of the link between node i and neighbor n, respectively, and pr_i and pr_n are protocol processing delays for a packet at node i and node n, respectively. The processing delay, generally speaking, is negligible compared to the sum of queuing, transmission, and propagation delays. Hence Equation (3.1) can be rewritten as

$$t_{in} \approx q_{in} + tx_{in} + pd_{in}. \tag{3.2}$$

In Equation (3.2), tx_{in} models the delay experienced because of the bandwidth of the link, while pd_{in} models the delay that a packet experiences while traveling between two nodes on a link. tx_{in} is dependent on the size of the packet, and the bandwidth of the link between nodes i and n, and pd_{in} is dependent on the distance between nodes i and n. For a particular link, neither of these delays changes with the traffic loads. However, q_{in} is directly dependent on the size of queue, and this in turn depends on the traffic loads. We approximate q_{in} as

$$q_{in} \approx \frac{ql_{in}}{b_{in}}, \tag{3.3}$$

where ql_{in} is the size of the queue (in bits) for neighbor n at node i, and b_{in} is the bandwidth of the link between node i and node n. During the initialization phase, we approximate the bandwidth and propagation delays of all the links of a node by transmitting hello packets. The *bee agents* estimate the queuing delay for a link by observing the size of the queue (ql_{in}) and using Equation (3.3). Then, by adding the transmission and propagation delay of the link to it, they approximate the delay that a packet will experience on the link for reaching neighbor n. Finally, they update the trip time t_{is} that a packet will take in reaching its source node s from current node i:

$$t_{is} \approx t_{in} + t_{ns}. \tag{3.4}$$

3.4.2 Goodness of a Neighbor

The goodness of a neighbor j of node l (l has N neighbors) for reaching a destination d is g_{jd}, defined as follows:

$$g_{jd} = \frac{\frac{1}{p_{jd}}(e^{-\frac{q_{jd}}{p_{jd}}}) + \frac{1}{q_{jd}}(1 - e^{-\frac{q_{jd}}{p_{jd}}})}{\sum_{k=1}^{N}(\frac{1}{p_{kd}}(e^{-\frac{q_{kd}}{p_{kd}}}) + \frac{1}{q_{kd}}(1 - e^{\frac{q_{kd}}{p_{kd}}}))}. \tag{3.5}$$

The fundamental motivation behind Equation (3.5) is to approximate the behavior of the real network. When the network is experiencing a heavy network traffic load, queuing delay plays the primary role in the delay of a link. In this case it is trivial to say that $q_{jd} \gg p_{jd}$ and we can see from Equation (3.5) that $g_{jd} \approx \frac{\frac{1}{q_{jd}}}{\sum_{k=1}^{N} \frac{1}{q_{kd}}}$. When the network is experiencing low network traffic, the propagation delay plays an important role in defining the delay of a link. As $q_{jd} \ll p_{jd}$, from Equation (3.5) we get $g_{jd} \approx \frac{\frac{1}{p_{jd}}}{\sum_{k=1}^{N} \frac{1}{p_{kd}}}$.

The plot of Equation (3.5) is shown in Figure 3.2(a). Once we started developing our engineering model of *BeeHive* inside the network stack of the Linux operating system as discussed in Section 1.3, we realized that the kernel library does not support mathematical functions like exponential, sine, square root, and cosine. We implemented the exponential function ourselves in the Linux kernel, but we observed that the processing complexity of the exponential is an order of magnitude higher than that of one of simple arithmetic functions. This motivated us to look for other forms of modeling the quality function represented in Equation (3.5). The basic challenge was that it should capture the network behavior as discussed in the previous paragraph but it should contain no exponentials. The reason for having no complex mathematical functions like exponential, sine, cosine, square root, etc., is that the quality calculation considerably increases the processing overhead of a routing algorithm; therefore, its processing complexity should be as small as possible. We tried different options and the following form gave similar results as those of the form discussed in Equation (3.5) (the graphical representation of Equation 3.6 is shown in Figure 3.2(b)):

$$g_{jd} = \frac{\frac{1}{p_{jd}+q_{jd}}}{\sum_{k=1}^{N} \left(\frac{1}{p_{kd}+q_{kd}}\right)}. \tag{3.6}$$

We then used our profiling framework (it will be introduced in Section 3.6) to measure the processing complexity (in cycles) of both forms. Table 3.1 summarizes the results. The results are a summary of ten independent runs, and in each run the goodness forms have been evaluated ten million times to get a more representative value. One can easily see that the new form is approximately ten times faster to compute than the one with exponentials. The form with the exponentials takes on average $1,100$ cycles, compared to 105 cycles taken by the non-exponential version. However, this simple form, as the results of our extensive experiments in Section 3.9 demonstrate, has not resulted in any performance degradation compared to the *BeeHive* algorithm presented in [273].

Equations	Max	Min	Av
3.5	1109	1093	1099
3.6	118	96	105

Table 3.1. Processing complexity of different forms

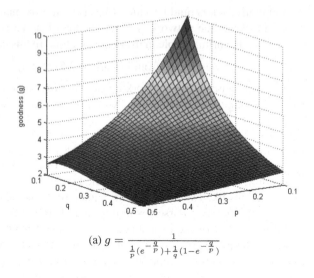

(a) $g = \dfrac{1}{\frac{1}{p}(e^{-\frac{q}{p}}) + \frac{1}{q}(1 - e^{-\frac{q}{p}})}$

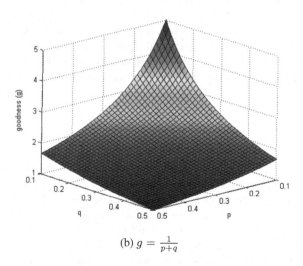

(b) $g = \dfrac{1}{p+q}$

Fig. 3.2. Goodness of a neighbor (different options)

3.4.3 Communication Paradigm of Agents

The forager bees in nature try to exploit different food sources based on their quality, and the distance to the flower site is communicated via a waggle dance. In order to use this bee model, we need to find a mathematical model which assigns an overall quality to neighbor nodes for reaching a destination. The overall quality of a neighbor node can be further represented as a function of delays (propagation, transmission, and queuing) of different paths leading from this node towards the destination. Such a model is realized with the help of a communication paradigm among different replicas of the same agent, which are launched from the same source node. The communication model, in which three paths exist between k and s, is explained in Figure 3.3. Node s launches three replicas of the same agent on three paths, and they arrive through different paths at node k. Each replica uses the estimation model described above to estimate the queuing delay and the propagation delay. The replica that arrived earlier is allowed to continue its exploration further while other replicas are killed. However, the other replicas do communicate their estimates to the replica that is allowed to continue the exploration. Using the communication paradigm, the replica calculates p_{ks} and q_{ks}, which incorporate the estimates of all replicas proportional to the quality of the path they traversed. Once this replica continues its exploration of the network, it tells the other nodes about the existence of a path from k to s through which a packet could reach s with a propagation delay of p_{ks} and queuing delay of q_{ks}. The other nodes forward data packets to node k based on the quality, which is a function of p_{ks} and q_{ks}. Once the data packet is at node k, it can take any one of three paths based on their quality, which is calculated based on the delay estimates of *bee agents*.

3.4.4 Packet-Switching Algorithm

We use *stochastic sampling with replacement* [98] for selecting a neighbor as a next hop towards a particular destination. This principle ensures that a neighbor j with goodness g_{jd} will be selected as the next hop with at least the probability $\frac{g_{jd}}{\sum_{k=1}^{N} g_{kd}}$; or more formally, the probability of taking j as a next hop towards destination d at node i is ϕ_{jd}^i, mathematically represented as

$$\phi_{jd}^i = \frac{g_{jd}}{\sum_{k=1}^{N} g_{kd}}. \tag{3.7}$$

We did not use any rescaling of the probabilities, as done by the authors of *AntNet* because this would increase the processing complexity of packet switching. Remember that any processing during the packet switching also lies on the critical path of a routing algorithm and hence has to be as efficient as possible. Our experiments suggest that the performance of *BeeHive* is the same as or better than the existing state-of-the-art routing algorithms, even without rescaling the probabilities. This shows that the trade-off between the quality of a path decision and its computing time as utilized in *BeeHive* does not harm its performance.

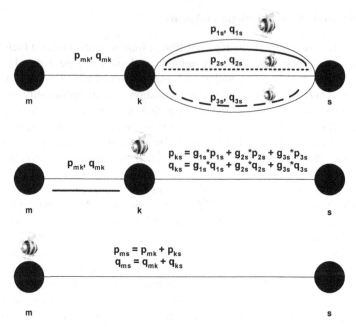

Fig. 3.3. Communication paradigm of *bee agents*

3.5 BeeHive Algorithm

In *BeeHive*, each node i maintains three types of routing tables: *Intra Foraging Zone* (IFZ), *Inter Foraging Region* (IFR), and *Foraging Region Membership* (FRM). The *Intra Foraging Zone* routing table R_i is organized as a matrix of size $(|D(i)| \times |N(i)|)$, where $D(i)$ is the set of destinations in the *foraging zone* of node i, and $N(i)$ is the set of only those neighbors of i which lie on a path towards the destination. Each entry P_{jd} is a pair of queuing delay and propagation delay (p_{jd}, q_{jd}) that a packet will experience in reaching destination d via neighbor j. Table 3.2 shows an example of R_i. In the *Inter Foraging Region* routing table, the queuing delay and propagation delay values for reaching the *representative node* of each *foraging region* through the neighbors of a node are stored. The structure of the *Inter Foraging Region* routing table is similar to the one shown in Table 3.2 where the destination entry is replaced by the representative node of the region containing the destination node. The *Foraging Region Membership* routing table provides the mapping of known destinations to the *representative node* of their *foraging region*. In this way we eliminate the need to maintain O($N \times D$) (where D is total number of nodes in a network) entries in a routing table and save a considerable amount of router memory needed to store it.

Consider Figure 3.1. The total number of entries in all routing tables maintained by Node 10 using the *BeeHive* algorithm is 14. Algorithms 1 and 2 describe the important features of the *BeeHive* algorithm. The important symbols used in Algorithms 1 and 2 are tabulated in Table 3.3.

R_i	$D_1(i)$	$D_2(i)$	$\ldots$	$D_d(i)$
$N_1(i)$	(p_{11}, q_{11})	(p_{12}, q_{12})	$\cdots$	(p_{1d}, q_{1d})
$\vdots$	$\vdots$	$\vdots$	$\ddots$	$\vdots$
$N_n(i)$	(p_{n1}, q_{n1})	(p_{n2}, q_{n2})	$\cdots$	(p_{nd}, q_{nd})

Table 3.2. *Intra foraging zone* routing table

i	Current node
n	Successor node of i
p	Predecessor node of i
s	source node of a packet
d	destination node of a packet
b_s^v	Bee agent of ID v launched by s
b_s^{xv}	Replica x of b_s^v
h_s	Hop limit for bee agents of s
$hops$	Current number of hops
D_{sd}	Data packet launched by s towards d
Fr_c	Foraging region containing node c
FZ_c	Foraging zone of node c
z	Representative node of Fr_s
w	Representative node of Fr_d
qd	queuing delay estimate of a bee agent from p to s
pd	propagation delay estimate of a bee agent from p to s
Short_Limit	Hop limit for short-distance bee agents (7 hops)
Long_Limit	Hop limit for long-distance bee agents
t	current time
t_{end}	Time to end simulation
Δt	Generation interval for bee agents (1 sec)
Δh	Generation interval for hello packets
b_{ip}	Estimated bandwidth of link from i to p
p_{ip}	Estimated propagation delay from i to p
l_{ip}	Size of normal queue at i for p (bits)
n_i	number of packets received at node i
N	number of neighbors of a node
%	Modulo operator
=	Logical comparison
Packet_Limit	packet limit for launching bee agents

Table 3.3. Symbols used in the *BeeHive* algorithm

Algorithm 1 *Bee launching and processing algorithms*

---

procedure launchBeeAgents(s,n_i,n)

    if t % $\Delta t = 0$ or n_i % Packet_Limit = 0 **then**

        if s is representative node of foraging region **then**

            $h_s \leftarrow$ Long_Limit, $\{b_s^v$ is a *long-distance bee agent*$\}$

        else

            $h_s \leftarrow$ Short_Limit, $\{b_s^v$ is a *short-distance bee agent*$\}$

        end if

        if $v = 0$ **then**

            $Fr_s \leftarrow s$ {claim to be the representative node}

        end if

        for $x \leftarrow 1$ to N **do**

            create a replica b_s^{xv} of bee agent b_s^v

            find address of neighbor at index x

            launch replica b_s^{xv} to neighbor at index x

            x++

        end for

    end if

end procedure

procedure manageRegions(s,Fr_s,hops,Fr_i)

    if $Fr_s = s$ && hops $< h_s$ **then**

        if $Fr_s < Fr_i$ **then**

            $Fr_i \leftarrow Fr_s$ {withdraw in favor of lower IP node as a representative node}

            {If the current representative node of i has joined another foraging region}

        else if $Fr_i = Fr_s$ && $Fr_s \neq s$ **then**

            $Fr_i \leftarrow i$ {claim to be the representative node}

        end if

    end if

end procedure

procedure processBeeAgents(b_s^{xv},i)

    if b_s^{xv} already visited i or its hop limit reached **then**

        kill b_s^{xv} {avoid loops}

    else if b_s^{xv} is inside FZ_s **then**

        $qd \leftarrow qd + \frac{l_{ip}}{b_{ip}}$ and $pd \leftarrow pd + p_{ip}$

        update IFZ entries $q_{ps} \leftarrow qd$ and $p_{ps} \leftarrow pd$

        update $qd \leftarrow \sum_{k \in N(i)} (q_{ks} \times g_{ks})$

        update $pd \leftarrow \sum_{k \in N(i)} (p_{ks} \times g_{ks})$

    else

        $qd \leftarrow qd + \frac{l_{ip}}{b_{ip}}$ and $pd \leftarrow pd + p_{ip}$

        update IFR entries $q_{pz} \leftarrow qd$ and $p_{pz} \leftarrow pd$

        update $qd \leftarrow \sum_{k \in N(i)} (q_{kz} \times g_{kz})$

        update $pd \leftarrow \sum_{k \in N(i)} (p_{kz} \times g_{kz})$

    end if

    if b_s^{jv} already reached i $\{\forall j \neq x\}$ **then**

        kill b_s^{xv}

    else

        use priority queues to forward b_s^{xv} to all neighbors of i except p

    end if

end procedure

Algorithm 2 *Packet-Switching and neighbor maintenance algorithms*

---

procedure switchDataPackets(D_{sd},i)

  if d is within FZ_i **then**

    consult IFZ of node i to find delays to node d

    calculate g_{kd}, $\forall k \in N(i)$

  else

    consult FRM of node i to find node w

    consult IFR of node i to find delays to node w

    calculate g_{kw}, $\forall k \in N(i)$

  end if

  probabilistically select a neighbor n ($n \neq p$) as per goodness

  enqueue data packet D_{sd} in normal queue for neighbor n

end procedure

procedure manageNeighbors(i)

  if t % $\Delta h = 0$ **then**

    send a hello packet to all neighbors

    if time out before a response from neighbor {4th time} **then**

      neighbor is down

      update the routing table at i

      launch special bees to inform other nodes in FZ_i

    end if

  end if

end procedure

3.6 The Performance Evaluation Framework for Nature-Inspired Routing Algorithms

We now introduce our performance evaluation framework that we used for an unbiased evaluation of the algorithms presented in Section 3.7. We used the guidelines suggested by Hinningbottom in [114] as well as our discussions with the Cisco network engineers in our system management group for defining the relevant performance parameters of our framework. A prototype version of the performance evaluation framework was introduced in [268]. Its conceptual block diagram is illustrated in Figure 3.4. The framework consists of two input modules: the *topology generator* and the *traffic generator*. The topology generator generates a topology of a given number of nodes and links and assigns a buffer capacity to a router, and the traffic generator enables different traffic patterns through a variation of different input parameters. The traffic generator can generate session-oriented traffic in which all packets of a session have the same destination. This type of traffic tests the behavior of a routing algorithm under congestion. For injecting dynamically changing traffic patterns, we have defined two states: *uniform* and *weighted*. Each state lasts ten seconds and then a transition to another state occurs. In a *Uniform* state (U) a destination

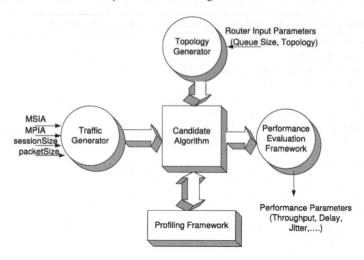

Fig. 3.4. Performance evaluation framework

is selected from a uniform distribution, while in a *Weighted* state (W) a destination selected in the previous *Uniform* state is favored over the other destinations.

$d_u(U) := \theta_d, (0 \le \theta_d \le 1).$

$d_w(W) := \theta_s(d_u) + (1 - \theta_s)\theta_{d'}, (0 \le \theta_{d'} \le 1).$

θ_d represents the probability of selecting node d as the destination of a packet in a session. In a U state, $\theta_d = 1/D$, where D is number of nodes in the network. During a W state, we are using $\theta_s = 0.4$; this means that we favor by 40% the destination from the previous *Uniform* state to be also the destination during the current *Weighted* state. The mechanism above ensures that the traffic patterns can be dynamically generated in an arbitrary fashion in order to represent a decent subset of traffic patterns in real networks. The session-oriented traffic is shaped through the variables *sessionSize, MSIA, and MPIA*. The size of a session is given by the sessionSize variable. MSIA is the mean of session inter-arrival times at a node, and MPIA is the mean of packet inter-arrival times. The session inter-arrival and packet inter-arrival times are taken from negative exponential distributions with mean MSIA and MPIA, respectively. The input parameters and their symbolic representations are shown in Table 3.4.

In sessionless traffic, the destination of each packet is selected from a uniform distribution. This traffic pattern, under low load, simulates static network conditions. Generally, researchers use either session-oriented or sessionless traffic, though we believe that a *good multi-path routing algorithm should be able not only to avoid congestion but also to be competitive under static network loads.*

MSIA	Mean of sessions inter-arrival times (sec)
MPIA	Mean of packets inter-arrival times (sec)
$sessionSize$	The size of a session in bits
D	The number of nodes in a network
l_t	The number of bi-directional links in a network
β_c	Buffer capacity (packets) of routers
δ_l	The size of a packet in bytes

Table 3.4. Input parameter symbols used in the chapter

This profiling framework has been developed to determine the processing complexity of the agents and the data packets for each routing algorithm. An ideal routing algorithm should be able to route data packets as quickly as possible, and the amount of time it spends in processing the agents should be as small as possible. The framework is based on the *rdtsc* machine instruction supported by Pentium architectures [50] that provides cycle-level profiling accuracy. We decided to report the complexity of an algorithm in cycles because this parameter is independent of the frequency of a processor. It is used as a standard parameter by the embedded systems community to report the complexity of an algorithm running on a hardware system.

Our framework keeps on logging the relevant parameters during the simulation, and it finally stores them into a data file. It measures a number of parameters that provide a comprehensive insight into the behavior of an algorithm over a wide range of states of the operating environment. In this way we can evaluate each algorithm in an unbiased manner. The output parameters from the framework and their symbolic representations are shown in Table 3.5.

T_{av}	Average throughput (Mbits/sec)
P_d	Packet delivery ratio (%)
P_{drop}	Packet drop ratio (%)
P_{loop}	Percentage of packets that followed a cyclic path
S_c	Session completion ratio (%)
t_d	Average packet delay (msec)
t_{90d}	90th percentile of packet delays (msec)
S_d	Average session delay (msec)
S_{90d}	90th percentile of session delays (msec)
J_d	Average jitter value (msec)
J_{90d}	90th percentile of jitter times (msec)
q_{av}	Average queuing delay (msec)
A_a	Average agent processing cycles
D_a	Average data processing cycles
A_t	Total agent processing cycles (in billions) per node
D_t	Total agent processing cycles (in billions) per node
R_o	Control overhead
S_o	Suboptimal overhead
h_i^{sd}	hops packet i took to reach from node s to node d
h_o^{sd}	minimum hops needed to reach from node s to node d
h_{av}	Average hops count of the data packets

Table 3.5. Output parameter symbols used in the chapter

Average throughput. Throughput is a measure of how much traffic is successfully received at the intended destination in a unit interval of time [114]. A routing protocol should try to maximize this value.

Packet delay. For all algorithms, we report the average packet delay and the 90th percentile of the packet delays. A good algorithm should be able to deliver packets with minimum delay and with minimum standard deviation of delays. In the rest of the chapter we use the term packet delay for the 90th percentile of packet delays, for brevity.

Session delay. Our network engineers suggested that in the case of session-oriented traffic, the most important parameter is the time needed to complete a session. On the application layer at the destination node one only gets the packets after all packets of a session have been received in the correct order. The packet delay factors out this waiting time and hence favors multi-path routing algorithms, which in general deliver packets out of order but with smaller delays.

Session completion ratio. The percentage of sessions that are able to complete without any support from transport layer protocols. For example, if only one packet in a session is dropped due to congestion or because the time-to-live (TTL) value becomes zero, we report the session as an incomplete one. We believe that this parameter indicates the drop pattern of packets in the face of congestion. Our results substantiate that it is more difficult to maximize this parameter than throughput.

Packet delivery ratio. This measure tells us how much of the data packets are successfully delivered at their destination. Under high loads a 1% improvement in packet delivery ratio at times may mean about a few hundred thousand more data packets delivered at their destination. Again, one cannot observe this improvement via throughput values only.

Packet loop ratio. The percentage of data packets that followed a cyclic path. A cyclic path represents an error in a routing algorithm and should be reported, but we do not have to kill such packets. The motivation for this approach is that a good stochastic routing algorithm must be able to quickly recover from such looping, and if a packet happens to loop infinitely, then it is dropped once its TTL value becomes 0.

Jitter. The jitter is defined as the difference in arrival times of two subsequent packets of the same session, sent from the same node, at their destination node. Let packet p_i^{sd} and p_{i+1}^{sd} be two subsequent packets sent from node s to node d and let t_i and t_{i+1} be their arrival times at destination node d. Now the jitter can be defined as $J_d = t_{i+1} - t_i$ if $t_{i+1} > t_i$, else $J_d = 0$. The jitter is an important parameter in Quality of Service (QoS) routing, especially for streaming multimedia applications. The subsequent packets in video streams should arrive with an acceptable jitter in order to avoid flickering in relaying the video.

Average agent processing cycles. This parameter defines the processing complexity of an agent, say ant or bee. The results show a tendency that can be easily reproduced from one run to another within an acceptable variance. This parameter gives an insight into the complexity of computations and actions an agent performs at a node.

Average data processing cycles. This parameter defines the processing complexity of switching a data packet to its next hop. It creates a considerable overhead for a routing algorithm, and should therefore be as small as possible.

Total agent processing cycles per node. This parameter provides information about the total number of cycles (in billions) that a node spends in processing the agents. We will see the effect of accumulating small differences in average agent processing cycles over a longer period of time.

Total data processing cycles per node. This parameter provides information about the total number of cycles (in billions) that a node spends in processing data packets. We will also see the effect of accumulating small differences in average data processing cycles over a longer period of time.

Control overhead. The ratio of the bandwidth occupied by the routing/control packets to the total available bandwidth in the network [62]. Generally, the authors report this parameter in order to show the control overhead of their routing algorithm.

Suboptimal overhead. This metric was introduced in [210], in the context of MANETs. We believe that it is equally relevant in fixed networks. It is defined as *"The difference between the bandwidth consumed when transmitting data packets from all the sources to destinations and the bandwidth that would have been consumed should the data packets have followed the shortest hop count path."* Formally, we define the parameter as

$$S_o = \frac{\sum_{d=1}^{D} \sum_{s=1}^{D} \sum_{i=1}^{K} (h_i^{sd} - h_o^{sd}) \times L_i^{sd}}{B_t}, \quad s \neq d, \tag{3.8}$$

where D is total number of nodes in the network, K is the total number of packets generated, L_i^{sd} is the length of packet i (in bits) from source s to destination d, and B_t is the total bandwidth of the network. We report this parameter because it implicitly includes the additional bandwidth consumed by data packets when they follow cyclic paths.

3.7 Routing Algorithms Used for Comparison

The focus of our research is on adaptive (flexible) routing algorithms, but we will use *OSPF* in our comparative simulation for the sake of comprehensiveness.

3.7.1 AntNet

AntNet has been described in detail in Chapter 2. In rest of the chapter, we use the name *AntNet-CL* for the version of *AntNet* in which a Forward_Ant agent uses the same queues that data packets use, while *AntNet-CO* corresponds to the version of *AntNet* in which a Forward_Ant agent utilizes priority queues and uses the estimation model described in [65] to estimate its trip time to the neighbor node. In the rest of the chapter, if not otherwise specified, we will refer to *AntNet-CO* as *AntNet* for the sake of the brevity.

3.7.2 DGA

DGA has been described in detail in Chapter 2. The poor performance of *DGA* under high traffic loads, as reported in [273], led us to investigate the problem in more detail. Our research revealed that launching half of the population of agents during the initialization phase resulted in approximately 50% control overhead, which is not tolerable. Therefore, through rigorous testing we empirically reached a value of 12 to 16 ant agents that should be launched during initialization phase, which reduced the control overhead to about 5% without any significant performance degradation. Our experiments also suggest that in this way we have significantly improved the performance parameters of *DGA* compared to those of the original algorithm.

Note that in contrast to the above-mentioned algorithms, the *bee agents*, as discussed in Section 3.1, need not be equipped with a stack to perform their duties. Moreover, our agent model requires only forward-moving agents, and they utilize an estimation model to calculate the trip time from their source to a given node. This model eliminates the need for global clock synchronization among routers, and it is expected that for very large networks routing information can be disseminated quickly with a rather small overhead compared to *AntNet*. Our agent model does not require storing the average trip time, the variance of trip times, and the best trip time for each destination at a node in order to determine the goodness of a neighbor with respect to a particular destination. Last but not least, *BeeHive* works with significantly smaller routing tables compared to *AntNet*.

3.7.3 OSPF

OSPF has been described in Chapter 2. In our implementation we simply take the fixed propagation costs and do not allow them to be changed by the network administrators. As a result, our implementation makes *OSPF* a single-path non-adaptive routing protocol. Such approach was also taken by the authors of *AntNet*.

3.7.4 Daemon

The daemon algorithm is an ideal algorithm in which all routers can have access to a single global instance of the network. They can update the costs of the links of the network based on their current queue lengths. Each router has instant access to the queue length of all other routers, and it applies the Dijkstra algorithm for each routing decision to find the shortest path from the current node to the destination node. The router updates the cost of the link from node i to node j (C_{ij}) after queuing the data packet D_{sd} in its network buffer interface by using the formula $C_{ij} = \frac{L_j^{sd} + l_{ij}}{b_{ij}} + p_{ij}$, where L_j^{sd} is the length of D_{sd} (in bits). The other parameters are defined in Table 3.3. The algorithm cannot be implemented in real routers because of the control overhead required to transmit changes in the queue lengths, and the processing overhead of running Dijkstra's algorithm for each routing decision. Nevertheless, the algorithm provides a benchmark for the comparison of algorithms.

3.8 Simulation Environment for BeeHive

In order to evaluate our algorithm *BeeHive* in comparison with *AntNet*, *DGA*, *OSPF*, and *Daemon*, we implemented all of them in the OMNeT++ simulator [255]. For *OSPF* we implemented a link-state routing that implements the deterministic Dijkstra Algorithm [69]. For *AntNet* and *DGA* we used the same parameters that were reported by the authors in [62] and [147], respectively. *BeeHive*, *OSPF*, and *Daemon* algorithms were given 30 seconds to initialize the routing tables. In comparison *DGA* and *AntNet* were given 50 seconds to initialize the routing tables for the experiments reported in this chapter. Our simulation server was a Fujitsu Siemens machine with a Pentium 4 processor of 3 GHz having 1 Giga byte of main memory and 30 Giga byte of secondary memory. We tested the algorithms on the three network instances simpleNet, NTTNet, and Node150. All reported values for an experiment are an average (μ) of the values obtained from ten independent repetitions (each lasting 1,000 seconds). The performance values of ten independent repetitions varied in the interval ($\mu \pm 5\%\mu$) for a confidence level of 95%.

3.8.1 simpleNet

simpleNet is a small network designed by the authors of [62] to study relevant aspects of a routing algorithm. We designed the experiments on simpleNet in order to study the effect of distributing network traffic loads on multiple paths, an important feature of *BeeHive* and *AntNet*. simpleNet is composed of eight nodes and nine bidirectional links each of 10 Mbits/sec bandwidth and 1 msec propagation delay. The topology is shown in Figure 3.5.

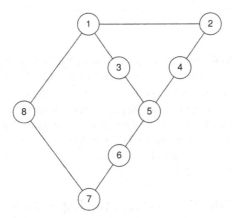

Fig. 3.5. simpleNet

3.8.2 NTTNet

The next network instance that we used in our simulation framework is the Japanese Internet Backbone (NTTNet). It has 57 nodes and 81 bidirectional links. The Link bandwidth is 6 Mbits/sec, while propagation delays range from 1 to 5 msec. Moreover NTTNet it is a non-balanced oblong network with a low degree of connectivity. Hence it puts strong demands on the routing protocols because in a narrowly shaped network, once a packet is forwarded in a wrong direction, it might not have a chance to be routed to the correct destination. NTTNet is shown in Figure 3.6. The advantages associated with the design options adopted in *BeeHive* are expected to become apparent from the topology of this and larger size.

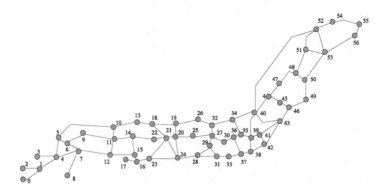

Fig. 3.6. NTTNet

3.8.3 Node150

Our next network, Node150, has 150 nodes and 200 bidirectional links. The link bandwidth is uniformly distributed between 6 and 10 Mbits/sec and the propagation delay is uniformly distributed between 1 and 5 msec. The topology was generated using the BRITE (Boston University Representative Internet Topology Generator) software developed by Medina at the University of Boston. Interested reader will find detailed information about the BRITE topology generator in [158, 156, 157]. The Node150 topology is shown in Figure 3.7.

3.9 Discussion of the Results from the Experiments

3.9.1 Congestion Avoidance Behavior

The purpose of the experiments discussed in this subsection was to study the congestion control behavior of the routing algorithms. The nearly saturated traffic load was

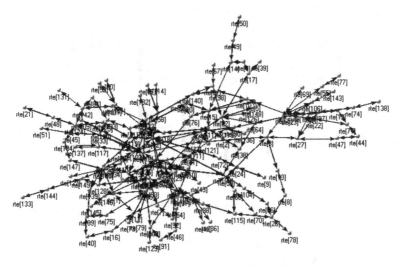

Fig. 3.7. Node150: captured with the OMNeT++ plotter

created by gradually decreasing the value of MSIA from 8.6 sec to 1.6 sec (remember that MSIA and MPIA values are taken from negative exponential distributions). *The performance values of the algorithms in the following bar charts are represented from left to right for the algorithmic legends from top to bottom respectively.*

simpleNet

The performance values of all of the algorithms were approximately the same once we performed the experiments by changing MSIA from 8.6 sec to 1.6 sec, MPIA = 0.005 sec, sessionSize = 2,130,000 bits, δ_l = 512 bytes, and β_c = 1,000 packets. Therefore, we designed a series of special experiments in which we enabled the traffic generators only at Nodes 1 and 7 (see Figure 3.5). We further ensured that all the sessions originating at Node 1 had Node 7 as the destination and viceversa, in order to saturate the queues on selected paths. We increased the sessionSize to 26,000,000 bits and decreased MSIA from 8.6 sec to 2.6 sec while other parameters remained the same as above. These traffic conditions did generate a challenging traffic pattern and showed the advantage of dynamic routing algorithms *BeeHive, AntNet,* and *Daemon* over classical non-adaptive algorithms like *OSPF.* Remember that *Daemon* has complete knowledge about the topology and queue lengths of all routers in it and chooses the shortest path towards the destination, while *BeeHive* and *AntNet* do a stochastic spread of the data packets based on the local information, collected by *bee* or *ant agents* respectively. Figures 3.8, 3.9, 3.10, and 3.11 show the important parameters obtained from the experiments.

Note that as we move close to saturated traffic loads, the packet delivery ratio of *OSPF* and *DGA* starts trailing significantly behind compared to that of *BeeHive, AntNet,* and *Daemon* algorithms. However, Figure 3.9(b) shows that the session completion ratio of *OSPF, DGA,* and *AntNet* is significantly smaller than that of *BeeHive*

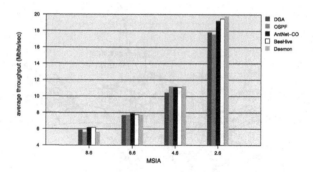

(a) Average throughput

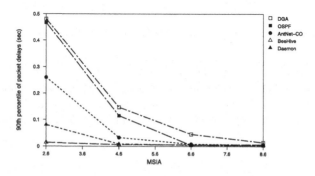

(b) 90th percentile of packet delays

Fig. 3.8. Congestion control behavior in simpleNet (throughput and packet delay)

and *Daemon* algorithms. This confirms our expectation (see Section 3.6): *during congestion, a small difference in packet delivery ratio results in a significantly larger difference in session completion ratio.* Note that *BeeHive* and *Daemon* are able to maintain higher throughput, smaller packet delay, and higher session completion ratio than the other algorithms under all the simulated scenarios.

Figure 3.10 shows that the control overhead and suboptimal overhead of *BeeHive* and *AntNet* are approximately the same. The suboptimal overhead of *Daemon* is the largest because it tries to distribute the packets on all available paths based on its global information about the network; as a result, the average hop count increases to about three hops. The additional fractional increase in h_{av} at extreme saturating loads translates to a significant increase in suboptimal overhead. The control overhead of *DGA* decreases with an increase in traffic load because it employs a genetic algorithm. Remember that in *DGA*, ant agents use the same buffers as data packets,

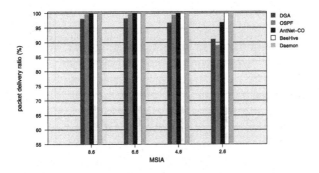

(a) Packet delivery ratio

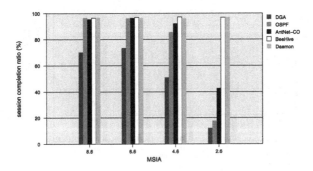

(b) Session completion ratio

Fig. 3.9. Congestion control behavior in simpleNet (packet delivery ratio and session completion ratio)

and next generations of agents are launched once the node receives four agents from the previous generation. Consequently, more agents will be traveling on the network if their trip time is smaller, and this happens in a small topology or under low traffic loads, or both. This explains the sharp increase in the control overhead with an increase in the network traffic.

Figure 3.11 shows the parameters that provide insight into the processing complexity of the agents and the efficiency with which a data packet can be switched. The number of cycles that a node spends in processing *bee agents* is smaller than that for *ant agents* in *AntNet*, and the packet-switching complexity of *BeeHive* is significantly smaller than that of *AntNet*. These are the benefits of employing the simple *bee agents*, and of maintaining small routing tables (see Figure 3.1). We have collected all other important parameters in Table 3.6. The processing complexity of

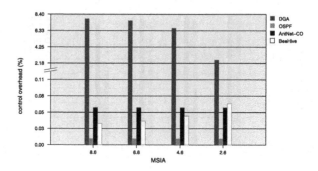

(a) Control overhead (%)

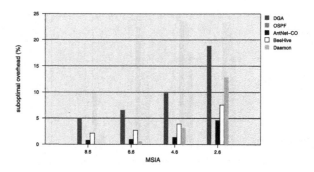

(b) Suboptimal overhead (%)

Fig. 3.10. Congestion control behavior in simpleNet (control and suboptimal overhead)

OSPF directly depends on how efficiently it has been realized. Therefore, to avoid any controversy, we do not report it. Similarly, it makes no sense to report the processing complexity of a benchmark-only *Daemon* algorithm, because it cannot be realized in a real-world router.

NTTNet

We continued our study of congestion control behavior on NTTNet, a relatively complex network topology. We enabled the traffic generators on all the nodes with the following parameters: MPIA = 0.005 sec, sessionSize = 2,130,000 bits, δ_l = 512 bytes, and β_c = 1,000 packets. We decreased the values of MSIA from 8.6 sec to 1.6 sec.

MSIA	Algorithm	t_d	S_d	S_{90d}	P_{loop}	J_d	J_{90d}	q_{av}	h_{av}	A_a	D_a
8.6	DGA	5.8	31736	32255	3.13	2	6	0.15	3.49	27787	4444
	OSPF	3	31732	32243	0	2	5.1	1	2	-	-
	AntNet-CL	3	31738	32262	0.01	2	5.2	0	2.24	69925	7244
	AntNet-CO	3	31719	32219	0.01	2	5.1	0	2.23	71791	7232
	BeeHive	3.75	31741	32250	0	2.7	6.2	0.03	2.61	38141	4077
	Daemon	2.96	31732	32242	0	2	5.1	0.07	2	-	-
6.6	DGA	9.43	31737	32247	2.72	1.6	7.2	0.63	3.51	27960	4186
	OSPF	3	31735	32242	0	1.2	3.9	0.5	2	-	-
	AntNet-CL	3.4	31736	32241	0.01	1.3	3.9	0.1	2.23	70183	7141
	AntNet-CO	3.4	31737	32244	0.01	1.4	3.9	0.2	2.22	71779	7112
	BeeHive	3.78	31724	32235	0	2	5.2	0.03	2.62	38898	4064
	Daemon	3.23	31734	32241	0	1.2	3.9	0.11	2.12	-	-
4.6	DGA	31.62	31753	32267	4.81	1.7	14.4	3.79	3.68	28153	3854
	OSPF	21.2	31742	32249	0	1	2.2	9.5	2	-	-
	AntNet-CL	7.7	31733	32249	0.02	1	4.5	1.8	2.25	70390	6952
	AntNet-CO	6.6	31744	32261	0.01	1	4.6	1.6	2.22	72353	6950
	BeeHive	3.89	31738	32256	0	2	4	0.07	2.62	38657	3979
	Daemon	4.6	31741	32246	0	1	2.3	0.43	2.5	-	-
2.6	DGA	187.74	31739	32297	2.08	1.7	24.6	38.19	3.89	28397	3284
	OSPF	224.5	31768	32300	0	0.8	1	110.9	2	-	-
	AntNet-CL	79.4	31762	32302	0.14	2.2	25.6	31.1	2.45	71802	6727
	AntNet-CO	87.1	31750	32268	0.13	2.1	25.2	34.4	2.43	73728	6715
	BeeHive	5.51	31728	32242	0	1	4.6	0.64	2.69	39250	3789
	Daemon	32.52	31762	32279	0	0.8	1	8.87	3.15	-	-

Table 3.6. Performance parameters for congestion control behavior in simpleNet

MSIA	Algorithm	T_d	S_d	S_{90d}	P_{loop}	J_d	J_{90d}	q_{av}	h_{av}	A_a	D_a
8.6	DGA	469.31	2803	3531	30.44	102.2	699.8	29.97	15.97	52448	8341
	OSPF	38.8	2634	2813	0	5	11	9.2	6.9	-	-
	AntNet-CL	32.6	2630	2791	4.38	6.9	53.3	1	9	75622	14655
	AntNet-CO	32	2630	2792	4.39	6.9	42.8	0.9	8.97	79515	14679
	BeeHive	25.28	2622	2773	2.45	6.7	20.9	0.35	7.74	25109	10825
	Daemon	20.74	2617	2765	0	4.9	10.9	0.1	6.62	-	-
6.6	DGA	587.24	2825	3578	27.61	114.1	784.8	44.42	14.86	50968	8176
	OSPF	75.1	2664	2907	0	5	11	15.8	6.88	-	-
	AntNet-CL	38.6	2637	2818	3.61	7	54.1	1.8	8.92	77933	14745
	AntNet-CO	38.3	2636	2811	3.65	6.9	49.8	1.5	8.82	81876	14819
	BeeHive	26.03	2622	2773	2.56	6.7	20.4	0.4	7.86	25889	11288
	Daemon	22.02	2620	2769	0	4.9	10.9	0.26	6.53	-	-
4.6	DGA	738.29	2850	3582	23.17	113.9	801.8	65.91	13.26	50606	8121
	OSPF	254.3	2736	3108	0	4	10	49.7	6.8	-	-
	AntNet-CL	69.4	2669	2917	2.64	9.2	79.3	5.1	8.76	84837	14977
	AntNet-CO	72.5	2673	2933	2.88	9.3	80.9	5.6	8.7	88373	14974
	BeeHive	30.1	2629	2788	2.5	7	28.6	0.95	7.81	26624	11558
	Daemon	21.51	2620	2768	0	4	10	0.2	6.65	-	-
2.6	DGA	1078.9	2883	3559	15.03	99.3	747.1	121.19	10.45	53028	8183
	OSPF	751.3	2677	2962	0	4	9.9	149.8	6.21	-	-
	AntNet-CL	235.4	2824	3303	2.65	26.6	182.9	23.9	8.82	104061	15220
	AntNet-CO	235.7	2825	3310	2.76	26.1	182.9	23.9	8.83	107924	15249
	BeeHive	125.26	2755	3143	3.79	27.9	162.5	12.67	8.06	29733	11702
	Daemon	45.83	2647	2836	0	5	13	3.56	6.99	-	-
1.6	DGA	1363.84	2861	3382	8.32	69.6	610	190.59	8.1	55964	8317
	OSPF	896.6	2695	2989	0	3	8	212.9	5.53	-	-
	AntNet-CL	987.1	2946	3550	3.03	82	448.3	113.9	8.59	117603	15876
	AntNet-CO	1001.8	2985	3637	3.38	64.4	356.8	112.3	8.77	138650	15824
	BeeHive	721.16	2854	3391	4.48	116.3	551.2	92.19	7.79	108788	11742
	Daemon	666.2	2880	3372	0	10	42	87.64	7.36	-	-

Table 3.7. Performance parameters for congestion control behavior experiments in NTTNet

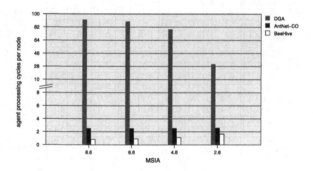

(a) Total agent processing cycles per node (in billions)

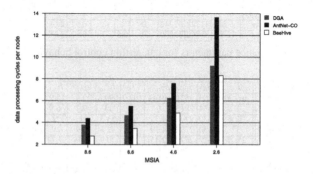

(b) Total data processing cycles per node (in billions)

Fig. 3.11. Congestion control behavior in simpleNet (agent and data processing complexity)

Figures 3.12 and 3.13 show the important performance parameters. One can easily conclude from the figures that *BeeHive*, *AntNet*, and *Daemon* algorithms scale nicely with an increase in the traffic load. The throughput, packet delay, and packet delivery ratio of *BeeHive* and *AntNet* are close to those of *Daemon*. However, the session completion ratio, of all algorithms, as expected, degrades significantly at MSIA = 1.6 sec. Note that *OSPF* and *DGA* are unable to cope with the congestion, but the performance of *DGA* is the worst among those of all algorithms.

From Figure 3.14 it is clear that *BeeHive* has significantly smaller control and suboptimal overhead compared to *AntNet* under all conditions. *OSPF* has the smallest control overhead and suboptimal overhead, but then one should try to complete the picture with other parameters such as throughput, packet delay, and packet delivery ratio, shown in Figures 3.12 and 3.13. An important conclusion from Figure 3.14 is that the suboptimal overhead is an order of magnitude higher than the control over-

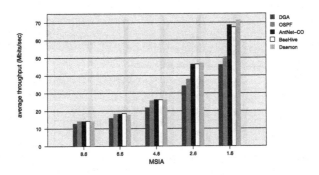

(a) Average throughput

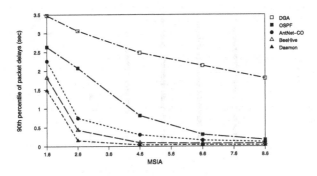

(b) 90th percentile of packet delays

Fig. 3.12. Congestion control behavior in NTTNet (throughput and packet delay)

head, and hence should be taken into account during the design of an adaptive routing algorithm. This parameter can be optimized if an algorithm brings more packets to a destination, in fewer hops, and with a small p_{loop} value. This parameter, to our knowledge, has received little attention in the nature-inspired routing community.

Figure 3.15 shows the agent processing and data packet processing complexities. The simple design of *bee agents*, which is a consequence of no stack processing, no complex mathematical formula evaluation, and only forward-moving agents, now starts showing its benefits because one can see an order of magnitude difference in the total time that a node spends in processing the agents. The processing of the agents in *AntNet* takes approximately 20 billion cycles compared to the five billion cycles taken by the processing of *bee agents*. One can easily note the sharp increase in the processing complexity of *bee agents* to 18 billion cycles at MSIA = 1.6 sec. This is a consequence of a significant increase in the average processing complexity

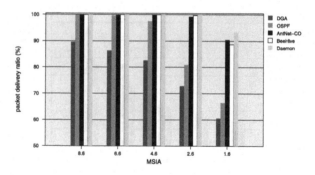

(a) Packet delivery ratio

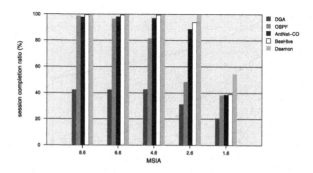

(b) session completion ratio

Fig. 3.13. Congestion control behavior in NTTNet (packet delivery ratio and session completion ratio)

of *bee agents* from 30,000 cycles to 108,788 cycles (see Table 3.7). The increase in the average cycle count appears to be counterintuitive as the actions that agents take remain the same as in the previous cases. We investigated the problem and it appeared that under saturated conditions the event-handling mechanism of OMNeT++ is time-consuming, especially if a packet needs to be flooded. All of the packets except *bee agents* follow point-to-point traversal of the network; therefore, the average processing complexity for *ant agents* and *data packets* remains approximately the same. Our conclusion is that under extremely saturated conditions, especially for MSIA = 2.0 sec and less, the results of the processing complexity might not be adequate because of the event-handling mechanism of OMNeT++. Except from MSIA = 1.6 sec, the average processing complexity of *bee agents* is always between 20,000

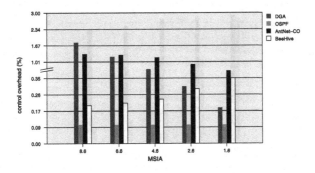

(a) Control overhead (%)

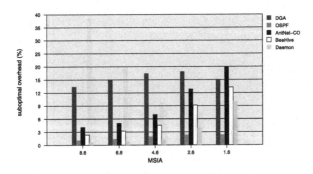

(b) Suboptimal overhead (%)

Fig. 3.14. Congestion control behavior in NTTNet (control and suboptimal overhead)

to 35,000 cycles, which is reasonably acceptable. The other important parameters are collected in Table 3.7.

Note that *DGA* has the same behavior as observed in the simpleNet topology: the control overhead significantly decreases as the network traffic load is increased by decreasing the MSIA. The performance of *DGA* is again the worst. However, our improvements in *DGA* significantly improved its performance compared to the original *DGA* (see Section 3.7). The performance of the original *DGA* is reported in [273].

Node150

The time required to simulate the algorithms on a Node150 network increased exponentially. Therefore, we decided to select the algorithms performing best on the

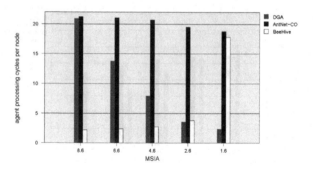

(a) Total agent processing cycles per node (in billions)

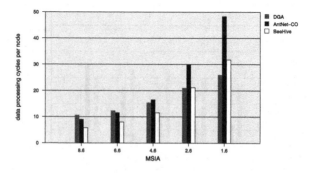

(b) Total data processing cycles per node (in billions)

Fig. 3.15. Congestion control behavior in NTTNet (agent and data processing complexity)

smaller topologies for simulation on the Node150 network. *AntNet-CO* is designed for quick spreading of the routing information; therefore, we did not simulate its counterpart *AntNet-CL* on a Node150 network. The poor performance of *DGA* on small topologies made it an obviously weaker candidate that could be easily dropped. The time needed to simulate *Daemon* on NTTNet was significantly greater than that of the other algorithms; therefore, we conducted only one set of experiments for *Daemon*. We also think that it makes sense to drop *Daemon* from our short list as the algorithm is more or less used as a benchmark, and it is impossible to implement it on any real network because of its communication and processing complexity. We did not drop *OSPF* because it is a widely used routing algorithm in the Internet and we want to always take it as a reference point (the authors of *DGA* did not compare their algorithm with *OSPF* in [147]). We believe that *AntNet-CO* and *BeeHive* are two clear winners from the experiments on smaller topologies and hence it makes perfect sense to compare their performance with one another. Figures 3.16, 3.17,

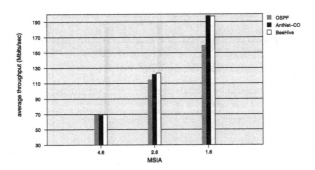

(a) Average throughput

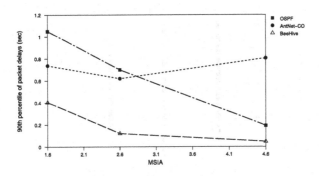

(b) 90th percentile of packet delays

Fig. 3.16. Congestion control behavior in Node150 (throughput and packet delay)

MSIA	Algorithm	T_d	S_d	S_{90d}	P_{loop}	J_d	J_{90d}	q_{av}	h_{av}	A_a	D_a
4.6	OSPF	48.8	2638	2820	0	4	10	16.2	5.4	-	-
	AntNet-CO	111.6	2716	3395	2.46	13.4	188	12.6	8.86	110235	22808
	BeeHive	22.12	2621	2769	0.59	6	15	0.3	5.63	34090	13648
2.6	OSPF	235.6	2663	2888	0	4	9.2	70.6	5.34	-	-
	AntNet-CO	109.2	2710	3134	1.73	14.6	146.6	11.6	8.12	112666	23120
	BeeHive	36.77	2642	2817	0.6	7	35.8	2.88	5.62	34399	13399
1.6	OSPF	384	2668	2891	0	3	8	107.6	5.18	-	-
	AntNet-CO	223.2	2848	3324	1.66	30	221.4	24.8	7.9	119439	23894
	BeeHive	129.38	2732	3016	0.62	18.2	113.6	19.38	5.64	47451	13212

Table 3.8. Performance parameters for congestion control behavior in Node150

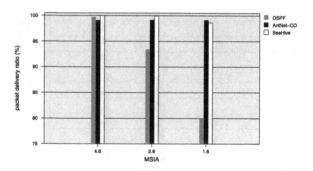

(a) Packet delivery ratio

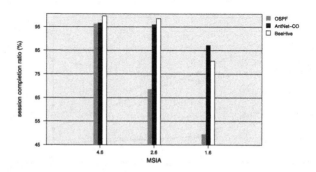

(b) session completion ratio

Fig. 3.17. Congestion control behavior in Node150 (packet delivery ratio and session completion ratio)

3.18 and, Figure 3.19 show the same behavior of *AntNet* and *BeeHive* as in the previous congestion control experiments. Both algorithms are able to deliver more packets and complete more sessions with an increase in the network traffic load compared with *OSPF*. Note that both the packet delay and the session delay of *BeeHive* are the smallest among the three algorithms. As expected, *OSPF* is not able to scale to increasing network traffic.

The benefit of collecting routing information in small *foraging zones* around a node in *BeeHive* becomes more apparent as one looks at Figure 3.18(a). The control overhead of *BeeHive* is significantly smaller than that of *AntNet*, and the feature of routing packets in fewer hops is manifesting its benefits in Figure 3.18(b). This significant difference in suboptimal overhead is due to the facts that a smaller number of data packets in *BeeHive* follow cyclic paths (see Table 3.8), and that *BeeHive*

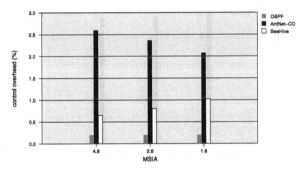

(a) Control overhead (%)

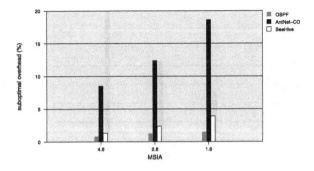

(b) Suboptimal overhead (%)

Fig. 3.18. Congestion control behavior in Node150 (control and suboptimal overhead)

delivers data packets at their destination in fewer hops. One can easily conclude that the suboptimal overhead of *BeeHive* is now approaching that of *OSPF*.

Figure 3.19 shows the processing complexity for agents and data packets. The simple behavior of *bee agents*, as discussed previously, is now showing significant benefits. The total time that a node spends in processing *bee agents* is approximately one fifth of its time processing *ant agents*. The reasons for a significantly smaller suboptimal overhead of *BeeHive*, discussed in the previous paragraph, are also valid for the smaller packet-switching complexity of *BeeHive* compared to *AntNet*.

Note that Node150 has links of 6–10 Mbits/sec bandwidth. As a result, the saturation of queue buffers at MSIA = 1.6 sec is not as visible as it was in the NTTNet experiments. Therefore, once we look at the average processing complexity of *bee agents*, it is around 47,000 cycles. This further strengthens our previous findings that an exponential increase in the average bee agent processing complexity in NTTNet

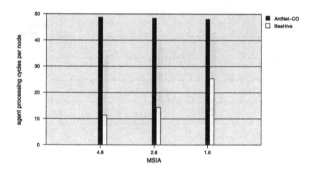

(a) Total agent processing cycles per node (in billions)

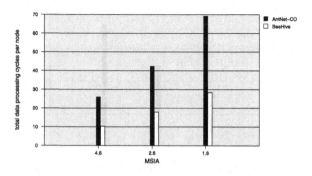

(b) Total data processing cycles per node (in billions)

Fig. 3.19. Congestion control behavior in Node150 (agent and data processing complexity)

at MSIA = 1.6 sec stems from specifics of the OMNeT++ simulator, and not from the *BeeHive* algorithm. *BeeHive* has clearly manifested its advantages on a Node150 network over *AntNet*, and we believe that the benefits will be even more apparent on larger topologies. We have collected other important performance parameters from these experiments in Table 3.8.

3.9.2 Queue Management Behavior

The purpose of these sets of experiments was to study the queue control behavior of the routing algorithms. We believe that it is important to know how the algorithms scale to different sizes of queue buffers. This is important for two reasons: one, to get an idea about an optimal queue buffer size for achieving the best performance, and two, to investigate the benefits that multi-path routing algorithms can bring on networks that have small devices like Personal Digital Assistants (PDAs) with limited main memory. Moreover, achieving better performance with small buffer capacities is a desirable property of any routing algorithm. During these experiments we kept MSIA = 2.6 sec, MPIA = 0.005 sec, sessionSize = 2,130,000 bits, δ_l = 512 bytes and varied the buffer capacity β_c from 50 packets to 4,000 packets.

simpleNet

One cannot see a significant difference between performance values in the simpleNet topology. Therefore, we skip the results.

NTTNet

The behavior of the algorithms during these experiments is summarized in Figures 3.20 and 3.21. One can see from Figure 3.20(a) that *BeeHive* is able to deliver about 2% more packets than *AntNet* for smaller buffer capacities; however, *AntNet* is able to catch up with *BeeHive* at a queue size of 1,000 packets. Note that 2% more packet delivery ratio compared to *AntNet* at β_c = 50 results in about 15% more sessions completed (see Figure 3.21(a)), which is a significant improvement and shows the superiority of *BeeHive* over *AntNet* for low buffer capacities. Figure 3.20(b) shows the real shortcoming of classical non-adaptive algorithms like *OSPF*, where an increase in buffer capacity does not result in any significant increase in packet delivery ratio or session completion ratio. This shows that *OSPF* is unable to manage higher traffic loads because of its lack of queue management. The performance of *DGA* improves with an increase in buffer capacity but is far inferior to *BeeHive* or *AntNet*. *BeeHive* has better scalability compared to all other algorithms except *Daemon*. On the other hand the *Daemon* algorithm has the best performance among all the algorithms. Another important observation from Table 3.9 is that the jitter value significantly increases for *BeeHive* and *AntNet* with an increase in buffer capacity. The reason is obviously that both algorithms stochastically spread data packets on multiple paths. As a result, subsequent packets might follow different paths. The difference in arrival time at destination increases with an increase in buffer capacity, and this connects the increasing jitter behavior with an increase in buffer capacity. The average agent processing complexity (A_a) of *bee agents* sharply increases at smaller buffer capacities, and the reason is similar to that for some OMNeT++-related behavior discussed in the previous Section 3.9.1.

We have collected other important performance parameters in Table 3.9.

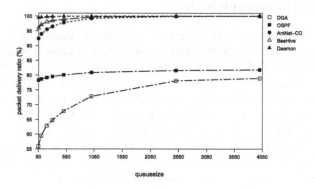

(a) Packet delivery ratio (%)

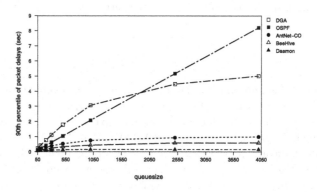

(b) 90th percentile of packet delays (sec)

Fig. 3.20. Queue management/control behavior of algorithms (packet delivery ratio and packet delay)

Node150

We decided to skip the buffer capacity experiments for Node150 network because experiments conducted on the NTTNet network provided good insight into this behavior. As we have already discussed, small buffer capacities create a more relevant problem on mobile devices like PDAs. All modern routers can easily support a queue length of 1,000 or more packets because of advancements in VLSI technology that lead to cost-effective production of memory chips.

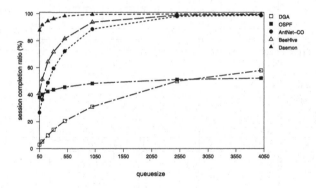

(a) Session completion ratio (%)

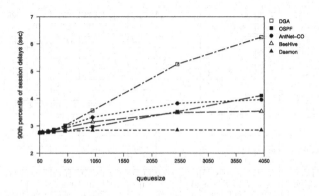

(b) 90th percentile of session delays (sec)

Fig. 3.21. Queue management/control behavior of algorithms (session completion ratio and session delay)

3.9.3 Hot Spots

The purpose of these experiments was to study the behavior of the algorithms in scenarios in which one node starts attracting bursts of traffic for a short period of time. The situation is quite common when some broadcasting channel breaks some news and everybody starts accessing its Website; then this Website is acting as a hot spot in the network. A good adaptive routing algorithm should be able to manage and cope with hot spots in the networks.

β_C	Algorithm	T_{av}	t_d	S_d	R_o	S_o	P_{loop}	J_d	J_{90d}	q_{av}	h_{av}	A_a	D_a
50	DGA	26.14	83.34	2602	2.51	12.74	11.48	11.9	62.9	6.2	9.58	56073	8884
	OSPF	36.75	48.3	2611	0.1	2.19	0	4	9	9	6.11	-	-
	AntNet-CL	43.5	48.7	2613	0.89	9.8	1.52	6	22.6	3	8.17	95167	15178
	AntNet-CO	43.26	49.8	2613	0.96	9.79	1.43	6.2	22.6	3	8.19	103831	15220
	BeeHive	45	38.43	2616	0.29	8.2	3.02	9	24.6	2.01	7.87	92375	11974
	Daemon	46.86	26.43	2622	-	3.38	0	4	10	0.85	6.87	-	-
100	DGA	27.84	147.1	2608	1.7	14.51	12.75	18.6	112.4	13.12	10.03	54950	8602
	OSPF	36.92	83.2	2612	0.1	2.21	0	4	9	16	6.13	-	-
	AntNet-CL	43.96	69.7	2621	0.89	10.41	1.74	8	36.8	5.8	8.31	97459	15200
	AntNet-CO	43.79	70.6	2623	0.95	10.46	1.76	8	37.2	5.8	8.32	105144	15230
	BeeHive	45.25	48.47	2622	0.29	8.3	3.15	11	38.3	3.3	7.88	75369	11970
	Daemon	46.92	28.59	2624	-	3.43	0	4	10	1.16	6.88	-	-
200	DGA	29.49	268.39	2627	1.06	15.81	13.57	29.5	201.8	26.85	10.25	56699	8422
	OSPF	37.15	154.6	2616	0.1	2.26	0	4	9	30.9	6.16	-	-
	AntNet-CL	44.7	106.1	2641	0.89	11.29	2.01	11.9	65.3	9.8	8.5	100047	15246
	AntNet-CO	44.6	108.2	2642	0.93	11.36	2.06	12	64.4	9.9	8.54	106533	15229
	BeeHive	45.93	66.46	2637	0.29	8.63	3.47	14.9	63.9	5.56	7.92	63308	11917
	Daemon	46.99	32.48	2627	-	3.57	0	4	11	1.71	6.91	-	-
300	DGA	30.28	380.92	2647	0.78	16.49	13.97	40.9	292.4	39.66	10.38	56091	8360
	OSPF	37.33	227.5	2621	0.1	2.28	0	4	9	45.8	6.18	-	-
	AntNet-CL	45.19	134.7	2664	0.89	11.79	2.18	14.9	87.3	13	8.61	101292	15285
	AntNet-CO	45.03	137.2	2666	0.92	11.89	2.29	15.2	88.6	13.2	8.65	107807	15235
	BeeHive	46.18	81.21	2655	0.29	8.74	3.48	18	86.3	7.41	7.95	56092	11893
	Daemon	47.03	36.46	2631	-	3.84	0	5	11	2.24	6.97	-	-
500	DGA	31.75	601.35	2701	0.52	17.26	14.24	60.1	456.3	64.92	10.44	54618	8318
	OSPF	37.59	377	2635	0.1	2.32	0	4	9	75.7	6.2	-	-
	AntNet-CL	45.82	177.4	2712	0.89	12.33	2.37	19.7	125.8	17.9	8.71	102764	15275
	AntNet-CO	45.8	179.8	2715	0.91	12.41	2.53	19.9	125.1	17.9	8.74	107597	15215
	BeeHive	46.45	104.33	2686	0.29	8.98	3.66	22.8	123.3	10.26	8	42240	11855
	Daemon	47.07	42.75	2639	-	3.89	0	5	12	3.13	6.98	-	-
1000	DGA	34.23	1078.9	2883	0.31	18.1	15.03	99.3	747.1	121.19	10.45	53018	8193
	OSPF	37.96	751.3	2677	0.1	2.35	0	4	9.9	149.8	6.21	-	-
	AntNet-CL	46.36	235.4	2824	0.89	12.82	2.65	26.6	182.9	23.9	8.82	104088	15230
	AntNet-CO	46.3	235.7	2825	0.91	12.88	2.76	26.1	182.9	23.9	8.83	107259	15135
	BeeHive	46.42	125.26	2755	0.29	9.12	3.79	27.9	162.5	12.67	8.06	29752	11705
	Daemon	47.1	45.83	2647	-	3.93	0	5	13	3.56	6.99	-	-
2500	DGA	36.64	1708.99	3533	0.19	16.34	14.63	128	946.2	221.17	9.74	63386	8148
	OSPF	38.3	1869.9	2808	0.1	2.38	0	4	10	366.8	6.22	-	-
	AntNet-CL	46.72	290.1	2988	0.9	13.09	2.77	33.3	247.6	29.8	8.86	103685	15143
	AntNet-CO	46.7	276.5	2973	0.9	13.21	2.93	31	234.3	28.2	8.89	106544	15042
	BeeHive	46.58	152.38	2846	0.29	9.24	4	33.4	219.3	15.98	8.08	26875	11605
	Daemon	47.1	45.92	2649	-	3.93	0	5	13	3.57	6.99	-	-
4000	DGA	37.03	1949.31	3958	0.16	14.86	13.26	122.8	943.8	272.37	9.31	73649	8162
	OSPF	38.41	2960.8	2934	0.1	2.39	0	4	10	577.1	6.22	-	-
	AntNet-CL	46.92	289.3	3012	0.9	13.23	2.85	33.5	247.1	29.6	8.88	104007	15124
	AntNet-CO	46.87	289.3	3009	0.9	13.26	2.94	32.4	244.7	29.7	8.9	106661	14993
	BeeHive	46.95	151.38	2855	0.29	9.18	3.87	33.1	221.5	15.92	8.04	26719	11562
	Daemon	47.1	45.92	2649	-	3.93	0	5	13	3.57	6.99	-	-

Table 3.9. Performance parameters for different buffer capacities in NTTNet

simpleNet

All algorithms were able to cope with the hot spot traffic in simpleNet; therefore, we are skipping the results.

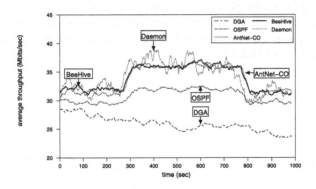

(a) Average throughput

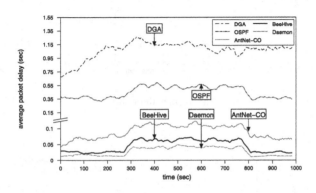

(b) Average packet delay

Fig. 3.22. Hot spot is Node 0 in NTTNet

NTTNet

We made Node 0 in Figure 3.6 act as a hot spot from 300 seconds to 800 seconds, which means that all nodes in the network sent data packets to this node with MPIA = 0.04 sec. This hot spot traffic was superimposed on a normal network traffic of MSIA = 3.6 sec, MPIA = 0.005 sec, and sessionSize = 2,130,000 bits. The other parameters are δ_l = 512 bytes and β_c = 1,000 packets. Figure 3.22 summarizes the results. It is clear from the figure that *BeeHive*, *AntNet*, and *Daemon* algorithms are able to cope with additional network traffic resulting from a hot spot. However, the packet delay of *BeeHive* is 50% less than that of *AntNet*. *OSPF* and *DGA* are unable to cope with hot spot traffic, and *DGA* has the worst performance among all algorithms. The other important parameters from the experiments are collected in Table 3.10. One can easily conclude that in almost all performance aspects *BeeHive* is better than *AntNet*. *Daemon*, as expected, has the best performance parameters (due to its access to the global network state).

Node150

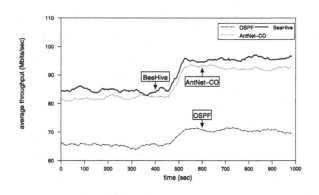

(a) Average throughput

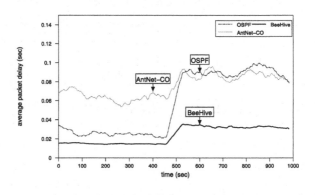

(b) Average packet delay

Fig. 3.23. Hot spot is Node 0 in Node150

All input parameters for hot spot experiments on Node150 remained the same as in the experiments conducted on NTTNet except for the start and end times. In Node150, the hot spot (Node 0) was active from 500 seconds to 1,000 seconds. From

Figure 3.23 it is evident that *BeeHive* is able to maintain higher throughput and a significantly smaller packet delay than *AntNet* throughout the experiment. Note that the time during which the hot spot is not active, packet delay of *AntNet* is approximately 20 msec greater than that of *OSPF*. The reason is that at MSIA = 3.6 sec no significant congestion resulted in the Node150 network. As a result, distributing packets on all possible paths, as *AntNet* does, is not a promising approach. We will explain this behavior of *AntNet* later under the sessionless traffic experiments. As expected, *BeeHive* and *AntNet* are again able to cope better with the hot spot traffic compared with *OSPF*. The other important parameters are collected in Table 3.10. Note that on this relatively bigger topology, performance parameters of *BeeHive* are significantly better than those of *AntNet*. The reason for this was explained during the discussion of congestion control experiments.

3.9.4 Router Crash Experiments

The purpose of these experiments is to study the fault-tolerant behavior of different algorithms. The experiments provide insight into how quickly an algorithm adapts its routes if a router crashes. Two features are of primary importance to handle fault tolerance in networks: one, a routing algorithm should be able to reroute packets on alternate paths toward their destination when an existing path is no more available, and two, once a router is repaired, its routing table should adapt quickly in order to start routing packets as quickly as possible. We do not report experiments for simpleNet, as one cannot see a significant difference in performance among different algorithms and consequently we lack the resources to complete all experiments on Node150; therefore, we conduct only important experiments on Node150. We dropped *OSPF* from the fault-tolerant experiments because it has significantly poor performance compared to *AntNet* and *BeeHive* without any router crash. We are including *DGA* for the sake of completeness; otherwise, we could have dropped it easily because it performed worst among all algorithms.

NTTNet

We report two experiments in which we analyzed the fault-tolerant behavior of the algorithms.

Experiment 1. In Experiment 1 the traffic generator parameters were MSIA = 4.6 sec, MPIA = 0.005 sec, sessionSize = 2,130,000 bits, δ_l = 512 bytes, and β_c = 1000 packets. In this experiment Nodes 20 and 43 (see Figure 3.6) crashed at 500 seconds and then remained down for the rest of the experiment. From Figure 3.6 it is clear that routers 20 and 43 are critical in NTTNet. One can easily conclude from Figure 3.24 that *BeeHive* and *Daemon* are able to maintain significantly higher throughput compared to *AntNet*. The packet delay of *Daemon* is significantly smaller compared to those of *BeeHive* and *AntNet*; however, *BeeHive* is able to maintain approximately half as much the packet delay as is *AntNet*. We have collected other important parameters in Table 3.10. Note that all performance parameters of *BeeHive* are significantly better than the *AntNet*. The packet delivery ratio of *BeeHive*

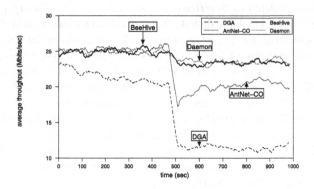

(a) Average throughput

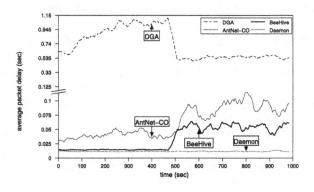

(b) Average packet delay

Fig. 3.24. Node 20 and Node 43 crashed at 500 seconds

is approximately the same as that of *Daemon*, which shows that *bee agents* are able to adapt routes once the routers crashed and *AntNet*, as expected, is not able to cope with the router crash problem.

Experiment 2. All of the traffic generator's parameters were the same in this experiment as in the previous experiment except the router crash times. In this experiment, Node 20 crashed at 300 seconds, Node 43 crashed at 500 seconds, and both routers were repaired at 800 seconds. The throughput and packet delay of algorithms are shown in Figure 3.25. The tendency of the algorithms is quite similar to that of the one shown in Figure 3.24. *BeeHive* is able to maintain higher throughput and lower packet delay compared to *AntNet*. All other important parameters are collected in Table 3.10. *Daemon* is the best performing algorithm among all algorithms

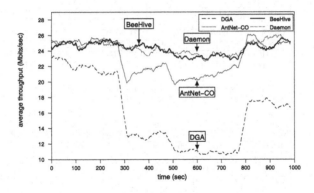

(a) Average throughput

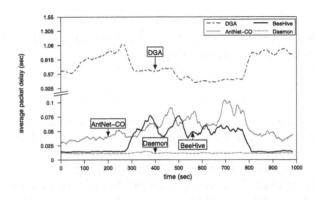

(b) Average packet delay

Fig. 3.25. Node 20 crashed at 300 seconds and Node 43 crashed at 500 seconds and both were repaired at 800 seconds

while *DGA* is the worst performing algorithm. Note that once Node 43 crashes, multiple paths still exist via Node 40 to the upper part of the network topology; therefore *BeeHive* is able to deliver more packets than *AntNet*.

3.9.5 Bursty Traffic Generator

The purpose of this set of experiments was to study the behavior of the algorithms under bursty network traffic. Such traffic consists of sudden bursts of packets followed by a long period of silence and inactivity (no traffic). Such a scenario is important because it investigates how quickly an algorithm can react to changes in network

Experiment	Algorithm	P_d	S_c	S_d	S_{90d}	R_o	S_o	P_{loop}	q_{av}	h_{av}	A_a	D_a
NTTNetHot	DGA	74.17	39.08	2865	3570	0.4	16.75	16.38	95.19	11.43	49484	8121
	OSPF	87.96	57.04	2721	3098	0.1	2.44	0	118.4	6.75	-	-
	AntNet-CL	98.06	93.18	2724	3052	1.71	20.07	2.51	19.9	9	105139	14740
	AntNet-CO	97.97	93.22	2727	3063	1.76	20.38	2.62	20.9	9.05	109717	14687
	BeeHive	98.29	96.09	2667	2898	0.27	7.09	2.98	10.84	8.21	31181	11632
	Daemon	98.22	97.11	2624	2777	-	2.43	0	8.63	6.98	-	-
Node150Hot	OSPF	95.19	89.69	2640	2825	0.2	0.84	0	5.47	5.4	-	-
	AntNet-CO	98.66	94.43	2709	3203	2.45	10.7	2.19	14.6	8.53	112412	22553
	BeeHive	99.55	97.73	2633	2803	0.73	2.17	0.62	7.54	5.77	36148	13606
NTTNetExp1	DGA	67.13	39.12	2795	3434	0.41	10.75	14.84	76.96	11.34	47548	7733
	AntNet-CL	92.02	79.54	2679	2960	1.63	10.76	2.12	9.1	8.21	82438	14572
	AntNet-CO	92.71	79.58	2690	2994	1.63	10.98	2.41	11.4	8.25	85599	14537
	BeeHive	99.67	92.48	2669	2960	0.22	5.02	7.19	6.03	8.09	27003	11447
	Daemon	100	96.14	2620	2771	-	1.35	0	0.39	6.68	-	-
NTTNetExp2	DGA	64.57	39.07	2800	3455	0.3	9.32	12.82	85.71	10	45750	7628
	AntNet-CL	93.73	81.44	2686	2977	1.65	11.46	2.39	9.5	8.32	82947	14630
	AntNet-CO	94.17	81.54	2686	2974	1.68	11.29	2.59	9.6	8.3	85699	14557
	BeeHive	99.7	93.15	2669	2962	0.22	5.11	7.46	5.83	8.11	26998	11592
	Daemon	100	96.84	2621	2772	-	1.32	0	0.43	6.66	-	-

Table 3.10. Performance parameters for hot spot and router down experiments

traffic. We skip the results for simpleNet as the performance of all algorithms were approximately the same for bursty traffic.

NTTNet

One can generate bursty traffic patterns using our traffic generator by decreasing the value of MPIA. In the experiments we decreased the value of MPIA from 0.005 sec to 0.0005 sec, which made a session to finish in one tenth of the time (260 msec) compared to when MPIA = 0.005 sec. However, we kept the value of MSIA constant at 2.6 sec. This resulted in our sending 2,130,000 bits (size of a session) in 260 msec followed by an inactivity period of approximately two seconds. Other parameters were δ_l = 512 bytes and β_c = 1,000 packets. One can see in Figure 3.26 that all algorithms are able to deliver approximately the same number of packets at MPIA

MPIA	Algorithm	P_d	t_d	S_c	S_d	S_{90d}	R_o	S_o	P_{loop}	J_d	J_{90d}	h_{av}	A_a	D_a
0.005	DGA	72.81	1079	31.19	2883	3559	0.31	18.1	15.03	99.3	747.1	10.45	53028	8183
	OSPF	80.86	751	48.39	2677	2962	0.1	2.35	0	4	9.9	6.21	-	-
	AntNet-CL	99.17	235	88.18	2824	3303	0.89	12.82	2.65	26.6	182.9	8.82	104061	15220
	AntNet-CO	99.2	236	88.68	2825	3310	0.91	12.88	2.76	26.1	182.9	8.83	107924	15249
	BeeHive	99.69	125	93.78	2755	3143	0.29	9.12	3.79	27.9	162.5	8.06	29733	11702
	Daemon	99.99	46	99.51	2647	2836	-	3.93	0	5	13	6.99	-	-
0.0005	DGA	70.91	1322	37.23	1019	1901	0.28	16.5	14.19	103.1	772.8	10.14	52567	7562
	OSPF	79.92	910	51.01	721	1297	0.1	2.34	0	1	3	6.23	-	-
	AntNet-CL	96.68	655	76.39	1113	1968	0.77	12.4	3.02	62.4	358.6	8.77	111788	14600
	AntNet-CO	97.26	634	79.66	1111	1967	0.83	12.74	3.21	59.5	343.1	8.82	112325	14592
	BeeHive	98	474	83	989	1714	0.29	9.28	4.19	86.2	372.1	8.08	31041	10255
	Daemon	99.58	447	96.92	883	1488	-	13.18	0	14	75	8.86	-	-

Table 3.11. Performance parameters for bursty traffic generators in NTTNet

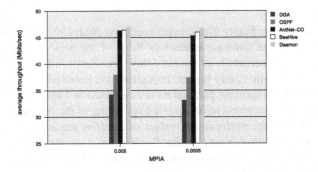

(a) Average throughput

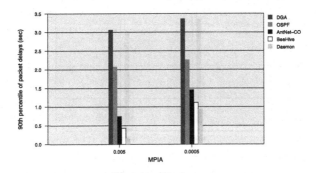

(b) 90th percentile of packet delays

Fig. 3.26. Bursty traffic behavior in NTTNet

= 0.0005 sec compared to MPIA = 0.005 sec but with a significantly greater packet delay.

The packet delay of *BeeHive* is quite close to *Daemon* and significantly lower than that of *AntNet*. All other important parameters are collected in Table 3.11. Note that all performance parameters of *BeeHive* are significantly better than those of *AntNet*. We again see the same tendency that *Daemon* is the best performing algorithm while *DGA* is the worst performing algorithm. The better performance of *Bee-Hive* is due to its quick spreading of routing information by utilizing only forward-moving agents (see Section 3.4).

Node150

We continued with the bursty traffic experiments on Node150. The traffic generator at a node received the same parameters as those of the previous experiment. The throughput and 90th percentile of packet delays are shown in Figure 3.27. *BeeHive* is able to maintain significantly higher throughput and lower packet delay compared to *AntNet*. All other important parameters are collected in Table 3.12. Note that the advantage of quick communication-oriented spreading of the routing information is now more visible as the performance values of *BeeHive* are an order of magnitude better than those of *AntNet*. Note also that *BeeHive* has a session delay of 1,000 msec, suboptimal overhead of 2.5%, and average hop count of 5.6 compared to 1,631 msec, 12.6%, and 8.2 hops for *AntNet*, respectively.

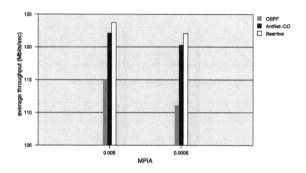

(a) Average throughput

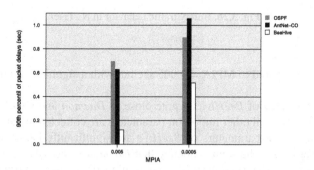

(b) 90th percentile of packet delays

Fig. 3.27. Bursty traffic behavior in Node150

3.9.6 Sessionless Network Traffic

MPIA	Algorithm	P_d	t_d	S_c	S_d	S_{90d}	R_o	S_o	P_{loop}	J_d	J_{90d}	h_{av}	A_a	D_a
0.005	OSPF	93.23	236	68.08	2663	2888	0.2	1.26	0	4	9.2	5.34	-	-
	AntNet-CO	99.16	109	95.89	2710	3134	2.36	12.42	1.73	14.6	146.6	8.12	112666	23120
	BeeHive	99.94	37	98.45	2642	2817	0.81	2.37	0.6	7	35.8	5.62	34399	13399
0.0005	OSPF	90.47	393	69.71	628	1002	0.2	1.21	0	1	2	5.32	-	-
	AntNet-CO	98.18	410	85.41	921	1631	2.25	12.62	1.97	43.2	291	8.21	117440	22950
	BeeHive	99	218	91.44	651	999	0.81	2.5	0.7	22.6	129	5.66	36870	11862

Table 3.12. Performance parameters for bursty traffic generators in Node150

The purpose of these experiments was twofold; first to test the algorithms in a domain where the performance of *OSPF* is the best, and second to gain confidence that our implementation of *DGA* is functionally correct. The authors of *DGA* only published their results under a sessionless network traffic with MPIA = 0.035 sec. The poor performance of *DGA* at higher loads, reported in the previous sections, required us to verify its functional correctness. One option is to use the same traffic patterns that the developers of *DGA* utilized and then compare the results. The improvements that we made in *DGA* to make it competitive with other algorithms might have introduced some undesired side effects. However, our results from such extensive studies further strengthened our belief that our improved version of *DGA* was still functionally similar to that of the original *DGA*. We skip reporting the results for simpleNet because the results do not show any significant difference in performance among the algorithms.

NTTNet

We used the sessionless traffic generator, in which the destination for each packet is chosen at random from a uniform distribution. We gradually decreased the load to a static level by increasing the MPIA from 0.005 sec to 1 sec. The idea was to test the feasibility of the algorithms under static conditions where *OSPF* is a state-of-the-art algorithm. At MPIA = 0.005 sec, i.e., under a significantly high load, *BeeHive* and *AntNet* completely outperformed *DGA* and *OSPF*, but as the load started to decrease, the packet delivery ratio of *OSPF* (see Figure 3.28) significantly improved. Note that *AntNet* drops about 2–3% of the packets at low loads. This came as a surprise to us. We investigated the problem and it appeared that stochastic distribution of packets over too many multiple paths, a desired property under high network traffic, is not a promising approach under static network loads. As a result, more data packets followed loops, as shown by the p_{loop} parameter in Table 3.13, and the packets that were dropped were those that had followed about 100 hops and had not arrived at the destination. The results in Figure 3.28 clearly demonstrate the superiority of *OSPF* over all other algorithms under low loads.

Closely monitor the performance values of *BeeHive* in comparison to those of *OSPF*. It appears that maintaining only those paths towards a destination whose quality is above a threshold value (bee behavior) results in striking a good compromise

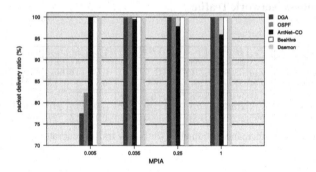

(a) Packet delivery ratio

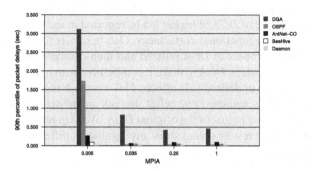

(b) 90th percentile of packet delays

Fig. 3.28. Sessionless network traffic in NTTNet

between *AntNet*, which maintains all possible multiple paths towards a destination, and *OSPF*, which maintains a single path towards the destination. All other important performance parameters are collected in Table 3.13. Our claim is verified by the performance parameters collected in Table 3.13. The performance of *BeeHive* is similar to or better than that of *AntNet* at higher loads (MPIA = 0.005 sec) and similar to that of *OSPF* under low loads. Compare the values of the p_{loop} and h_{av} parameters under a low network traffic load (MPIA = 1.0 sec).

Node150

We used the same traffic generator parameters as in the previous experiment on the Node150 network, and the tendency of the performance parameters were approximately the same. It is clear from Figure 3.29 that *AntNet* now drops approximately

MPIA	Algorithm	T_{av}	t_d	R_o	S_o	P_{loop}	q_{av}	h_{av}	P_{drop}	A_a	D_a
0.005	DGA	36.23	1046.57	0.31	23.21	17.37	112.55	11.65	22.33	48351	6913
	OSPF	38.46	627.6	0.1	2.4	0	112	6.23	17.66	-	-
	AntNet-CL	46.74	35.5	0.79	9.83	1.08	1	8.19	0.01	113576	15599
	AntNet-CO	46.74	35.3	0.78	9.45	1	1	8.11	0.01	118669	14525
	BeeHive	46.74	33.78	0.3	8.83	3.23	1.32	7.98	0.01	25614	11268
	Daemon	46.74	22.03	-	2.44	0	0.27	6.66	0	-	-
0.035	DGA	6.67	150.64	4.39	14.1	36.95	2.08	26.69	0.17	51028	9345
	OSPF	6.68	20	0.1	0.51	0	10	6.9	0	-	-
	AntNet-CL	6.68	24	0.75	1.27	0.93	0	8	0.02	116385	15868
	AntNet-CO	6.68	24	0.75	1.19	0.87	0	7.88	0.02	123341	15132
	BeeHive	6.68	23.03	0.19	1.16	2.66	0.01	7.84	0	25046	11144
	Daemon	6.68	19.83	-	0.29	0	0.01	6.58	0	-	-
0.25	DGA	0.93	87.67	4.98	2.17	35.25	0.04	28.67	0.01	51629	12036
	OSPF	0.94	20	0.1	0.07	0	12	6.9	0	-	-
	AntNet-CL	0.93	25.5	0.74	0.24	3.15	0	8.6	0.11	115419	16285
	AntNet-CO	0.93	25.3	0.74	0.23	3.15	0	8.5	0.11	122168	16097
	BeeHive	0.93	22.48	0.17	0.15	1.76	0	7.66	0	25409	11130
	Daemon	0.94	19.76	-	0.04	0	0	6.58	0	-	-
1	DGA	0.23	86.45	5.04	0.54	32	0.02	28.56	0.06	49657	13465
	OSPF	0.23	20	0.1	0.02	0	12	6.9	0	-	-
	AntNet-CL	0.23	29.9	0.72	0.1	8.77	0	10.08	0.33	116211	16990
	AntNet-CO	0.23	30	0.72	0.1	8.72	0	10	0.34	123457	16959
	BeeHive	0.23	22.09	0.17	0.04	1.21	0	7.5	0	25508	11178
	Daemon	0.23	19.74	-	0.01	0	0	6.58	0	-	-

Table 3.13. Performance parameters for sessionless traffic in NTTNet

5–6% of the packets and has a significantly higher packet delay compared to *BeeHive* and *OSPF* under low loads. All other performance parameters are collected in Table 3.14. The performance values of *BeeHive* are again quite close to those of *AntNet* under high loads and quite close to *OSPF* under low loads. Now compare the p_{loop} and h_{av} values of all algorithms. One can see that about 19% of the packets enter into loops in *AntNet* compared to 0.7% for *BeeHive* for MPIA = 1.0 sec. Consequently, the average hop count increases to 24 hops for *AntNet*. These results show that for large topologies and for extremely low network traffic, routing tables in *AntNet* do not converge; as a result, 5–6 % of packets keep on looping in the networks. We believe that a threshold-based exploitation of paths, such as the bee behavior, has been instrumental for *BeeHive* in getting a performance quite close to *OSPF* under low loads.

MPIA	Algorithm	T_{av}	t_d	R_o	S_o	P_{loop}	q_{av}	h_{av}	P_{drop}	A_a	D_a
0.005	OSPF	118.13	224.4	0.2	1.33	0	73	5.36	3.95	-	-
	AntNet-CO	122.98	31	2.09	10.57	1.14	0	7.65	0.02	119621	24701
	BeeHive	123.03	22.64	0.81	2.7	0.91	0.32	5.7	0	31311	13813
0.035	OSPF	17.57	19	0.2	0.21	0	8	5.4	0	-	-
	AntNet-CO	17.5	33	2.08	2.03	2.14	0	8.57	0.38	116840	27320
	BeeHive	17.57	20.76	0.5	0.38	0.89	0.01	5.69	0	31029	14422
0.25	OSPF	2.46	19	0.2	0.03	0	8	5.4	0	-	-
	AntNet-CO	2.39	55.2	2.05	0.74	8.36	0	14.47	3.03	114867	29997
	BeeHive	2.46	20.49	0.46	0.05	0.78	0	5.65	0	31102	14366
1	OSPF	0.62	18.8	0.2	0.01	0	8	5.4	0	-	-
	AntNet-CO	0.57	91.2	2	0.36	19.16	0	23.87	6.89	112407	31462
	BeeHive	0.62	20.33	0.45	0.01	0.7	0	5.61	0	31421	14830

Table 3.14. Performance parameters for sessionless traffic in Node150

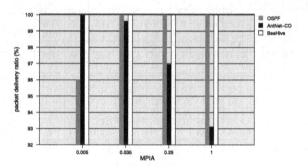

(a) Packet delivery ratio

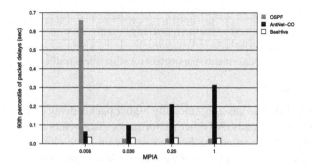

(b) 90th percentile of packet delays

Fig. 3.29. Sessionless network traffic in Node150

3.9.7 Size of Routing Table

The size of routing tables utilized by *AntNet*, *BeeHive*, and *OSPF* is shown in Figure 3.30. The benefits of the *foraging zone* and *foraging region* concepts become clear for bigger and complex topologies. For NTTNet, *AntNet* has 162 entries on average, in comparison to 78 and 57 for *BeeHive* and *OSPF* respectively. For Node150 *AntNet*, *BeeHive*, and *OSPF* have 400, 194, and 150 entries respectively in the routing table. The number of entries for *BeeHive* are a sum of the number of entries for all three routing tables maintained by *BeeHive*.

Figure 3.30 demonstrates the clear advantage of the way route discovery and maintenance is done in *BeeHive*. The concepts of *foraging zone* and *foraging region* result not only in smaller routing tables but also in a smaller routing overhead compared to *AntNet*. We believe that for topologies larger than 150 nodes, *BeeHive* will maintain significantly smaller routing tables compared to *AntNet*.

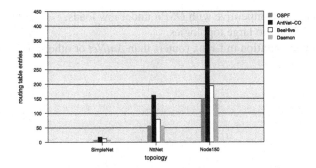

Fig. 3.30. Size of routing table

3.10 Summary

A honeybee colony is able to optimize its stockpiles of nectar, pollen, and water through an intelligent allocation of labor among different specialists, who communicate with each other using a sophisticated communication protocol that consists of *signals* and *cues* in continuously changing internal and external environments. The dance language and foraging behavior of honeybees inspired us to develop a dynamic, simple, efficient, robust, flexible, and scalable multi-path routing algorithm. The algorithm does not need any global information such as the structure of the topology and the cost of links among routers; rather, it works with the local information that a *short-distance bee agent* collects in a *foraging zone*. BeeHive does not utilize an optimized clustering algorithm to avoid the overhead of routing through cluster heads. It works without the need for global clock synchronization; this not only simplifies its installation on real routers but also enhances fault tolerance. In contrast to *AntNet*, our algorithm utilizes only forward-moving *bee agents* that help in disseminating the state of the network to the routers in realtime. The *bee agents* take less than 1% of the available bandwidth but provide significant enhancements in throughput and packet delay over other state-of-the-art approaches.

We implemented two state-of-the-art nature-inspired algorithms (*AntNet* and *DGA*) for the OMNeT++ simulator and then compared our *BeeHive* algorithm with them. Through extensive simulations representing dynamically changing operating network environments, we have demonstrated that *BeeHive* achieves better or similar performance compared to *AntNet*. However, this enhancement in performance is achieved with following preferred features:

- Significantly smaller routing tables of the order of the size as in *OSPF*.
- Simple agents, which resulted in significantly less control overhead compared to *AntNet*.
- Simple agents also resulted in significantly less processing complexity compared to *AntNet*.
- No infinite loop problems, which *AntNet* has, under static network traffic.

- Performance is competitive with *OSPF* under low loads.
- Better scalability to large topologies.
- Easier implementation in Linux routers than *AntNet* or other protocols.

4

A Scalability Framework for Nature-Inspired Routing Algorithms

The major contribution of the work presented in this chapter is a comprehensive framework for the scalability analysis of distributed routing protocols. The framework models the productivity of a routing protocol on a number of performance values. The cost model consists of processing, communication, and resource costs for deploying a routing algorithm in an operational environment in real-world networks. The framework is general enough and can be easily utilized to investigate the scalability of an agent-based distributed software system. Finally, we studied the scalability behavior of two state-of-the-art nature-inspired agent-based routing protocols: AntNet and BeeHive. We also evaluated the scalability of OSPF, which is currently employed in the Internet. The results give valuable insight into the scaling capacity of agent-based routing algorithms. We believe that the work will motivate designers of routing protocols to consider scalability as an important metric in the design and development of state-of-the-art routing protocols and to empirically validate them according to this measure.

4.1 Introduction

A routing protocol has to be deployed on large-scale telecommunication networks. Therefore, not envisaging the repercussions of scalability on the performance of a routing protocol might lead to severe performance bottlenecks. Consequently, the Web applications which utilize the networking systems running such protocols might show poor responsiveness, resulting in customer dissatisfaction and loss of valuable revenues in a highly competitive market [230, 229]. Scalability means the ability of a routing protocol to efficiently transport the network traffic between any pair of the nodes with adequate quality of service, over a wide range of network configurations. The increased performance should be in proportion to additional costs [127]: processing, communication, and router's resources.

The major factors that might undermine the scalability of a routing protocol are: excessive consumption of the router's resources and the routing complexity [295].

In order to counter these problems, designers of the routing protocols have to be equipped with analytical and empirical tools to systematically investigate the scalability behavior of a routing algorithm. In this chapter, we present for a routing protocol a new scalability framework, which has the following new features:

- It provides a comprehensive cost model which incorporates the processing, communication, and router's resource costs.
- It utilizes a recently proposed concept of *total overhead* in [210] for calculating the communication costs rather than supporting the existing practice of taking only the control overhead.
- It provides a quality of service value which depends on a number of refined and processed parameters rather than only on delay.
- It utilizes a refined throughput metric for a routing algorithm rather than the raw throughput.
- It defines power and productivity metrics for a routing algorithm on the basis of the above-mentioned parameters, which provide significant insight into its benefit-to-cost ratio.
- It defines a scalability matrix, which facilitates the scalability analysis. The matrix enables a designer to study the scalability of a routing algorithm either across different topologies by fixing the network traffic load, or on the same topology by increasing the network traffic load.
- It uses a new, comprehensive empirical performance evaluation framework that collects the relevant performance values required for the scalability framework.

We believe that our new framework will enable the designers of routing protocols to establish their scalability in an early stage of *protocol engineering* [140]. Such a framework will be instrumental in practicing the principles of *Software Performance Engineering* (*SPE*), which also emphasizes the consideration of performance and scalability issues early in the design and architectural phase [230, 231, 289, 229], to rectify the deficiencies in a simulation environment. This will not only obviate the risk of a disaster once the algorithm is deployed on large-scale networks, but also avert the cost overruns due to tuning or redesigning the algorithm later in the protocol engineering cycle. Consequently, such a pragmatic protocol engineering cycle will be capable of reducing the time to market of a new protocol.

The framework is general enough to act as a guideline for analyzing the scalability of any agent-based network system. However, in this chapter, we limit our analysis to two state-of-the-art agent-based routing algorithms: *AntNet* and *BeeHive*. We will also analyze the scalability of a classical non-adaptive routing protocol,*OSPF*. We now briefly discuss the existing work on scalability analysis. We will point out considerable lack of concrete work on comprehensive scalability analysis for routing protocols. This should facilitate to emphasize our novel direction of work.

4.1.1 Existing Work on Scalability Analysis

Scalability analysis has received extensive treatment in the area of parallel computing. Here, the focus is to analyze the scalability of a parallel algorithm on massively parallel platforms. The important metrics are: *speedup, efficiency, and scalability* [243]. Speedup measures how the rate of doing work increases with an increase in the number of processors as compared to one processor. Efficiency is the work rate per processor, and scalability is defined as the ratio of efficiencies on two platforms. Ideally, efficiency and scalability metrics have a value of unity while speedup is k if k processors are added from one configuration to another. The interested reader will find detailed treatment of the scalability of parallel computing systems and algorithms in [99, 104, 227, 142, 211].

The work on analyzing the scalability of distributed systems is inspired by the work of Giessler et al. [97], in which they proposed a *power metric* for computer network systems. The metric is defined as $P = \frac{T_{av}}{tr}$, where T_{av} is the average throughput of the system and t_r is the ratio of the average packet delay to the minimum packet delay. Kleinrock extended this definition in [139] to combined loss and delay systems, and was able to define the optimal operating point at which the defined power is maximized. In [138], he discussed the effect of flow control procedures on the throughput in computer networks. Earlier work of Kuemmerle and Rudin presented in [143] is also of paramount importance because the authors compared the performance of circuit-switched networks with packet-switched networks, considering delay performance and usage cost. Our cost model utilizes some of the definitions presented in their work. In [197], Rosner concluded that packet-switched networks outperform circuit-switched networks with respect to both transmission utilization efficiencies and overall network costs. However, the major focus of these studies does not relate to comparing the efficiencies of routing protocols, which is the focus of our work.

In [125, 126, 127], Jogalekar and Woodsite somehow misinterpreted t_r (introduced in the last paragraph) for the average packet delay. As a result, they thought that the power is significantly influenced by smaller packet delays. Therefore, they modified the definition of power for distributed systems as

$$P(k) = \frac{\lambda(k)}{1 + \frac{T(k)}{T'}}, \tag{4.1}$$

where $\lambda(k)$ is defined as the throughput, $T(k)$ is the average response of the system, T' is an acceptable response time for the user, and k is the scalability parameter, i.e., number of nodes or users. The metric is useful in studying the scalability of distributed systems because the basic objective of scaling up a distributed system is to support more throughput for a fixed response time or to reduce the response time for a fixed throughput, or a combination of both [125]. In order to incorporate costs, the authors defined the performance of distributed systems as

$$F(k) = \frac{\lambda(k)}{(1 + \frac{T(k)}{T'}) \times C(k)}, \tag{4.2}$$

where C is the cost. Finally, they defined a p-scalability metric $\Psi_p(k_1, k_1')$ that defines the scalability of a distributed system from configuration k_1 to k_1' as

$$\Psi_p(k_1, k_1') = \frac{F(k_1')}{F(k_1)}. \tag{4.3}$$

In [126], the authors have shown their view of the scalability behavior of a distributed system as illustrated in Figure 4.1. The authors arbitrarily suggested that a

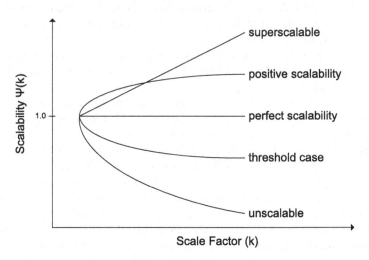

Fig. 4.1. The scalability behavior in different situations [126]

distributed system is scalable if $\Psi_p(k_1, k_1') > 0.8$, where 0.8 is a threshold value case to reflect an acceptable benefit-to-cost ratio. However, as might be expected, the focus of this work was on distributed systems and not on distributed algorithms or routing protocols.

To our knowledge, the work presented in [38, 37] is the first preliminary treatment of the scalability of an agent-based distributed routing protocol. The authors used equation (4.1) and argued that the throughput and delay of a routing protocol are functions of the number of links (L) and the average number of hops (h_{av}) needed to reach a destination in a particular topology. Therefore, the power of a routing algorithm, as described in [38, 37], becomes

$$P(k) = \frac{L}{h_{av}}. \tag{4.4}$$

Consequently, the power has now become a function of the topology rather than of the algorithm. In their cost model, they simply take the control overhead into consideration. Subsequently, they defined the performance of a routing protocol F as

$$F(k) = \frac{L}{C(k) \times h_{av}}, \tag{4.5}$$

where C is the control overhead. Finally, they defined a scalability metric $\Psi(k_1, k_1')$ for a routing protocol as

$$\Psi(k_1, k_1') = \frac{\frac{L_2}{C(k_1') \times h_{av2}}}{\frac{L_1}{C(k_1) \times h_{av1}}}, \tag{4.6}$$

where L_2, $C(k_1')$, and h_{av2} are the number of links, control overhead, and average hops, respectively, in topology k_1', and L_1, $C(k_1)$, and h_{av1} are the number of links, control overhead, and average hops, respectively, in topology k_1. They took these values from the work of Di Caro and Dorigo [62] and then concluded that *AntNet* is scalable from NSFNet to NTTNet because $\Psi(k_1, k_1') > 1.0$. However, their treatment completely lacked an empirical framework.

In [232], the authors have emphasized the need for an empirical simulation environment that incorporates a real scalable environment for performance evaluation of distributed algorithms. In [133], Katz and Yung have proposed a scalable protocol for authenticated key exchange. They defined four complexity metrics for a network protocol: *round, message, communication, and computational*. The *round complexity* is simply the number of rounds until the protocol terminates. The *message complexity* is the total number of messages sent (regardless of their length) by all parties in the course of protocol execution. The *communication complexity* is the total number of bits communicated throughout the execution of the protocol; this now includes the message length as well. *Computation complexity* is the maximum amount of computation done by any player in the protocol. Our cost model is based on some of their ideas as well.

In [52], Costa et al. have proposed a QoS routing algorithm, *Single Mixed Metric* (SMM), and have done its scalability analysis using a simulation environment. In [210], Santivanez et al. introduced a new metric called *total overhead* for a routing protocol that is defined as follows: *total overhead induced by a routing protocol is the difference between the amount of bandwidth actually consumed by the network running such a protocol minus the minimum traffic load that would have been required should the nodes have had full topology information a priori* [210]. This metric includes the control overhead, the bandwidth occupied by the control messages, and the suboptimal overhead, resulting from extra hops the packets took in excess of the ideal minimal hop number. However, the focus of this work was to analytically study the asymptotic scalability with respect to a scalable parameter. Nevertheless, the *total overhead* is a valuable concept in analyzing the communication cost of a routing protocol.

In [290], Woodside proposed a scalability metric for analyzing the scalability of mobile agent systems. The work is an extension of his earlier work in scalability of distributed systems reported in [125, 126, 127]. He evaluated the new model on a class of mobile agents, using basic and robust models for the workload and delay. However, he completely ignored the cost in his analysis.

4.1.2 Organization of the Chapter

We will introduce our scalability model in Section 4.2, and a description of the simulation environments is presented in Section 4.3. Section 4.4 provides a comprehensive description of the results obtained from extensive experiments on a set of

topologies varying in size and complexity. We will then discuss the empirical results obtained from our scalability model in Section 4.5 and then comment on the
scalability of the algorithms. Finally, we provide a summary of the chapter.

4.2 The Scalability Model for a Routing Algorithm

In a sense, a distributed routing algorithm is a distributed communication system
with the objective to optimize throughput and reduce the packet delay. However, the
algorithm has to achieve this objective with minimal cost. We could, therefore, define
the power of a routing algorithm as a function of a number of performance values,
which are collected by our comprehensive performance evaluation framework. In the
following subsections we will define the cost model, power model, and scalability
model of a routing algorithm.

4.2.1 Cost Model

A distributed routing algorithm is a piece of software that has many costs associated with it. The costs that we will consider in our scalability model are: *processing,
communication, and router resources.* We will ignore the costs related to the development, implementation, testing, installation, and maintenance of a routing algorithm
that arise in a network of real-world routers. Instead we will consider the cost of one
important resource of a router: memory. Now, we define a few parameters that will
influence our cost model:

- *Control processing ratio* C_p is the ratio of the number of cycles that a node spends
  in processing the control packets to the number of bits that have been delivered
  in the network.
- *Data processing ratio* C_d is the ratio of the number of cycles that a node spends
  in switching the data packets to the number of bits that have been delivered in
  the network.
- *Processing ratio* C_β is the ratio of the control processing ratio to the data processing ratio, i.e., $C_\beta = \frac{C_p}{C_d}$.
- *Total overhead* C_t is the sum of the control overhead and suboptimal overhead,
  which have been defined in Chapter 3.
- *Bandwidth ratio* C_w is the ratio of the total extra bits, generated due to the total
  overhead, to the total number of bits delivered at their destination.
- *Memory overhead* C_m is the cost associated with storing routing tables in the
  memory of a router.

C_p and C_d define the processing complexity with respect to the number of bits delivered at their destination, while C_β defines the relative processing overhead of control
packets with respect to the packet switching. The motivation for a similar parameter is justified in [143]. A routing algorithm should spend only a fraction of time
on processing the control packets as on packet switching, which is its actual task

[295]. A smaller ratio is definitely desirable for a routing protocol. C_t provides information about the extra bits, due to total overhead, that have been propagated in the network per time unit in relation to the total available bandwidth in the network. C_w is important at low network traffic loads because it is a function of the number of delivered bits. On the other hand, C_t is important for larger network traffic loads because it increases with an increase in the network traffic. We believe that together these complementary costs capture the communication costs of a routing algorithm.

The amount of memory needed to store the routing tables is directly proportional to the number of entries in the routing tables; therefore, our memory cost model is defined as

$$C_m = \frac{R}{D \times log(\frac{L}{h_{av}})},$$ (4.7)

where R is number of entries in the routing table, D total number of nodes in the network, L total number of links, and h_{av} is the average number of hops needed to reach a destination. In [38], Carrillo et al. have shown that the number of links between any (source, destination) pair is a function of the total number of links in the network and the average number of hops ($\frac{L}{h_{av}}$). The exact combination, however, depends on the routing algorithm and on the topology. Therefore, $\frac{R}{log(\frac{L}{h_{av}})}$ represents the number of entries needed to model the links between a node and other destinations. In this way, the cost associated with larger routing tables is defined as a function of the number of entries in the routing tables and the characteristics of a given topology. So the total cost C in our cost model is

$$C = 1 + C_\beta + e^{C_t} + C_w + C_m.$$ (4.8)

e^{C_t} is used to emphasize the influence of the total overhead if $C_t \geq 0.001$ because it is a vital parameter to modeling the communication cost of a routing algorithm. However, its value otherwise is relatively small by virtue of its definition. We have added a constant cost of 1 to take care of the costs related to operating system functions like context switching, interrupt processing, and network stack processing. We assume that these costs are the same for all routing protocols, which appears reasonable.

4.2.2 Power Model of an Algorithm

In [125, 126, 127], the authors, as discussed in Section 4.1.1, have defined the power of a distributed system as

$$P = \frac{T_{av}}{1 + \omega},$$ (4.9)

where ω corresponds to the ratio of average response time to acceptable response time. In the case of a routing protocol, ω becomes the ratio of average packet delay (t_d) to acceptable packet delay ($\frac{t_d}{t_{acc}}$). For our purpose, we defined $t_{acc} = K * t_{min}$, where t_{min} is the minimal packet delay achieved by any of the algorithms at MSIA = 4.6 sec and K is a constant (1.5 in our case) .

We would like to mention that even this new definition of power is unable to capture the influence of important performance values of a routing protocol, like jitter, session completion ratio, and the standard deviation of the packet delay distributions. Therefore, we now define new parameters that will enable us to have a comprehensive formula for the power of a routing algorithm. The session ratio ρ is defined as

$$\rho = (p_d)^{(2 \times (1 - S_c))}. \tag{4.10}$$

p_d and S_c correspond to packet delivery ratio and session completion ratio respectively, and they are defined in Chapter 3. ρ captures the effect of packet delivery ratio coupled to the session completion ratio. A good routing algorithm should have packet delivery ratio and session completion ratio values as large as possible. With this, we will define our *scaled throughput* T_{net} as a function of this performance value:

$$T_{net} = (T_{av})^{1 + \frac{\rho}{a}}. \tag{4.11}$$

The value of ρ has been scaled down by a constant value a (currently 10).

ζ is the ratio of the 90th percentile of the packet delays to the average delay (t_{90d}/t_d). A smaller value indicates that the algorithm has been able to deliver the majority of the packets within small deviation from the average delay. In order to incorporate ζ into our power formula, we will define κ

$$\kappa = 1 - e^{-\frac{\omega}{\zeta}}. \tag{4.12}$$

This equation gives weight to ζ only if its value is in proportion to the value of ω. Otherwise, if $\omega << \zeta$, the influence of ζ is de-emphasized. Similarly, the *jitter ratio* χ is defined as the ratio of the average jitter (J_d) to the acceptable jitter J_{acc} (J_{acc} is set to 30 msec). We now define a variable σ with the motivation that χ could only influence the power of a routing algorithm if its value is significantly greater than that of J_{acc} and comparable with the sum of ζ and ω:

$$\sigma = 1 - e^{-\frac{\chi}{(\omega + \zeta)}}. \tag{4.13}$$

We define a quality of service value, Υ, as

$$\Upsilon = 1 + \omega + \kappa + \sigma. \tag{4.14}$$

Now we define the power of a distributed routing algorithm γ as

$$\gamma = \frac{T_{net}}{\Upsilon}. \tag{4.15}$$

We believe that equation 4.15 models the power of a distributed routing algorithm as a function of a number of relevant performance values. Finally, we define the productivity Γ of a routing algorithm as

$$\Gamma = \frac{\gamma}{C}. \tag{4.16}$$

4.2.3 Scalability Metric for a Routing Algorithm

Here, we define our scalability metric in the same spirit as that of [126, 127, 38]. First, we define the scalability value Ω of a distributed algorithm with respect to a number of scalability parameters $k_1, k_2, \ldots k_n$ as

$$\Omega(k_1, k_2, \ldots k_n) = \frac{\Gamma}{k_1 \times k_2 \cdots \times k_n}. \qquad (4.17)$$

Now we are interested in whether the algorithm is scalable at a new value k_1' compared to a scalable parameter k_1. We can reach this decision by defining a scalability metric $\Psi(k_1, k_1')$ as follows:

$$\Psi(k_1, k_1') = \frac{\Omega(k_1', k_2, \ldots, k_n)}{\Omega(k_1, k_2, \ldots, k_n)}. \qquad (4.18)$$

We now suggest the following classifications for defining the scalability of a routing protocol. The classification is based on our experience in evaluating the algorithms (*BeeHive, AntNet*, and *OSPF* on large topologies). We say that a routing algorithm is

- *perfectly scalable* if $\Psi(k_1, k_1') \geq 1$.
- *positively scalable* if $0.9 \leq \Psi(k_1, k_1') < 1$.
- *nearly scalable* if $0.8 \leq \Psi(k_1, k_1') < 0.9$.
- *marginally scalable* if $0.7 \leq \Psi(k_1, k_1') < 0.8$.
- *not scalable* if $\Psi(k_1, k_1') < 0.7$.

4.3 Simulation Environment for Scalability Analysis

The simulation environment has been already introduced in Chapter 3. We tested the algorithms on six network instances: simpleNet, NTTNet, Node150, Node350, Node650, and Node1050.

4.3.1 simpleNet

simpleNet is defined in Chapter 3, and we will refer to this topology with the symbol n8 in the rest of the chapter.

4.3.2 NTTNet

NTTNet is defined in Chapter 3, and we will refer to this topology with the symbol n57 in the rest of the chapter.

4.3.3 Node150

Node150 is defined in Chapter 3, and we will refer to this topology with the symbol n150 in the rest of the chapter.

4.3.4 Node350

Node350 is a 350-node network with 464 bidirectional links. The link bandwidth is uniformly distributed between 7 and 14 Mbits/sec, and the propagation delay is uniformly distributed between 1 and 5 msec. The topology was generated using the BRITE software. The Node350 topology is shown in Figure 4.2. We will refer to this topology with the symbol n350 in the rest of the chapter.

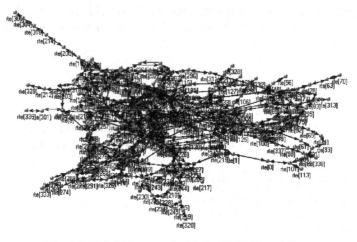

Fig. 4.2. Node350 captured with the OMNeT++ plotter

4.3.5 Node650

Node650 is a 650-node network with 785 bidirectional links. The link bandwidth is uniformly distributed between 15 and 20 Mbits/sec, and the propagation delay is uniformly distributed between 1 and 6 msec. The topology was again generated using the BRITE software. The Node650 topology is shown in Figure 4.3. We will refer to this topology with the symbol n650 in the rest of the chapter.

4.3.6 Node1050

Node1050 is a 1050-node network with 1,295 bidirectional links. The link bandwidth is uniformly distributed between 11 and 20 Mbits/sec, and the propagation delay is uniformly distributed between 1 and 5 msec. The topology was generated using the BRITE software. The Node1050 topology is shown in Figure 4.4. We will refer to this topology with the symbol n1050 in the rest of the chapter.

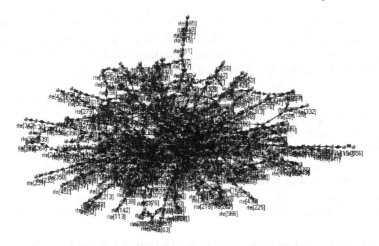

Fig. 4.3. Node650 captured with the OMNeT++ plotter

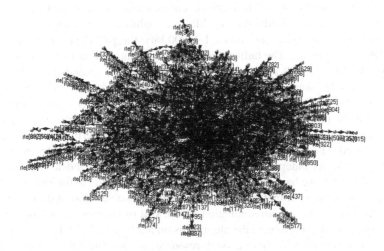

Fig. 4.4. Node1050 captured with the OMNeT++ plotter

4.4 Discussion of the Results from the Experiments

In this section we discuss the behavior of the algorithms based on important performance values, obtained from extensive experiments in OMNeT++, like *throughput, packet delay, packet delivery ratio, session completion ratio, control overhead,
suboptimal overhead, agent-processing complexity, packet-switching complexity,* and
size of the routing tables. The objective of this exercise is to study the behavior of the
algorithms with respect to the above-mentioned performance values in large topologies. However, we would like to mention that our primary objective is to focus on
the scalability behavior of *AntNet* and *BeeHive*, and not on the congestion avoidance

behavior as in Chapter 3. The reason for not investigating the congestion control behavior in large topologies is that we do not have the computational resources to simulate congested load scenarios for them. Nevertheless, we have included *OSPF* in
our comparison as a benchmark because it is a state-of-the-art algorithm for normal
or static traffic loads. Note that *OSPF* is a single-path routing algorithm; therefore, it
cannot scale with an increase in the network traffic load as suggested by the results
presented in Chapter 3 and by Di Caro and Dorigo in [62].

The congestion state for large topologies with greater bandwidths could not be
reached with the same parameters for the traffic generator as in the smaller topologies. The major emphasis of the work is on answering the question: *can nature-
inspired stochastic routing algorithms, like* AntNet *and* BeeHive, *competently perform routing in large topologies?* The answer will be of great significance to the
networking community in general and to the nature-inspired routing community in
particular.

We gradually decreased the value of MSIA from 4.6 sec to 1.6 sec in all of our
experiments. As discussed earlier, MSIA = 2.6 sec or less can cause a network congestion in n57, but for n650 this might be a normal load. This is due to the fact that
the links in n57 have a bandwidth of 6 Mbits/sec while in n650 they are on average
19 Mbits/sec. The session size was 2,130,000 bits, the packet size was 512 bytes,
and MPIA = 0.005 sec. The buffer size for storing data packets in routers was limited to 1,000 packets. *BeeHive* and *OSPF* were given 30 seconds to initialize the
routing tables. In comparison, *AntNet* was given 500 seconds to initialize the routing
tables, as done by Di Caro and Dorigo in [62]. In all the reported experiments, the
bee generation interval is 1 second, and the short-distance limit is seven hops. The
performance values are obtained from 1,000 seconds of experiments unless otherwise specified. They are an average of the values obtained from ten independent runs
for n8 and n57, five independent runs for n150, and three independent runs for n350,
n650, and n1050. We have to reduce the independent runs due to limited availability
of high-performance computers. Even then, the burdensome effort of the testing took
more than six months on our simulation server. Out of six months, it took approximately two months to evaluate the algorithms on n650 and n1050. Even our current
simulation server does not have enough resources to simulate either congested network traffic load in larger topologies, or topologies greater than 1,050 nodes. The
worst scenario is with n1050 and MSIA = 1.6 sec, in which case we could only
simulate 150 seconds of the network traffic; therefore, the results from this scenario
are provided to illustrate a tendency. We are actively pursuing our efforts to explore
opportunities to simulate even congested loads in large topologies for 1,000 seconds.

4.4.1 Throughput and Packet Delivery Ratio

Figures 4.5 and 4.6 show the throughput and the packet delivery ratio of the algorithms as the size of the topology is increased from eight nodes to 1,050 nodes. Each
figure consists of three subfigures which show the behavior of the algorithms at a
particular MSIA value. Figure 4.5 shows that all of the three algorithms are able to
maintain approximately the same throughput upto n150. As the topology grows to

350 nodes or more, the throughput of *AntNet* significantly starts trailing that the other algorithms. The reader has to correlate the throughput with the packet delivery ratio in Figure 4.6 to get a comprehensive picture. The larger throughput values in Figure 4.5 might mislead with the conclusion that the behavior of *BeeHive* and *OSPF* are the same. Figure 4.6(a) shows that at MSIA = 4.6 sec, three algorithms are able to deliver approximately all the packets upto n150. Beyond topology of this size the packet delivery ratio of *AntNet* dropped from 99% to about 70%, 40%, and 30% in n350, n650, and n1050 respectively. *OSPF* is able to deliver all packets except in n57 and n1050. In Figure 4.6(b) we can observe a similar tendency for *AntNet* at MSIA = 2.6 sec as in the previous case. *OSPF* starts significantly trailing *BeeHive* in n57 and n1050. n57 appears to be a complex topology with a low degree of connectivity compared with other topologies; therefore, the performance of *OSPF* significantly degrades in this instance. The same tendency for the packet delivery ratio of *AntNet* and *OSPF* can be seen in Figure 4.6(c) at MSIA = 1.6 sec. The packet delivery ratio of *OSPF* is significantly lower compared with that of both *AntNet* and *BeeHive* in n57. However, as might be expected, the packet delivery ratio of *AntNet* significantly degrades after n150. *BeeHive*, at MSIA = 1.6 sec, is able to maintain a significantly higher packet delivery ratio than *OSPF* in all topologies, however, due to scaling problems in Figure 4.5, the same difference is not obvious in the throughput.

We can easily conclude from this series of experiments that *BeeHive*, as far as throughput is concerned, is able to scale to larger topologies. However, *AntNet*, in comparison, scales well till the n150 topology but its throughput significantly deteriorates in n350 or larger topologies. The other performance values of interest are collected in Tables 4.1, 4.2, and 4.3.

Topology	Algorithm	t_d	P_{loop}	q_{av}	h_{av}	P_{drop}	S_c	S_d	S_{90d}	J_d	J_{90d}	A_a	D_a
n8	OSPF	2.99	0	1	1.92	0	99.7	2602	2749	4	9.99	-	-
	AntNet-Co	2.99	0.003	0	2.18	0	99.7	2601	2747	4	9.99	70555	6894
	BeeHive	3.26	0	0.019	2.29	0	99.7	2603	2751	4.99	10.9	17920	4107
n57	OSPF	254	0	49.6	6.8	2.51	81.4	2736	3108	4	9.99	-	-
	AntNet-Co	72.4	2.87	5.59	8.69	0.195	96.9	2673	2933	9.29	80.8	88373	14974
	BeeHive	30.1	2.5	0.946	7.81	0.029	99.4	2629	2788	7	28.6	26624	11558
n150	OSPF	48.8	0	16.2	5.4	0.29	96.4	2638	2820	4	10	-	-
	AntNet-Co	111	2.45	12.6	8.85	0.89	96.7	2716	3395	13.4	188	110235	22808
	BeeHive	22.1	0.594	0.3	5.62	0.003	99.6	2621	2769	6	15	34090	13648
n350	OSPF	31	0	8.99	7.66	0.099	98.3	2626	2781	4	10	-	-
	AntNet-Co	109	10.1	11	15.5	28.1	54.5	2660	3058	27.3	216	289068	34610
	BeeHive	25.4	0.257	0.17	7.94	0.014	99.5	2623	2771	5.33	14.6	33547	16590
n650	OSPF	33	0	0	7.42	0.03	99.4	2630	2779	4	10	-	-
	AntNet-Co	256	10	12	18.1	55.9	28.7	2752	3428	58.6	486	-	-
	BeeHive	34.8	0.099	0.06	7.72	0.006	99.5	2633	2782	5	13	-	-
n1050	OSPF	116	0	12	7.45	9.34	70.1	2634	2787	4	10	-	-
	AntNet-Co	320	6	18.5	15.2	67.2	16.7	2830	4074	30.5	329	-	-
	BeeHive	39.1	0.227	0.804	8.13	0.115	98.1	2640	2792	7	27	-	-

Table 4.1. Performance values for MSIA = 4.6 sec

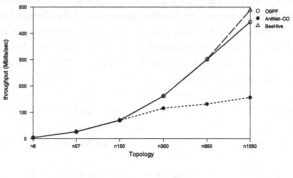

(a) MSIA = 4.6 sec

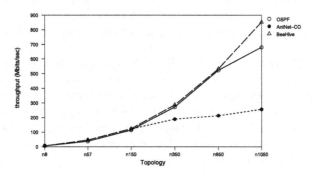

(b) MSIA = 2.6 sec

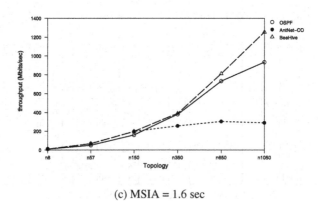

(c) MSIA = 1.6 sec

Fig. 4.5. Throughput (Mbits/sec))

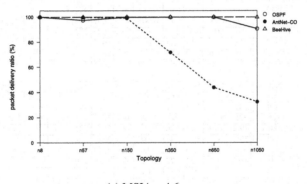

(a) MSIA = 4.6 sec

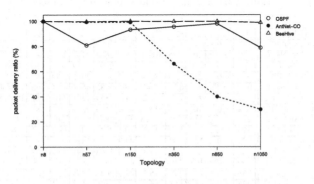

(b) MSIA = 2.6 sec

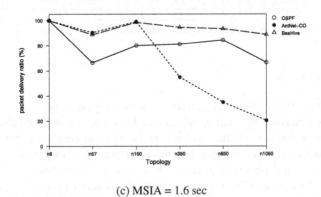

(c) MSIA = 1.6 sec

Fig. 4.6. Packet delivery ratio (%)

Topology	Algorithm	t_d	P_{loop}	q_{av}	h_{av}	P_{drop}	S_c	S_d	S_{90d}	J_d	J_{90d}	A_a	D_a
n8	OSPF	2.99	0	1	1.92	0	99.7	2601	2747	4	8.99	-	-
	AntNet-Co	2.99	0.006	0	2.17	0.001	99.6	2601	2747	4	9.49	70893	6879
	BeeHive	3.26	0	0.02	2.27	0	99.7	2602	2748	4	9.99	18107	4096
n57	OSPF	751	0	149	6.21	19	48.3	2677	2962	4	9.89	-	-
	AntNet-Co	235	2.76	23.9	8.83	0.778	88.6	2825	3310	26.1	182	107924	15249
	BeeHive	125	3.79	12.6	8.06	0.309	93.7	2755	3143	27.9	162	29733	11702
n150	OSPF	235	0	70.6	5.34	6.75	68	2663	2888	4	9.19	-	-
	AntNet-Co	109	1.73	11.5	8.12	0.829	95.8	2710	3133	14.6	146	112666	23120
	BeeHive	36.7	0.601	2.88	5.62	0.055	98.4	2642	2816	7	35.8	34399	13399
n350	OSPF	149	0	36.3	7.6	4.42	73.1	2649	2836	4	8.99	-	-
	AntNet-Co	336	9.32	32	15.5	33.8	44.9	2762	3124	92.3	638	292420	33429
	BeeHive	49.2	0.29	3.13	7.96	0.29	94.3	2647	2819	8	44.6	36446	16296
n650	OSPF	73.3	0	5.33	7.41	2.04	79.7	2641	2795	4	8.99	-	-
	AntNet-Co	316	7.82	19.6	16.3	59.9	24.6	2742	3382	63	492	-	-
	BeeHive	46.3	0.129	1.54	7.72	0.14	95.7	2645	2798	6	25	-	-
n1050	OSPF	160	0	19	7.37	21.1	54.5	2638	2796	4	9	-	-
	AntNet-Co	391	5.29	26	15.1	69.9	14.2	2766	3615	49	392	-	-
	BeeHive	80.9	0.183	5.87	8.2	1.158	77.5	2678	2866	16	82	-	-

Table 4.2. Performance values for MSIA = 2.6 sec

Topology	Algorithm	t_d	P_{loop}	q_{av}	h_{av}	P_{drop}	S_c	S_d	S_{90d}	J_d	J_{90d}	A_a	D_a
n8	OSPF	2.99	0	1	1.93	0	99.7	2602	2749	2.99	8	-	-
	AntNet-Co	2.99	0.005	0	2.18	0.001	99.6	2601	2748	2.99	8	71209	6852
	BeeHive	3.3	0	0.04	2.28	0	99.7	2602	2749	4	8.99	18086	4087
n57	OSPF	896	0	212	5.53	33.4	38.2	2695	2989	2.99	8	-	-
	AntNet-Co	1001	3.37	112	8.77	9.44	38.5	2985	3637	64.4	356	138650	15824
	BeeHive	721	4.47	92.1	7.79	11.2	39	2854	3391	116	551	108788	11742
n150	OSPF	384	0	107	5.18	19.9	49.4	2668	2891	3	8	-	-
	AntNet-Co	223	1.66	24.8	7.9	0.813	87.3	2847	3324	30	221	119439	23894
	BeeHive	129	0.622	19.3	5.64	1.42	80.1	2732	3016	18.2	113	47451	13212
n350	OSPF	334	0	72.3	7.41	18.8	43.2	2666	2870	3	8	-	-
	AntNet-Co	930	6.4	82	13.3	44.8	8.83	2968	3491	154	869	280192	32586
	BeeHive	150	0.477	16.6	7.96	6.28	68.9	2679	2888	18.3	109	46559	16266
n650	OSPF	158	0	17.3	7.31	15.5	48.5	2640	2795	3	8	-	-
	AntNet-Co	425	6.26	35.9	14.9	64.7	15.2	2746	3182	78.3	521	-	-
	BeeHive	138	0.116	13.6	7.69	6.56	48.4	2654	2819	14	81	-	-
n1050	OSPF	215	0	30	7.25	33.3	38.7	2641	2802	3	8	-	-
	AntNet-Co	579	4.8	49	14.8	78.7	4.08	2785	3365	117	614	-	-
	BeeHive	197	0.084	20.6	8.19	11	34.9	2676	2869	29.5	157	-	-

Table 4.3. Performance values for MSIA = 1.6 sec

4.4.2 Packet Delay

The 90th percentile of the packet delays is another important performance value for studying the behavior of a routing algorithm. Figure 4.7 shows the behavior of the packet delay as the size of the topology increases from eight to 1,050 nodes. The values of packet delays are plotted in three different figures for MSIA values of 4.6 sec, 2.6 sec, and 1.6 sec respectively. One observation is quite obvious: the n57 topology again appears to be more challenging than the other ones. The packet delay of *AntNet*, as shown in Figure 4.7(a), is better than that of *OSPF* in n57 topology at MSIA = 4.6 sec. The packet delay of *AntNet* then keeps on rising in larger topologies and reaches 1,800 msec in n1050, which is significantly higher than that of *OSPF* and *BeeHive*. A similar trend for the packet delay for *AntNet* is observed at MSIA = 2.6 sec and MSIA = 1.6 sec in Figures 4.7(b) and 4.7(c), respectively. The packet delay of *AntNet* becomes significantly greater in n350 and larger topologies. At MSIA =

4.6 sec, *OSPF* has the largest packet delay in n57, and then it gradually decreases to about 400 msec in n1050. Figure 4.7(b) shows that the packet delay of *OSPF* is highest again in n57, and then it gradually drops to 500 msec. A similar trend in packet delay for *OSPF* is observed at MSIA = 1.6 sec. However, at this MSIA value, the difference in the packet delay of *OSPF* compared with other algorithms is smaller. Please note that *BeeHive* has the greatest packet delivery ratio and the smallest packet delay in all the scenarios.

One can find an important conclusion from the experiments: *if the network is in a congestion state, then* BeeHive *is able to do excellent load balancing, as shown by the performance values in n57. Otherwise, if the network is not in congestion, then* BeeHive *is able to utilize a shortest path like* OSPF. This adaptive behavior is the result of the exploring of multiple paths in parallel by a swarm of replicas of *bee agents* in a small zone around its launching node. Routing the data packets only on those paths that have a quality value above a certain threshold appears to be a good compromise between maintaining just a single path, as *OSPF* does, and trying to maintain all possible paths, as *AntNet* tries to do. Moreover, exploring and managing the paths in a deterministic way, and then distributing the data packets in a stochastic manner on these paths, provides an excellent mix of stochastic elements with deterministic elements. Please note that *OSPF* explores the paths and routes data packets on them in a deterministic manner, while *AntNet* does both functions in a stochastic manner.

4.4.3 Control Overhead and Suboptimal Overhead

In this subsection we discuss the cost overhead associated with the transmission of the agents and the extra bandwidth consumed by data packets by virtue of their taking more hops than in the ideal case. The benefits associated with the stackless design of *bee agents*, and of gathering the routing information in a local *foraging zone*, are clearly exhibited in Figure 4.8. Remember that the *bee agents* have a fixed size of 48 bytes. The tendency of control overhead is approximately similar at different values of MSIA: *AntNet* has an overhead comparable to that of *OSPF* and *BeeHive* up to n150. Then it sharply increases in larger topologies. The obvious reasons are: first, *ant agents* are equipped with a stack, and its size increases with an increase in hops, and second, the stochastic exploration embodies a greater chance of running into loops, causing a significant waste of bandwidth. Note that the control overhead of *BeeHive* at MSIA values of 4.6 sec (see Figure 4.8(a)) and 2.6 sec (see Figure 4.8(b)) is less than or equal to 1%, while at MSIA = 1.6 sec (see Figure 4.8(c)) in n1050 it increases to about 2%. However, the control overhead is comparable to *OSPF*. Figure 4.9 shows the suboptimal overhead of the algorithms. Figures 4.9(a), 4.9(b), 4.9(c) show a similar tendency about the suboptimal overhead. Two observations are apparent: one, the suboptimal overhead increases with an increase in the network traffic load, and two, the suboptimal overhead is significantly higher than the control overhead, especially in case of n57. One can easily conclude from Figure 4.9 that the suboptimal overhead of *OSPF* is the smallest. This is due to the transporting of the data packets on the shortest paths; however, in congestion scenarios *OSPF*

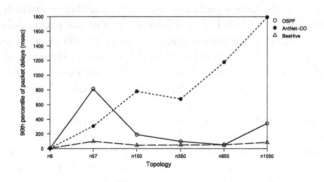

(a) MSIA = 4.6 sec

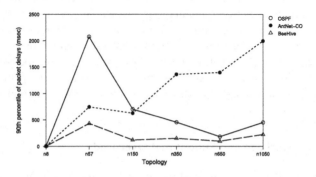

(b) MSIA = 2.6 sec

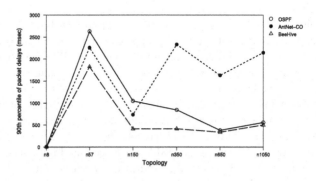

(c) MSIA = 1.6 sec

Fig. 4.7. Packet delay (msec))

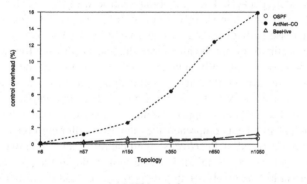

(a) MSIA = 4.6 sec

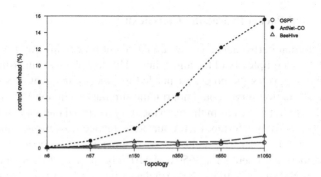

(b) MSIA = 2.6 sec

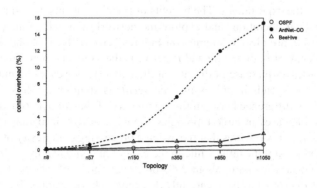

(c) MSIA = 1.6 sec

Fig. 4.8. Routing overhead (%)

delivers fewer data packets. The suboptimal overhead of *AntNet* sharply increases after n57 till n350, and then sharply drops at n650 and n1050. The reason for the significant increase is that the data packets routed by *AntNet* take more hops (h_{av}) to reach their destination, and the number of data packets which follow cyclic paths (p_{loop}) also increase significantly in larger topologies. However, the sharp decline in the suboptimal overhead in n650 and n1050 is due to the fact that only 40% or less of the data packets are delivered (see Figure 4.6) at their destination. Remember that the packets which have been dropped are not included in the suboptimal overhead; therefore, one has to take a comprehensive view by combining Figures 4.9 and 4.6. The suboptimal overhead of *BeeHive* is significantly smaller than that of *AntNet* in all topologies, and except for n57, it is of the same order as that of *OSPF*. *BeeHive* is able to maintain a relatively small suboptimal overhead even though its packet delivery ratio is the highest among the three algorithms.

4.4.4 Agent and Packet Processing Complexity

The next important performance values are the agent processing complexity and the packet-switching complexity of the algorithms. The complexity is measured in terms of the number of cycles the processor needed to process an agent or switch a data packet. Our simulation server could not run the simulations in n650 and n1050 without the need to switch pages from the main memory to virtual memory and vice versa. Therefore, disk activity interfered with our performance framework, and it effected both agent and data processing complexity parameters. Consequently, we decided not to report these parameters for n650 and n1050 because of the uncertainty caused by the page replacement activity. Nevertheless, the trend obtained from n650 and n1050 was a simple extrapolation of the cycle complexity values in small topologies.

Figure 4.10 shows the number of cycles that a node spends in processing the agents during the experiments. The benefits of stackless agents, which perform simple mathematical operations and explore the network in a deterministic fashion in a *foraging zone*, are clearly inferable in Figures 4.10(a), 4.10(b), and 4.10(c). The general tendency of both *AntNet* and *BeeHive* is the same: an increasing processing complexity with an increase in the size of the network; however, the number of cycles that a node spends in processing *bee agents* is significantly less than that for *ant agents*. The difference between the complexity of the *ant agents* and *bee agents* significantly increases in large topologies. Looking at the A_a parameter in Tables 4.1, 4.2, and 4.3, it is obvious that the average processing complexity of a *bee agent* does not increase significantly, while in the case of an *ant agent* this value increases from 50,000 cycles for n8 to about 290,000 cycles for n350. The packet-switching algorithm of *BeeHive* has been carefully designed to have features like no rescaling of the goodness values, and little search time due to smaller routing tables. *BeeHive* only maintains those paths whose quality is above a threshold; therefore, few packets enter into loops. Consequently, all of these factors ensure that a node spends a significantly smaller number of cycles in a packet-switching algorithm, which is its main task in comparison to *AntNet*. Figures 4.11(a), 4.11(b), and 4.11(c) show the same tendency: the packet-switching complexity of both algorithms is comparable

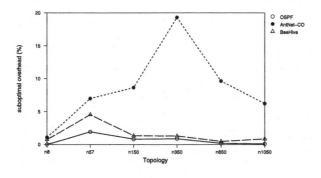

(a) MSIA = 4.6 sec

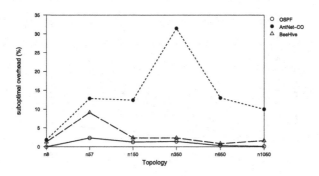

(b) MSIA = 2.6 sec

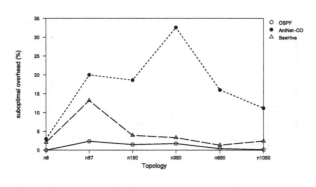

(c) MSIA = 1.6 sec

Fig. 4.9. Suboptimal overhead (%)

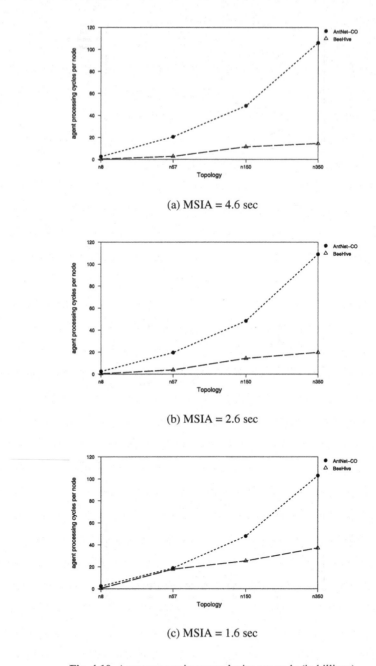

(a) MSIA = 4.6 sec

(b) MSIA = 2.6 sec

(c) MSIA = 1.6 sec

Fig. 4.10. Agent processing complexity per node (in billions)

in small topologies like n8, and then the difference between them starts increasing in large topologies like n150 and n350.

We did not report the processing complexity of *OSPF* because the value depends on how efficiently the Dijkstra algorithm is implemented. In order to avoid any controversy, we ignore the processing costs for *OSPF*, i.e., $C_\beta = 0$, during the scalability analysis.

4.4.5 Routing Table Size

The benefits of dividing the network into *foraging regions* and *foraging zones* are clearly exhibited in Figure 4.12. The size of the routing table of *BeeHive* is significantly smaller than that of *AntNet* in larger topologies. Note that the size of the *BeeHive* routing table includes entries in all three routing tables, and its size is approximately the same as the one for *OSPF*. Running *AntNet* requires, on average, the same number of entries in the routing tables of a node as the number of the links in the network. Note that n57, n150, n350, n650, and n1050 have 162, 400, 928, 1,570, and 2,590 bidirectional links, respectively. Figure 4.12 shows clearly that *BeeHive* is able to maintain the best performance in all topologies under all network traffic loads, but with significantly smaller routing tables.

4.4.6 Investigation of the Behavior of *AntNet*

The poor performance of *AntNet* in large topologies came as a surprise to us because its performance values are significantly better in n8, n57, and n150. We are in close contact with Di Caro at IDSIA, one of the developers of *AntNet*. He has been kind enough to audit the source code of our implementation of *AntNet*. We decided to investigate the reason for this anomalous behavior, and our performance evaluation framework collected a number of auxiliary parameters that provided us with a valuable insight into the behavior of *AntNet*. After some initial brainstorming we developed two hypotheses for this behavior of *AntNet*:

- **H3:** the *ant agents* could not manage the traffic load in large topologies.
- **H4:** the exploration of routes and their maintenance is flawed.

We conducted experiments for MSIA = 10.6 sec and MSIA = 20.6 sec in n350, n650, and n1050, and even at these small network loads the performance of *AntNet* did not improve. Moreover, we noted that if a single path routing algorithm like *OSPF* can manage the load at MSIA = 4.6 sec, then a multiple-path routing algorithm like *AntNet* should be able to manage it as well. Subsequently, we focused our investigation on H4 and we collected different parameters from the performance evaluation framework.

Figure 4.13(a) shows the percentage of *ant agents* that actually managed to return to their source node. It is interesting to note that the percentage of *ant agents* that return to their source node decreases from 80% in n8 to 40%, 20%, 30%, 10%, and 5% in n57, n150, n350, n650, and n1050, respectively. Figure 4.13(b) shows the number of *ant agents* that were launched per node in the network and how many of

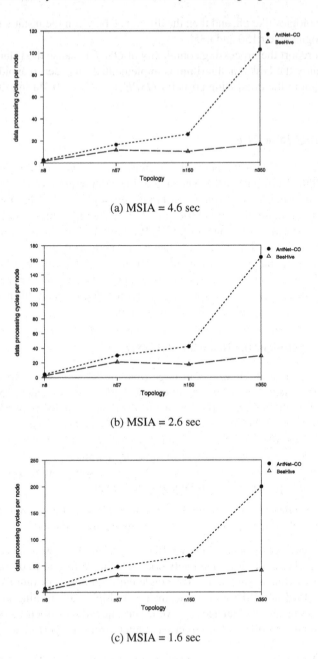

(a) MSIA = 4.6 sec

(b) MSIA = 2.6 sec

(c) MSIA = 1.6 sec

Fig. 4.11. Packet switching complexity per node (in billions)

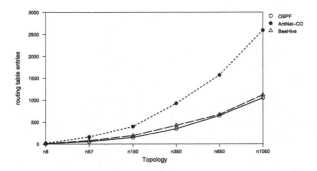

Fig. 4.12. The size of the routing table

them actually returned to their source node. Figure 4.13(b) again shows the trend that as the size of the network increases, only a small fraction of the *ant agents* are able to return to their nodes. Moreover, each node in *AntNet*, as suggested by Figure 4.13(b), launches about four *ant agents* per second. Even if all these *ant agents* arrive back at their source node (which of course is not the case), a node in the n650 network would be able to launch an *ant agent* for a destination after approximately every 150 seconds, assuming that all destinations receive the same amount of traffic. The sampling period for a destination increases further to 250 seconds in n1050. This shortcoming coupled with the point-to-point mode of transmission for *ant agents* further degrades the route discovery and management process, especially in large topologies.

Figure 4.14(a) demonstrates the disadvantage of stochastic exploration that allows *ant agents* to loop. The hop values are for the complete journey of an ant, both forward and backward. The average hop count of the received *ant agents* is of the order of the nodes in the network. Consequently, the lifetime of the *ant agents*, as shown in Figure 4.14(b), sharply increases with an increase in the size of the network. It is obvious that the stochastic network exploration coupled with forward and backward journeys, as expected, is not a promising approach for large networks. We set the TTL value of *ant agents* to $(1.5 \times D)$; therefore, it is highly unlikely that the *ant agents* are dropped because their TTL value has reached 0. Figure 4.14(a) clearly supports this argument as the hop values of the *ant agents* are well below the allowable limit. Our investigation further revealed that the *ant agents* got deleted because they followed cyclic paths. Di Caro and Dorigo in [62] kill all those *ant agents* that spend more than half of their life in a cyclic path. The motivation behind this feature in *AntNet* as explained in [62] is: *such an* ant agent *is carrying an old and misleading memory of the network state and it is counterproductive to use it to update the routing tables.* However, our investigation revealed a side effect of this feature: once an *ant agent* follows a cyclic path during the early stage of its network exploration, the condition to kill it is nearly always met because it spends more than half its life in the cycle; as a result, it is killed. A substantial percentage of the *ant agents* that got killed were those that followed a cyclic path by starting at their source node or within few

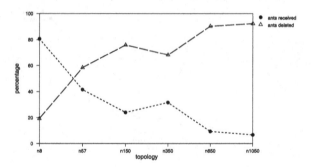

(a) Percentage of ant agents received and deleted

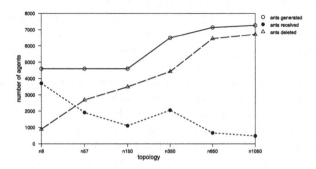

(b) Number of ant agents sent, received, and deleted per node

Fig. 4.13. The behavior of AntNet (agents sent, received, and deleted)

hops of their source node. Consequently, the rate at which the routes leading to different nodes are sampled is significantly reduced. This phenomenon, however, does not manifest itself as a serious bottleneck in relatively small topologies like n8, n57, and n150, where still a reasonable number of *ant agents* manage to return to their source nodes, and hence update the routing tables.

4.5 Towards an Empirically Founded Scalability Model for Routing Protocols

Our comprehensive performance evaluation framework introduced in Chapter 3 proved its usefulness in validating our scalability model because it calculated the performance values utilized by the scalability model. We developed a scalability an-

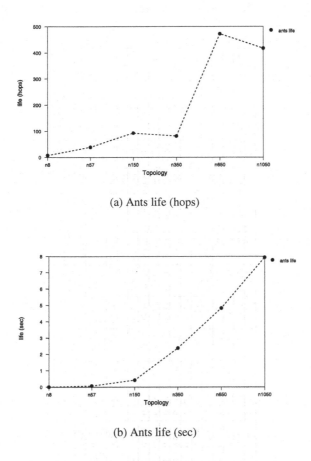

(a) Ants life (hops)

(b) Ants life (sec)

Fig. 4.14. The behavior of AntNet (agent life in hops and sec)

alyzer tool that retrieves the relevant performance values for calculating the values of different scalability parameters from the database, where the results of different experiments were stored. Values of relevant parameters are cataloged in Tables 4.4, 4.5, and 4.6 for *BeeHive*, *AntNet*, and *OSPF* algorithms, respectively. One can summarize from Table 4.4, when compared with similar Tables 4.5 and 4.6 for *AntNet* and *OSPF*, respectively, that *BeeHive* has power and productivity comparable to those of *OSPF* and *AntNet* in n8. As expected, *OSPF* has the best power and productivity values in this topology. However, in n57 *OSPF* has the worst power among all algorithms. This is due to its poor performance values: small throughput and significantly high packet delays (see Section 4.4). *BeeHive* has the best power and productivity in n57 at MSIA = 4.6 sec and MSIA = 2.6 sec. At MSIA = 1.6 sec, *BeeHive* and *OSPF* have similar productivity values, while the productivity of *AntNet* is the smallest. *AntNet* is trailing *BeeHive* because it achieves similar performance as *BeeHive*, but at

Topology	MSIA	C_p	C_d	C_β	C_w	C_t	C_m	C	ζ	ω	κ	χ	σ	ρ	Υ	T_{net}	γ	Γ
n8	4.6	0.688	3.3	0.208	0.365	0.007	0.672	3.25	1.53	0.815	0.412	0.166	0.068	0.999	2.29	4.29	1.86	0.574
	2.6	0.493	3.27	0.15	0.361	0.013	0.672	3.19	1.56	0.815	0.406	0.133	0.054	0.999	2.27	7.93	3.48	1.089
	1.6	0.39	3.28	0.118	0.363	0.021	0.672	3.17	1.63	0.825	0.395	0.133	0.052	0.999	2.27	13.2	5.84	1.83
n57	4.6	5.9	24.8	0.237	1.76	0.047	0.428	4.47	3.24	0.43	0.123	0.233	0.061	0.999	1.61	36.4	22.5	5.04
	2.6	4.65	25.9	0.179	1.96	0.094	0.429	4.67	3.45	1.78	0.403	0.929	0.162	0.999	3.35	68	20.3	4.34
	1.6	15	26.8	0.561	1.96	0.135	0.426	5.09	2.53	10.3	0.982	3.86	0.26	0.863	12.5	96.8	7.71	1.51
n150	4.6	24.5	22	1.111	0.862	0.019	0.296	4.29	2.07	0.442	0.191	0.2	0.076	0.999	1.71	106	62.3	14.5
	2.6	17.4	21.6	0.805	0.781	0.031	0.296	3.91	3.26	0.734	0.201	0.233	0.056	0.999	1.99	199	99.9	25.5
	1.6	19.2	21.5	0.892	0.762	0.049	0.296	4	3.2	2.58	0.553	0.606	0.099	0.994	4.23	333	78.6	19.6
n350	4.6	31	36.2	0.857	1.222	0.018	0.241	4.34	1.94	0.508	0.23	0.177	0.069	0.999	1.8	269	149	34.3
	2.6	23.9	35.7	0.672	1.128	0.03	0.235	4.06	3.06	0.984	0.274	0.266	0.063	0.999	2.32	505	217	53.5
	1.6	33.5	37.1	0.903	1.19	0.043	0.252	4.39	2.76	3	0.662	0.61	0.1	0.965	4.76	691	145	33
n650	4.6	0	0	0	1.149	0.011	0.192	3.35	1.494	0.696	0.372	0.166	0.073	0.999	2.14	534	249	74.4
	2.6	0	0	0	0.944	0.016	0.192	3.15	2.11	0.925	0.354	0.2	0.063	0.999	2.34	996	425	134
	1.6	0	0	0	0.894	0.023	0.192	3.11	2.44	2.76	0.677	0.466	0.085	0.931	4.52	1514	334	107
n1050	4.6	0	0	0	1.76	0.02	0.181	3.96	2.18	0.782	0.3	0.233	0.075	0.999	2.15	908	420	106
	2.6	0	0	0	1.53	0.031	0.181	3.75	2.76	1.61	0.442	0.533	0.114	0.994	3.17	1668	525	140
	1.6	0	0	0	1.496	0.044	0.18	3.72	2.54	3.94	0.787	0.983	0.14	0.856	5.86	2312	394	105

Table 4.4. BeeHive

Topology	MSIA	C_p	C_d	C_β	C_w	C_t	C_m	C	ζ	ω	κ	χ	σ	ρ	Υ	T_{net}	γ	Γ
n8	4.6	5.3	5.35	0.99	0.285	0.011	1	4.28	1.66	0.747	0.361	0.133	0.053	0.999	2.16	4.24	1.96	0.457
	2.6	3.03	5.33	0.568	0.27	0.019	1	3.85	1.66	0.747	0.361	0.133	0.053	0.999	2.16	7.89	3.64	0.945
	1.6	1.87	5.32	0.352	0.259	0.03		3.64	1.66	0.747	0.361	0.099	0.04	0.999	2.14	13.4	6.24	1.71
n57	4.6	44.5	35.7	1.246	2.99	0.163	0.883	7.29	4.22	1.034	0.217	0.309	0.057	0.999	2.3	36.6	15.8	2.17
	2.6	23.9	36.8	0.651	2.88	0.275	0.883	6.73	3.18	3.35	0.651	0.87	0.124	0.998	5.13	67.8	13.2	1.96
	1.6	15.6	40.1	0.388	2.91	0.412	0.883	6.7	2.25	14.3	0.998	2.14	0.121	0.883	16.4	99.6	6.06	0.905
n150	4.6	106	56.7	1.87	4.94	0.112	0.614	9.55	7.03	2.22	0.27	0.446	0.047	0.999	3.53	104	29.6	3.1
	2.6	59.6	52.1	1.142	3.66	0.147	0.614	7.58	5.77	2.18	0.314	0.486	0.059	0.999	3.55	195	54.9	7.24
	1.6	36.3	52.3	0.694	3.17	0.206	0.614	6.71	3.29	4.46	0.742	1	0.121	0.997	6.32	335	53	7.9
n350	4.6	322	312	1.031	11.7	0.257	0.547	15.5	6.21	2.18	0.296	0.91	0.102	0.738	3.57	164	46	2.95
	2.6	203	305	0.664	10.6	0.38	0.547	14.3	4.05	6.72	0.809	3.07	0.248	0.632	8.77	263	29.9	2.09
	1.6	142	274	0.517	8.09	0.39	0.547	11.6	2.5	18.6	0.999	5.13	0.215	0.335	20.8	308	14.8	1.272
n650	4.6	0	0	0	25.7	0.22	0.447	28.4	4.6	5.12	0.671	1.95	0.182	0.309	6.97	153	22	0.774
	2.6	0	0	0	18.3	0.252	0.447	21	4.42	6.32	0.76	2.1	0.177	0.25	8.25	243	29.4	1.402
	1.6	0	0	0	14.2	0.28	0.447	17	3.83	8.5	0.891	2.61	0.19	0.168	10.5	333	31.5	1.85
n1050	4.6	0	0	0	29.4	0.221	0.419	32	5.6	6.4	0.68	1.016	0.081	0.153	8.16	169	20.7	0.647
	2.6	0	0	0	20.9	0.256	0.419	23.6	5.1	7.82	0.784	1.63	0.118	0.125	9.72	275	28.3	1.196
	1.6	0	0	0	19.3	0.266	0.419	22	3.7	11.5	0.956	3.9	0.225	0.047	13.7	296	21.5	0.976

Table 4.5. AntNet

Topology	MSIA	C_p	C_d	C_β	C_w	C_t	C_m	C	ζ	ω	κ	χ	σ	ρ	Υ	T_{net}	γ	Γ
n8	4.6	0	0	0	0.048	0.001	0.444	2.49	1.337	0.747	0.428	0.133	0.061	0.999	2.23	4.24	1.89	0.76
	2.6	0	0	0	0.027	0.001	0.444	2.47	1.337	0.747	0.428	0.133	0.061	0.999	2.23	7.95	3.55	1.438
	1.6	0	0	0	0.016	0.001	0.444	2.46	1.337	0.747	0.428	0.099	0.046	1	2.22	13.4	6.04	2.45
n57	4.6	0	0	0	0.781	0.02	0.31	3.11	3.2	3.62	0.677	0.133	0.019	0.99	5.32	35.5	6.68	2.14
	2.6	0	0	0	0.636	0.024	0.31	2.97	2.76	10.7	0.979	0.133	0.009	0.802	12.7	50.7	3.98	1.341
	1.6	0	0	0	0.482	0.025	0.31	2.81	2.93	12.8	0.987	0.099	0.006	0.602	14.7	63.8	4.31	1.53
n150	4.6	0	0	0	0.469	0.01	0.23	2.71	3.89	0.976	0.221	0.133	0.027	0.999	2.22	105	47.5	17.5
	2.6	0	0	0	0.4	0.015	0.23	2.64	3	4.7	0.791	0.133	0.017	0.955	6.5	179	27.5	10.4
	1.6	0	0	0	0.33	0.017	0.23	2.57	2.73	7.68	0.939	0.1	0.009	0.797	9.62	239	24.9	9.65
n350	4.6	0	0	0	0.83	0.012	0.206	3.04	3.12	0.62	0.179	0.133	0.034	0.999	1.83	269	146	48.1
	2.6	0	0	0	0.703	0.018	0.206	2.92	3.05	2.98	0.623	0.133	0.021	0.975	4.62	471	102	34.8
	1.6	0	0	0	0.598	0.021	0.206	2.82	2.52	6.68	0.928	0.1	0.01	0.788	8.61	605	70.2	24.8
n650	4.6	0	0	0	0.755	0.007	0.185	2.94	1.55	0.66	0.345	0.133	0.058	0.999	2.06	532	258	87.5
	2.6	0	0	0	0.514	0.008	0.185	2.7	2.49	1.466	0.444	0.133	0.033	0.991	2.94	972	330	122
	1.6	0	0	0	0.403	0.009	0.185	2.59	2.39	3.16	0.732	0.1	0.017	0.839	4.9	1275	259	99.9
n1050	4.6	0	0	0	0.762	0.008	0.169	2.94	2.96	2.31	0.542	0.133	0.024	0.942	3.88	788	202	69
	2.6	0	0	0	0.515	0.008	0.169	2.69	2.81	3.2	0.678	0.133	0.021	0.805	4.9	1149	234	87
	1.6	0	0	0	0.388	0.008	0.169	2.56	2.59	4.3	0.809	0.1	0.014	0.607	6.12	1415	231	90

Table 4.6. OSPF

significantly higher communication, processing, and memory costs. Moreover, n57, as mentioned before, is a challenging topology; therefore, congestion can already result at smaller network traffic loads.

n150 appears to be a simple topology with a high degree of connectivity; therefore, the power and productivity of all algorithms are significantly higher in this topology compared to n57. In n150, *BeeHive* has the best power and productivity for all network traffic loads except at MSIA = 4.6 sec, where *OSPF* has the best productivity. However, as might be expected from the experiments reported in Section 4.4, the power of *OSPF* starts trailing compared to *AntNet* at MSIA = 2.6 sec, and at MSIA = 1.6 sec under high loads. Yet, it achieves a higher productivity than *AntNet* because of its better benefit-to-cost ratio.

In the n350 topology, *OSPF* has the best productivity among all algorithms at MSIA = 4.6 sec because the network traffic load at this value cannot cause a congestion for this instance of network; therefore, it achieves a similar power as *BeeHive*, but with lower costs. Consequently, its productivity is significantly higher than that of *BeeHive*. However, as the network traffic load is increased by taking the MSIA values of 2.6 sec and 1.6 sec, the power and productivity of *OSPF* significantly start trailing those of *BeeHive*. Note that the power and productivity of *AntNet* in n350 have significantly degraded and are the lowest among all algorithms. This behavior is due to its poor performance as described in Section 4.4.

The power and productivity of the algorithms show the same tendency in the n650 topology. The power and productivity of *BeeHive* are significantly greater among all algorithms at MSIA = 2.6 sec and MSIA = 1.6 sec; however, at MSIA = 4.6, sec values are slightly smaller than those of *OSPF*. The power and productivity of *AntNet* are further decreased due to its poor performance. Finally, *BeeHive* has again the best power and performance among all algorithms for all traffic patterns in n1050. Both values are significantly higher than those of *OSPF*. Since n1050 is a relatively difficult topology with a rather low degree of connectivity, the power and productivity of *AntNet* are further deteriorated.

One has to be careful in analyzing the scalability of a routing algorithm based on the scalability metric Ψ introduced in Section 4.2. One drawback of Ψ is that it simply takes the ratio of the productivity values in two configurations and then the ratio of the scalability parameters with respect to which the scalability has to be studied. The ratio factors out the influence of the productivity value itself; e.g., if an algorithm A has productivity values of 0.9 and 1.2 in two configurations and another algorithm B has productivity values of 10 and 15, then both algorithms are termed as perfectly scalable although the algorithm A has smaller productivity values in both states. Therefore, one has to always complete the picture by looking both at Ψ and Γ.

4.5.1 Scalability Matrix and Scalability Analysis

In order to comprehensively analyze the scalability of the algorithms, we have developed a *scalability matrix* for each algorithm. The scalability matrix is populated under the following rules:

- The topologies are listed in the first column and the first row. Within each topology, different traffic patterns, modeled by changing the values of MSIA, are listed.
- In this matrix, we are interested in studying the scalability either by keeping the same topology and changing the traffic patterns, or by keeping the same traffic patterns but changing the topology.
- An initial configuration is defined by selecting a topology in the first column followed by choosing an MSIA value. A final configuration is defined by selecting a topology from the first row and then choosing the MSIA value.
- A transition from an initial configuration to a final configuration is only allowed in the matrix if the following three conditions are met:
  - The topology of the final configuration is extending the current configuration (in terms of nodes and/or links).
  - The MSIA value of the final configuration is smaller than or equal to the one in current configuration.
  - The MSIA value of the final configuration is equal to the initial configuration if the topologies are different.
- If a transition is allowed, then the corresponding value is simply a ratio, $\frac{\Omega_2}{\Omega_1}$, where Ω_2 and Ω_1 are scalability values in the final and initial configurations, respectively (they are defined in Section 4.2). Basically, we are interested in studying how the algorithm scales for the same traffic patterns once new nodes are added, or how the algorithm scales to additional injection of traffic load but for a fixed topology.
- An invalid transition is marked by a "$\times$."

Let us take the example of the scalability matrix of *BeeHive* illustrated in Table 4.7. If we want to study the scalability of *BeeHive* from n8, MSIA = 4.6 sec, to n350, MSIA = 4.6 sec, then the corresponding entry is 1.38, which shows that the algorithm is perfectly scalable for these configurations. However, the scalability metric from n57, MSIA = 4.6 sec, to n57, MSIA = 2.6 sec, is 0.48, which is an indication that the algorithm is not scalable for these configurations. Similarly, the following transitions, for example, are not allowed: n8, MSIA = 4.6 sec, to n57, MSIA = 2.6 sec, and n57, MSIA = 1.6 sec, to n57, MSIA = 2.6 sec. We now discuss the scalability of each algorithm separately.

4.5.2 Scalability Analysis of BeeHive

One can easily conclude from Table 4.7 that *BeeHive* is perfectly scalable at MSIA = 4.6 sec if n8 is considered as the reference topology. *BeeHive* is also perfectly scalable at MSIA = 2.6 sec except in n57, where it is not scalable, and in n1050, where it is positively scalable. However, at MSIA = 1.6 sec, the algorithm is not scalable except for n650, where it is marginally scalable. The algorithm is also perfectly scalable to an additional network traffic load in the n8 topology.

The shortcoming of the scalability metric, as discussed before, manifests itself if one takes n57 as the reference. *BeeHive* appears to be perfectly scalable for all network traffic loads, including MSIA = 1.6 sec, which is counterintuitive. The reason is obvious: the productivity of *BeeHive* is significantly smaller in n57 at MSIA = 1.6 sec; therefore, the scalability metric significantly increases.

If we take n150 as the reference topology, then the *BeeHive* algorithm is perfectly scalable at MSIA = 4.6 sec. In n150, the algorithm is nearly scalable, perfectly scalable, and marginally scalable for n350, n650, and n1050, respectively, at MSIA = 2.6 sec. A similar tendency is seen at MSIA = 1.6 sec. The algorithm is able to perfectly scale to the additional traffic injected at MSIA = 2.6 sec but not at MSIA = 1.6.

The algorithm is nearly scalable or better if n350 is taken as the reference topology for all configurations except one, in which additional load is injected in n350 at MSIA = 1.6 sec. The algorithm also does not scale at MSIA = 2.6 sec and MSIA = 1.6 sec if n650 is taken as the reference topology. However, at MSIA = 4.6 sec it is nearly scalable. In n650, the algorithm is perfectly scalable to an additional injected load at MSIA = 2.6 sec; however, it is not scalable at MSIA = 1.6 sec. In n1050 the algorithm does not scale to the additional traffic load at MSIA = 1.6 sec.

BeeHive is nearly scalable or better in most of the cases with MSIA = 4.6 sec and MSIA = 2.6 sec; however, for MSIA = 1.6 sec it is not scalable.

4.5.3 Scalability Analysis of AntNet

The power of *AntNet* has significantly deteriorated in n350 and onwards; therefore, the scalability metric is of little significance. However, in n1050 the algorithm appears to perfectly scale to the additional injected traffic load at MSIA = 2.6 sec because the corresponding Ψ value is 1.038. However, this is due to the productivity values of 0.647 and 1.196 at MSIA = 4.6 sec and 2.6 sec, respectively (see Table 4.5). Consequently, their ratio leads to a value of 1.038. One can easily conclude from the scalability metrics of *AntNet*, as shown in Table 4.8, that the algorithm is not scalable for most of the configurations.

4.5.4 Scalability Analysis of OSPF

The scalability matrix for *OSPF* is illustrated in Table 4.9. As expected, the general trend is that most of the time *OSPF* is not scalable for MSIA = 1.6 sec, either across different topologies or within the same topology. Moreover, the scalability metrics are generally inferior to those of *BeeHive*. The exceptions to this generalization are the cases where the productivity of *OSPF* is extremely low in one topology and significantly improves in an another topology. One such anomaly is observed at MSIA = 2.6 sec with n57 as the reference topology. One might conclude that in this case *OSPF* is perfectly scalable and its scalability metrics are also of higher value compared to *BeeHive*. This is due to the poor productivity of *OSPF* in n57, as cataloged in Table 4.6. The productivity of *OSPF* at MSIA = 2.6 sec increases from 1.34 in n57 to 10.4, 34.8, 122, and 87 in n150, n350, n650, and n1050, respectively. In comparison, the productivity of *BeeHive* (see Table 4.4) increases from 4.34 in n57 to 25.5, 53.5, 134, and 140 in n150, n350, n650, and n1050, respectively. The complete

	n8·4.6	n8·2.6	n8·1.6	n57·4.6	n57·2.6	n57·1.6	n150·4.6	n150·2.6	n150·1.6	n350·4.6	n350·2.6	n350·1.6	n650·4.6	n650·2.6	n650·1.6	n1050·4.6	n1050·2.6	n1050·1.6
n8 · 4.6	1	1.087	1.139	1.254	×	×	1.365	×	×	1.387	×	×	1.61	×	×	1.42	×	×
n8 · 2.6	×	1	1.048	×	0.563	×	×	1.249	×	×	1.119	×	×	1.52	×	×	0.978	×
n8 · 1.6	×	×	1	×	×	0.114	×	×	0.565	×	×	0.458	×	×	0.713	×	×	0.43
n57 · 4.6	×	×	×	1	0.488	0.104	1.088	×	×	1.106	×	×	1.288	×	×	1.131	×	×
n57 · 2.6	×	×	×	×	1	0.213	×	2.21	×	×	1.98	×	×	2.7	×	×	1.73	×
n57 · 1.6	×	×	×	×	×	1	×	×	4.92	×	×	3.98	×	×	6.21	×	×	3.74
n150 · 4.6	×	×	×	×	×	×	1	0.994	0.471	1.016	×	×	1.183	×	×	1.039	×	×
n150 · 2.6	×	×	×	×	×	×	×	1	0.474	×	0.896	×	×	1.221	×	×	0.783	×
n150 · 1.6	×	×	×	×	×	×	×	×	1	×	×	0.81	×	×	1.261	×	×	0.761
n350 · 4.6	×	×	×	×	×	×	×	×	×	1	0.877	0.376	1.165	×	×	1.023	×	×
n350 · 2.6	×	×	×	×	×	×	×	×	×	×	1	0.429	×	1.362	×	×	0.874	×
n350 · 1.6	×	×	×	×	×	×	×	×	×	×	×	1	×	×	1.55	×	×	0.939
n650 · 4.6	×	×	×	×	×	×	×	×	×	×	×	×	1	1.025	0.503	0.878	×	×
n650 · 2.6	×	×	×	×	×	×	×	×	×	×	×	×	×	1	0.49	×	0.641	×
n650 · 1.6	×	×	×	×	×	×	×	×	×	×	×	×	×	×	1	×	×	0.603
n1050 · 4.6	×	×	×	×	×	×	×	×	×	×	×	×	×	×	×	1	0.749	0.345
n1050 · 2.6	×	×	×	×	×	×	×	×	×	×	×	×	×	×	×	×	1	0.461
n1050 · 1.6	×	×	×	×	×	×	×	×	×	×	×	×	×	×	×	×	×	1

Table 4.7. Scalability Matrix for BeeHive

		n8			n57			n150			n350			n650			n1050		
		4.6	2.6	1.6	4.6	2.6	1.6	4.6	2.6	1.6	4.6	2.6	1.6	4.6	2.6	1.6	4.6	2.6	1.6
n8	4.6	1	1.177	1.309	0.668	×	×	0.364	×	×	0.148	×	×	0.02	×	×	0.01	×	×
	2.6	×	1	1.112	×	0.291	×	×	0.408	×	×	0.05	×	×	0.018	×	×	0.009	×
	1.6	×	×	1	×	×	0.074	×	×	0.246	×	×	0.016	×	×	0.013	×	×	0.004
n57	4.6	×	×	×	1	0.513	0.145	0.544	×	×	0.222	×	×	0.031	×	×	0.016	×	×
	2.6	×	×	×	×	1	0.283	×	1.402	×	×	0.173	×	×	0.062	×	×	0.032	×
	1.6	×	×	×	×	×	1	×	×	3.31	×	×	0.228	×	×	0.179	×	×	0.057
n150	4.6	×	×	×	×	×	×	1	1.32	0.885	0.407	×	×	0.057	×	×	0.029	×	×
	2.6	×	×	×	×	×	×	×	1	0.67	×	0.123	×	×	0.044	×	×	0.023	×
	1.6	×	×	×	×	×	×	×	×	1	×	×	0.069	×	×	0.054	×	×	0.017
n350	4.6	×	×	×	×	×	×	×	×	×	1	0.401	0.149	0.14	×	×	0.073	×	×
	2.6	×	×	×	×	×	×	×	×	×	×	1	0.373	×	0.359	×	×	0.189	×
	1.6	×	×	×	×	×	×	×	×	×	×	×	1	×	×	0.783	×	×	0.253
n650	4.6	×	×	×	×	×	×	×	×	×	×	×	×	1	1.024	0.833	0.52	×	×
	2.6	×	×	×	×	×	×	×	×	×	×	×	×	×	1	0.813	×	0.527	×
	1.6	×	×	×	×	×	×	×	×	×	×	×	×	×	×	1	×	×	0.323
n1050	4.6	×	×	×	×	×	×	×	×	×	×	×	×	×	×	×	1	1.038	0.517
	2.6	×	×	×	×	×	×	×	×	×	×	×	×	×	×	×	×	1	0.498
	1.6	×	×	×	×	×	×	×	×	×	×	×	×	×	×	×	×	×	1

Table 4.8. Scalability Matrix for AntNet

picture is that *BeeHive* has a higher productivity than *OSPF* in all cases, but its scalability metric is less than that of *OSPF* because its productivity in n57 is significantly higher compared to *OSPF*. Right now, we are pursuing different options to take care of such anomalous behavior in our definition of the scalability metric.

For now, we propose a principle: *a reference topology for analyzing the comparative scalability behavior of routing protocols must be chosen in such a fashion that all candidate algorithms should have similar productivity values in it.* This principle is crucial to follow in the current scalability framework because it takes care of the complex relationship between the performance of a routing protocol and the topology. Luckily, n8 is one such topology; therefore, the scalability analysis based on the performance values in this topology is completely unbiased because all three algorithms have similar productivity values for all traffic patterns (see Tables 4.4, 4.5, and 4.6).

4.6 Summary

We have empirically evaluated the behavior of two state-of-the-art nature-inspired routing algorithms and a single-path routing algorithm on a set of topologies ranging from eight to 1,050 nodes. The results unequivocally demonstrate that *BeeHive* is able to deliver superior performance under both high and low network traffic loads in all topologies. This is a result of considering scalability as an important factor in the design and development of our *bee agent* model. The true benefits of the algorithm are clearly visible in larger topologies. The performance of *AntNet*, however, significantly degrades in topologies of 350 or more nodes.

We also proposed power and productivity metrics for a distributed routing algorithm, which depend on a number of important performance values. The productivity metric provides unbiased insight into the behavior of a routing algorithm: at what cost does the algorithm improve its performance? Finally, we defined a scalability metric to reach a decision on the scalability of a routing algorithm and demonstrated that the *BeeHive algorithm scales better than the rest of the algorithms in a majority of the cases because of its superior benefit-to-cost ratio.*

The scalability metric suffers from a shortcoming: it takes the ratio of the productivity values in two configurations, and as a result it simply factors out the magnitude of the productivity. Therefore, one has to cross-examine it with the productivity value for a comprehensive scalability analysis. In our future work, we want to rectify this shortcoming. Our idea is that the scalability metric should not be just a simple ratio of the two productivity values, but it should be a function of their magnitude as well. We think that this can be achieved by defining a productivity metric for each topology as suggested in [210] and then modeling the scalability metric of the algorithm based on all these factors. Moreover, we would like to modify our simulation model in such a manner that we are able to run simulations under congested network traffic loads in large topologies.

	n8			n57			n150			n350			n650			n1050		
	4.6	2.6	1.6	4.6	2.6	1.6	4.6	2.6	1.6	4.6	2.6	1.6	4.6	2.6	1.6	4.6	2.6	1.6
n8 4.6	1																	
2.6	1.07	1																
1.6	1.127	1.053	1															
n57 4.6	0.397	x	x	1														
2.6	x	0.131	x	0.353	1													
1.6	x	0.087	0.249	0.705		1												
n150 4.6	1.237	x	3.11	x			1											
2.6	x	0.387	x	2.95	x		0.334	1										
1.6	x	x	0.21	x	2.39		0.191	0.572	1									
n350 4.6	1.451	x	3.65	x	1.172		x			1								
2.6	x	0.557	x	4.25	x		1.44	x	0.411		1							
1.6	x	0.231	x	2.63	x		1.099	0.179	0.436			1						
n650 4.6	1.423	x	3.58	x	1.15		x	0.981	x				1					
2.6	1.046	x	7.98	x	2.7		x	1.87	x	0.786				1				
1.6	x	0.501	x	5.7	x		2.38	x	2.16	0.396	0.504				1			
n1050 4.6	0.69	x	1.73	x	0.557		x	0.475	x	0.484	x					1		
2.6	x	0.462	x	3.52	x		1.193	x	0.828	x	0.441		x	0.716			1	
1.6	x	x	0.279	x	3.18		x	1.328	x	1.208	x	0.557	0.456	0.636				1

Table 4.9. Scalability Matrix for OSPF

We believe that the scalability framework presented in this chapter will help designers of routing protocols to consider scalability as an important factor in their design and development, adding empirical validation as a corrective measure at an early stage.

5

BeeHive in Real Networks of Linux Routers

The major contribution of the work presented in this chapter is a Natural Engineering *approach that we developed to design and implement nature-inspired routing protocols in Linux routers. The approach helped us in developing a* natural routing framework *inside the network stack of the Linux kernel. The* natural routing framework *was instrumental in our realizing* BeeHive *inside the Linux kernel. We developed a protocol verification system to compare the performance of the algorithms from a simulated network with that of a real network. We performed extensive experiments in both environments to show that the performance of* BeeHive *is significantly better than that of* OSPF. *The work is a quantum leap because we have laid the ground for empirically refuting the deeply rooted notion held by engineers in the telecommunication industry that nature-inspired routing protocols are not economically viable because they cannot be implemented with existing resources. The work, we believe, will also be instrumental in bridging the gap between different communities through the process of cross-fertilization of their design doctrines.*

5.1 Introduction

The work presented in this chapter was undertaken in order to respond to attitudes and prejudices from the classical networking community. This community strongly held the belief that nature-inspired routing protocols, though intriguing in character, cannot provide benefits in real networks as they are only evaluated in simulations. The criticism is that these algorithms do not pay attention to engineering principles and constraints. Consequently, their installation on real-world routers will require additional hardware and software resources. As a result, the benefit-to-cost ratio of the algorithms will render them economically nonviable to the telecommunication market where cutthroat competition exists between different router vendors. To date, according to our knowledge, the lack of any empirical work showing the benefits of such algorithms in real networks has certainly helped to maintain and strengthen this attitude.

We agree with the engineers working in the telecommunication industry about the *current lack of an engineering vision during the design and development of nature-inspired routing protocols*. *AntNet* had been proposed eight years ago and *DGA* was proposed four years ago, and as of today, these significant and novel algorithms have not been implemented into real networks. So, there is not yet any commitment to accept nature-inspired routing algorithms in real networks. Given this, effort is needed to overcome the above-mentioned attitude widely spread in industry.

With this in mind, we placed a strong emphasis on designing and developing an engineering approach during the design and development of our nature-inspired protocol. As a particular imperative, we decided that *the algorithm must be realizable in existing real-world routers, without the need of additional hardware or software resources to make it economically viable. The algorithm should not require any software components in the simulation that are not available in the real-world networks.* We hold this as a basic principle for *Natural Engineering* (see Chapter 1). As a result, this will lead to:

- Developing economically viable nature-inspired networking systems, an aspect so far neglected by the nature-inspired routing community.
- Ensuring that the algorithm developers avoid making unrealistic assumptions about the environment of real networks. In particular, algorithms must not utilize any features that are not available in real-world routers.
- Taking into account resource constraints in the real world, a fact that has so far received very little attention in the nature-inspired routing community.
- Demonstrating that nature-inspired routing protocols can truly deliver performance results in real networks similar to those obtained in a simulation environment. This will be instrumental for refuting the suspicion held in the commercial world. At any rate, *it will still contribute to a radical directional shift of design work in the nature-inspired routing community.*
- Providing for cross-fertilization of ideas and paradigms from fundamentally different design approaches. This opens a good perspective for developing state-of-the-art networking systems for complex networks of the new millennium.

The major contribution of the work is the engineering approach mentioned above that we realized while developing our nature-inspired routing algorithm, *BeeHive*, to operate in packet-switched telecommunication networks. We have implemented *BeeHive* inside the network stack of the Linux operating system in a real-world router. Using our comprehensive performance evaluation framework, we compared its performance with *OSPF*, in a real network of eight Linux routers, which were part of our network systems lab of LS III at Technical University of Dortmund. The results obtained from the extensive experiments reveal that our engineering approach has paid its dividend because *BeeHive* outperforms *OSPF* in a real network of Linux routers even in those aspects that had been considered critical by the network engineering community.

5.1.1 Organization of the Chapter

Section 5.2 will provide an overview of different design approaches available for implementing a routing algorithm in a real Linux router. The section will also introduce the important engineering issues that the designer of a nature-inspired routing protocol should consider during the life cycle of a protocol. We will introduce our *natural routing framework* in Section 5.3 by highlighting its algorithmic-independent and the algorithmic-dependent components. The reader will appreciate that the framework is general enough to realize the relevant features of a nature-inspired routing protocol. Section 5.4 introduces our protocol verification framework that we developed to comprehensively compare and evaluate an algorithm both in a simulation environment and in a real network of Linux machines. We will introduce the motivation behind our experiment design in Section 5.5 and then discuss the results. Finally, we will provide a summary of the chapter.

5.2 Engineering of Nature-Inspired Routing Protocols

The engineering approach is of paramount importance for the realization of a routing protocol in a real network router [140]. Our extensive research into this intriguing domain has unfolded the true challenges that one faces in undertaking this strenuous task. We followed the basic software engineering design principle: *a good software organization emphasizes structures and components that are easily reusable. The reusability is achieved by decoupling their design from the implementation-level details* [25, 40]. Such an approach enables a designer to concentrate on top-level design issues and pay little attention to the implementation details at an initial stage of the protocol development. Combined with our *Natural Engineering* approach, this helped in accomplishing the challenging task of creating a *natural routing framework* inside the network stack of the Linux kernel. We addressed the following issues, in order:

1. Structural design of a routing framework.
2. Structural semantics of the network stack.
3. System design issues.

5.2.1 Structural Design of a Routing Framework

The performance of a routing framework depends on its *structure* and the cost of certain operating system (OS) functions, such as context switching and interrupt handling [46]. The *structure* of a framework defines the partitioning of its important components between the user space and the kernel space [137]. Consequently, one should carefully make this crucial decision because a thoughtless or imperfect partitioning of the components between the user space and the kernel space leads to frequent context switching, resulting in a poor performance [137, 48]. A *structure* should be designed in such a manner that it carefully optimizes access to the OS functions like process scheduling, memory allocation and management, and interrupt handling. The extensive studies done by Kay and Pasquale [134] indicate that the cost of processing an interrupt, copying a packet from the kernel space to the user space, and context switching overheads is at least as expensive as that of the protocol processing. A good routing framework should take all these factors into account.

Thekkath et al. [248] have provided a good survey of the choices for partitioning the components of a protocol stack between the user space and the kernel space. Their major emphasis is on structuring a *protocol stack* and not a *routing framework*. Nevertheless, the design principles and choices are valid for a *routing framework* as well. According to Keshave [137], a designer has to make a multi-way trade-off among the following factors in order to reach at a suitable decision on the partitioning problem:

- *Software engineering considerations* are influenced by the difficulty level of writing, testing, and maintaining the code for the routing framework either in the user space or in the kernel space.
- *Customization* is the ability of a routing framework to satisfy the requirements of the applications.
- *Security* concerns stem from the fact that a user process can harm other user processes. One has to be careful in placing a service into an untrusted user space as opposed to a trusted kernel space.
- *Performance* is the ability of a routing framework to accomplish its task as quickly as possible. The interested reader will find details about performance enhancement techniques and systems in [165, 81, 267, 134, 120].

In the following we introduce three basic choices: *a monolithic routing framework in the kernel space, a monolithic routing framework in the user space, and a hybrid implementation in the kernel space and user space.* We considered the first one in our routing framework.

Monolithic routing framework in the kernel space

The most important components of a routing framework are: routing tables, route discovery and maintenance, and packet switching. In this approach, all these components are encapsulated inside a module within the Linux kernel. Figure 5.1 illustrates this structure. This routing framework has the highest security and the best performance; however, it is a challenging task to implement a module inside the kernel and then debug the kernel code. The reason is: the kernel space has only a basic set of data structures that are significantly limited in their functionality. The important libraries like math, Standard Template Library (STL), etc. available in the user space are nonexistent in the kernel space, prolonging the implementation and debug phases. Last but not least, the routing framework is strongly coupled with the kernel code of a particular kernel version, and porting it to new enhanced kernel versions may not be straightforward at times, which makes its source code difficult to maintain. The interested reader will find further details in [248, 267].

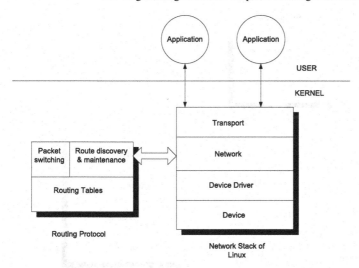

Fig. 5.1. Monolithic implementation in kernel space

Monolithic routing framework in user space

In this approach, all components of a routing framework are implemented as a single process in the user space [248, 258, 84]. Figure 5.2 illustrates this structure. A user space module provides the flexibility of using user space data structures and libraries, which significantly reduce the time to realize a routing framework in a router. The debugging in user space is significantly easier and straightforward compared to that in the kernel space. The routing framework is completely decoupled from the services/features of the kernel, which significantly simplifies its code maintenance. However, the performance of the routing framework is poor because for each arrival of a packet, two context switches are made: one from the kernel space to the user space to find the next hop leading toward the destination; another from the user space to the kernel space to queue the data packet at the network interface of the next hop. In the case of agents, an additional copy of the agent is to be made from the kernel space to the user space in order to carry out its detailed processing in the user space. Finally, the agent is copied back from the user space to the kernel space. If the CPU scheduler does not give enough preference to the routing framework process, then this scheduling latency degrades the overall performance of the algorithm. Consequently, all these factors significantly degrade the packet-switching rate of a network router. Last but not least, moving a routing framework into the user space raises serious security concerns as well. However, the solutions to the security concerns are discussed in [81, 85, 248].

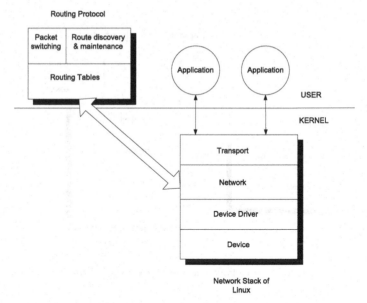

Fig. 5.2. Monolithic implementation in user space

Hybrid routing framework

In this approach, the control path is separated from the data path. Consequently, the route discovery and maintenance component is moved into the user space while the routing tables and the packet-switching components remain in the kernel space. The two modules communicate with one another via the /proc file system [149]. This approach combines the benefits of easy implementation in the user space for handling the complex behavior of agents (which requires special libraries and tools) with the efficient packet switching of data packets in the kernel space, without the need of two context switches for each routing decision. Figure 5.3 illustrates the approach. The frequency of agent-processing is significantly smaller compared to that of packet switching under high loads; therefore, the performance of the algorithm is not significantly degraded. The agent processing component is the most important one of a nature-inspired routing algorithm, and this approach completely decouples it from the corresponding kernel version; therefore; it becomes easier to maintain such a routing framework.

We decided to opt for the monolithic implementation in the kernel space after thoroughly reviewing the merits and demerits of the three approaches. The security and performance issues played a vital role in our decision-making process.

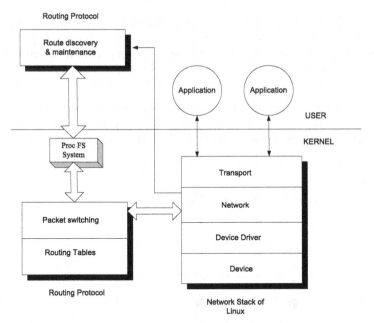

Fig. 5.3. Hybrid implementation

5.2.2 Structural Semantics of the Network Stack

The next factor that influences the design of a routing framework is the interface between different layers of a network protocol stack. Generally, three strategies are employed [137]:

1. The *single context* approach calls for a single non-preemptible thread of execution in which only one packet is processed at a time by our locking the access to the stack. For such a system the only option available is monolithic implementation in the kernel space.
2. The *tasks-oriented model* implements each layer as a task scheduled by a central task scheduler with a pointer to the buffer containing a packet. In this approach a packet is shepherded through the stack by a sequence of tasks that mutually schedule each other. Such an approach allows for Quality of Service (QoS) guarantees because a scheduler has the freedom to schedule high-priority tasks over low-priority tasks [5]. However, this approach has a higher latency because of frequent intervention of the scheduler during the journey of a packet through the stack. For such a system, any of the above-mentioned three strategies will work.
3. The *upcall* architecture associates a thread with each data packet, and it is responsible for all layers of the protocol processing. High-priority threads handle high-priority packets. Each layer's registers send and receive entry points to and from its lower layer. As a result, it can make a procedure call on the preregistered

entry point [47, 120] during the reception or transmission of the packets. In such a model, no data is copied between the layers, which enhances its performance. Linux has an upcall architecture; therefore, it is important to identify the entry points to and from the data link layer and the transport layer. For such a system, any of the above-mentioned three implementation strategies can be utilized.

The upcall system architecture facilitates conveniently structuring actions of a routing protocol in an asynchronous manner as responses to the five basic events shown in Figure 5.4 through entry and exit points. The NetFilter architecture in Linux

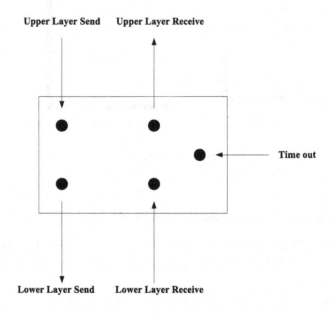

Fig. 5.4. Protocol block from [137]

[193, 1], based on the concept of Packet Filter [166], provides a simple mechanism for receiving and processing packets from the protocol stack in the network layer. This helps in realizing a routing framework as a protocol block as illustrated in Figure 5.4.

5.2.3 System Design Issues

A routing framework is part of a distributed system built from hardware and software resources [137]. The development of a routing framework, therefore, requires a thorough understanding of system design techniques with a high emphasis on computer networks. An interested reader will find a detailed treatment on operating systems in

[220, 246], on computer architecture in [112], and on network systems/protocols in Unix in [237, 236, 234, 293, 235]. As mentioned before, we followed an engineering approach, which tries to optimally balance the requirements for *constrained* and *unconstrained* resources during the design and development of a routing framework. We tried to account for the following resources: time, space, computation, and labor.

1. *Time:* Our routing framework must be simple enough so that it requires as little time of the processor as possible. The framework must not put any additional hardware constraints on the existing computers for its functionality. This will decrease its time to market as well.
2. *Space:* Like time, space is also a limited resource. Space constraints are expressed by limits on available memory, data packets in routers, routing tables, and other routing information obtained from the agents. A routing algorithm should try to utilize this resource as efficiently as possible. A carefully designed and efficiently implemented routing framework is a prerequisite for achieving this goal.
3. *Computation:* Our framework must be able to work with the low-end Pentium III series of processors that we have at our disposal in our LS III network systems lab. This will refute the popular notion held by the networking system community that Nature-inspired approaches are too complex to be realized with the existing hardware and software resources.
4. *Labor:* Our framework should be designed in such a fashion that we can efficiently implement it with the help of two full-time software developers working for eight months. The workload initially estimated was about 2,800 man- and machine-hours. Luckily, we were able to accomplish the task in approximately 3,000 man- and machine-hours. A well-planned design strategy, as discussed in the previous section, helped in our meeting important milestones and deadlines. In the first phase, we implemented and tested the *BeeHive* algorithm in a virtual network of virtual Linux machines [299], and in the second phase we made the final transition of realizing the algorithm in a real network of real Linux machines [105].

We agreed to work on a design model that is scalable: it can be easily adapted to new architectures, systems, and increasing network size. We expect that this feature will significantly reduce the cost of porting the framework to large heterogeneous systems.

5.3 Natural Routing Framework: Design and Implementation

We designed, developed, and implemented our *natural routing framework* with two layers of abstraction: algorithmic-independent and algorithmic-dependent. The algorithmic-independent framework consists of structures and services that facilitate the realization of a nature-inspired routing protocol inside the network stack of the Linux kernel by utilizing its interfaces and services. As mentioned in the previous section, typically a nature-inspired routing protocol consists of three components:

1. *Agents* which collect information about the state of the network and then store it at the nodes they visit.
2. *Routing tables* that act as the repository for storing the routing information collected by the agents. The tables also cater to information exchange, either directly, or indirectly through the environment. The nature-inspired routing protocols require specialized tables for their correct functioning.
3. *The Packet-switching* task is achieved by distributing data packets in a stochastic manner based on quality, which is calculated based on the information in the routing tables along the paths.

An algorithmic-independent framework consists of structures and features that contribute to implementing the above-mentioned components of a routing algorithm.

5.3.1 Algorithm-Independent Framework

The algorithm-independent framework consists of an algorithmic module inside the Linux kernel. The motivation for a monolithic implementation in the kernel space has been comprehensively substantiated in the previous section. The framework consists of three components: agent manager, agent processor, and packet processor (for both incoming and outgoing data packets). For these components, we now discuss in detail the implementation of our module.

Agent manager

The task of the agent manager is to periodically or aperiodically launch the agents which collect the routing information from the network. The agents can be launched either in a broadcast mode or in a point-to-point mode, depending on the design of the algorithm. The manager sleeps in an *interruptible* fashion after launching the agents. The *interruptibility* enables the manager to asynchronously/aperiodically launch the agents if an event of interest occurs.

Agent processor

The agent processor runs in the background as a *daemon* and is responsible for receiving and processing the agents. Once it receives an agent it executes the behavior component of the agent. Subsequently, if the agent needs to update the routing information in the routing tables, then it can request the agent processor, the only component authorized to update the routing table, to do it. Once the agent has executed its actions, the agent processor forwards it, if needed, to all neighbor nodes of the current node (if broadcast mode is selected) except the one from which it arrived. Otherwise, in the point-to-point mode, it will forward the agent to the next hop determined by the routing algorithm.

Packet processor

The packet processor is allowed to access the routing information in the routing tables but not to modify it. A packet processor can receive packets either in its *pre_input_hook*, which in turn is connected to the IP_PRE_ROUTING of the NetFilter, represented as "PRE Mangle" in Figure 5.5, or in its *output_hook*, which in turn is connected to IP_LOCAL_OUTPUT, represented as "OUT Mangle" in Figure 5.5. The packets passing through the current node are processed in the *pre_input_hook*. If the current node is the destination of the packet, then it is delivered to the transport layer; otherwise, its next hop interface is looked in the routing table and it is sent there. The packets that are generated at the current node are processed in the *output_hook*, and the routing decision is made in this hook.

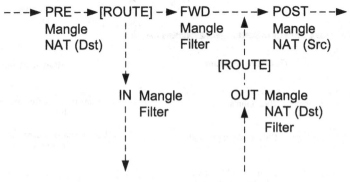

Fig. 5.5. NetFilter hooks

5.3.2 Algorithmic-Dependent BeeHive Module

The *BeeHive* module utilizes the services and interfaces of the above-mentioned algorithmic-independent routing framework. The tasks of processing and managing the information of the *bee agents* are accomplished by utilizing the agent manager and agent processor components of the framework. The packet-switching functionality is achieved through utilizing the services of the packet processor, and the routing information is maintained in routing tables especially designed and developed for multi-path stochastic routing algorithms. Figure 5.6 shows in detail the block diagram of the *BeeHive* module.

Bee processing and management

The important task of the *agent manager* is to periodically or aperiodically launch a *bee agent* as depicted in the procedure *launchBeeAgents(s,n_i,n)* in Algorithm 1 (see Chapter 3). The *bee agents* can also be asynchronously launched once a node

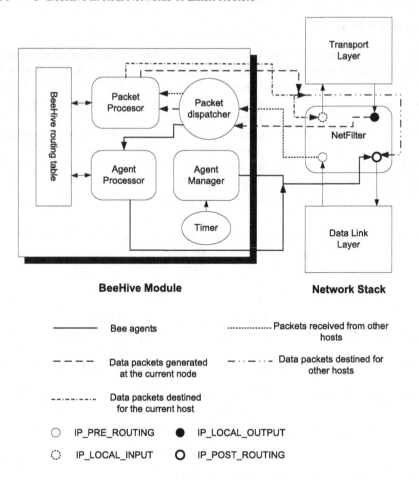

Fig. 5.6. Block diagram of BeeHive module

receives m (currently 240) packets; the paths leading to the destination node of frequent network traffic should be sampled and explored at a higher frequency than the others. The agent manager, once awake, does the following three tasks:

- It creates a *bee agent* and writes its own IP address into the source field. It assigns it a bee agent id and a time-to-live (TTL) value in term of the number of hops permitted.
- It retrieves the IP address of each valid network interface except the loopback adapter. It creates a replica of the *bee agent* for each valid IP interface and assigns it a distinct replica id.

- It increments the bee agent id.

The *bee agents* launched by the *agent managers* are received at other nodes through the *agent processors*. The *agent processor*, as discussed earlier, is a daemon process that listens on a predefined UDP port reserved for the transmission and reception of *bee agents*. The tasks of an *agent processor* are outlined in the *processBeeAgents()* method of Algorithm 1 (see Chapter 3). The concepts of *foraging regions* and *foraging zones* are not implemented because their benefits on a small topology are not obvious. The *agent processor* does the following tasks:

- If the time-to-live (TTL) timer has reached 0 or if the replica has already visited the same node, then it is killed (to avoid loops).
- At start-up time, the *agent processor* estimates the propagation delay of the transmission links connected to its neighbors. The received replica updates its estimate of the queuing delay by calculating the queue length of the interface of the neighbor from where it arrived at the current node. Similarly, it updates its estimate of the propagation delay.
- The received replica then requests the *agent processor* to update in the routing tables the estimates of the queuing and propagation delays of its source node for the neighbor from where it arrived. Please remember that only an *agent processor* has the permission to modify the routing tables.
- If available, the replica reads the estimates of the queuing and propagation delays of other replicas that are also launched from its source node. The replica updates its view of the network state based on the information communicated by the other replicas (see Figure 3.3).
- If a replica of the same generation (bee agent id) has already been received at the node then the *agent processor* kills the current replica.
- The *agent processor* retrieves the IP address of each valid network interface except the loopback adapter and the network interface from where the replica is received. The replica is broadcast to the retrieved IP addresses without modifying its replica or agent id.

Packet processor

The data packets are sent to the *packet processor* by the *packet dispatcher*, which is connected to NetFilter at two interfaces: *pre_input_hook* and *out_hook*. The packets generated by the applications running at the current host are received in the *out_hook*. The packets sent from the network interface card are received in the *pre_input_hook*. The general task of a packet processor is to access the information in the routing tables, calculate the quality of each neighbor that can be the next hop towards the destination, and then stochastically choose the next hop toward the destination.

Floating-point operations are not supported in the Linux kernel; therefore, one has to adapt the packet-switching algorithm such that it works with integers only. The simple solution is to rescale the probabilities in the range of 0 to 100 instead of 0 to 1. We simply rescale $g = \frac{K}{p+q}$, where K is a large integer value, p is the propagation delay, and q is the queuing delay. Finally, the integer probability θ_{jd} is $\theta_{jd} = \frac{g_{jd} \times 100}{\sum_{k=1}^{N} g_{kd}}$, where N is the number of neighbors of the current node.

The *packet processor* can receive a data packet from the network interface card through the *pre_input_hook* of the *packet dispatcher*. If the destination of the packet is the current node, then it is given to the upper layers via the *IP_LOCAL_INPUT* hook of NetFilter, represented by "IN Mangle" in Figure 5.5; otherwise, the function *b_route_input()* is called. This function accesses the bee routing table and applies the above-mentioned algorithm to find a next hop. The Linux kernel caches the routing tables for efficient routing of the data packets, under the assumption that the quality of the routes will not drastically change in small intervals of time. In the *b_route_input()* function, the complete processing relating to routing is done, and finally *ip_route_input()* of the network stack is called by giving it the next hop as the final destination. This simple circumvention makes Linux access its cached routing tables with the next hop as the final destination. As a result, the packet will be queued in the network interface of the next hop host determined by the *BeeHive* algorithm. The *packet processor* calls the *ip_route_output_flow_output()* function for processing the data packets that have been received through the *out_hook* of the *packet dispatcher*. These packets are generated at the current node. The processing related to routing is quite similar, as discussed in the *pre_input_hook* case.

Bee routing tables

The routing tables of the Linux kernel render themselves useless when it comes to supporting nature-inspired routing protocols because of the following reasons:

- Nature-inspired routing algorithms take a routing decision on a packet basis. Therefore the idea of caching based on session, destination IP, source IP, port, and type of service (TOS) values does not significantly help.
- Nature-inspired routing algorithms are multi-path routing algorithms. Consequently, it is possible to reach each destination host through multiple neighbors, and this information has to be incorporated into the routing tables.
- Nature-inspired routing protocols implicitly assume that the hosts are *polymorphic*. A polymorphic host has different IP addresses for each of its network cards, but all of these IP addresses represent the same host. *OSPF* does not care about this because it finds a single path to each IP address.

These shortcomings motivated us to design and develop a *BeeHive*-specific routing table that would not suffer from the above-mentioned limitations. The bee routing table consists of three other tables as shown in Figure 5.7: IP table, host table, and neighbor table. The IP table contains all the possible destination IP addresses. Each IP address is mapped to a host in the host table. If a host has multiple IP addresses then the corresponding IP entries point to the same host. Each host points to a neighbor table that contains all neighbors that could lead to the same destination. The tables are currently implemented with the help of linked list structures. The IP table consists of a linked list of rt_ip structures as shown in Figure 5.8. Each entry in the IP table is linked to an entry of the hosts table, which is a double linked

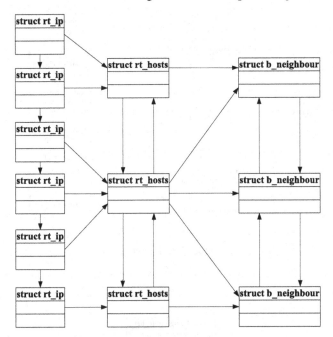

Fig. 5.7. Top-level routing table

list of rt_hosts structures (see Figure 5.8). The basic idea is that different rt_ip structures (IP addresses) of the same host point to the same rt_hosts structure, and an rt_hosts structure points to all its rt_ip structures. We maintain a double linked list of b_neighbour structures (as shown in Figure 5.9) that is only accessible through an entry of a linked list of rt_neighbour structures (as shown in Figure 5.10). Each entry in the double linked list of rt_hosts (host table) points to its own linked list of the rt_neighbour structures through which a packet can reach its destination node. In order to efficiently implement the routing table, the linked lists of rt_neighbour structures of different hosts point to their corresponding entries in the same double linked list of b_neigbhour structures. This significantly improves the fault management; if a neighbor node crashes then its corresponding b_neighbour entry is made invalid and this virtually removes it from the linked list of all rt_neighbour structures because an invalid entry is not considered in the routing process. Refer to Table 5.1 for the description of important symbols used in the rest of the chapter.

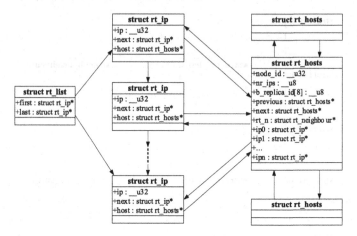

Fig. 5.8. IP table

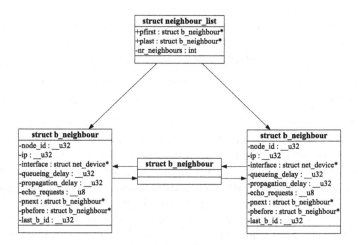

Fig. 5.9. Neighbor table

5.4 Protocol Verification Framework

We have designed and developed a comprehensive and sophisticated protocol verification framework for extensive performance evaluation, comparison, and verification of a routing protocol, both in a simulation environment and in real networks. The conceptual block diagram of the framework is illustrated in Figure 5.11.

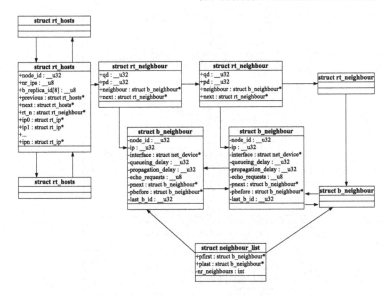

Fig. 5.10. Detailed bee routing table

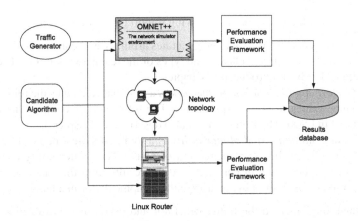

Fig. 5.11. Protocol verification framework

P_{gen}	Total number of packets generated
P_{rec}	Total number of packets received
S_{to}	Total number of sessions started
S_{co}	Total number of sessions completed
T_{av}	Average throughput (M/sec)
P_d	Packet delivery ratio (%)
sSize	size of a session (in bytes)
pSize	size of a packet (in bytes)
MSIA	Mean of session inter-arrival times (sec)
R_o	Control overhead (%)
MPIA	Means of packet inter-arrival times (sec)
S_o	Suboptimal overhead (%)
P_{loop}	Percentage of packets that followed a cyclic path
S_c	Session completion ratio (%)
t_d	Average packet delay (msec)
t_{90d}	90th percentile of packet delays (msec)
S_d	Average session delay (msec)
h_o^{sd}	minimum hops needed to reach from node s to node d
S_{90d}	90th percentile of session delays (msec)
h_i^{sd}	hops packet i took to reach from node s to node d
J_d	Average jitter value (msec)
h_{av}	Average hops count the data packets
J_{90d}	90th percentile of jitter times (msec)
β_c	Queue Length (in packets)
env	Environment
$algo$	Algorithm
$scen$	Scenario

Table 5.1. Symbols used in the chapter

The framework consists of the following important building blocks:

1. The *traffic generator* is a tool that generates arbitrary but realistic traffic patterns which represent different classes and types of the real network traffic. The traffic generator has been introduced in Chapter 3.
2. The *performance evaluation framework* is a tool that calculates the relevant performance values from the experiments. The framework is comprehensively introduced in [268]. The framework is also described in Chapter 3.
3. The *results database and plotter utility* is a tool that stores the most important parameters from the experiments in a database. The plotter utility consists of a set of scripts written in Perl that read the results from the database and then present them, as desired by the user, either graphically or in a table.

We have implemented the traffic generator and performance evaluation framework both in the network simulator OMNeT++ [255] and in an application layer of the Linux operating system network stack as illustrated in Figure 5.11. By assigning the same values to the above-mentioned parameters of the traffic generator, we expect to generate nearly similar traffic patterns in simulation and real network environments. In the next step of our protocol verification process, we utilize a topology generator that generates a topology of a given number of nodes and links. The links are fully characterized by their bandwidth, propagation delay, and bit error rate. During this process, we select a topology that we will use in our simulation environment and in the real network of Linux routers. The topology is known as simpleNet and it is shown in Figure 5.12. The detailed description of each Linux router machine that includes its processor, RAM, cache, hard disk, and network interface cards is cataloged in Table 5.2.

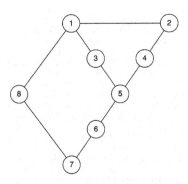

(a) simpleNet in OMNeT++

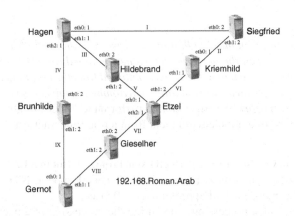

(b) simpleNet of Linux machines in LS III lab

Fig. 5.12. simpleNet topology: 8 routers, 9 bidirectional links each of 10 Mbits/sec

The protocol verification principle is: *if we generate the same traffic patterns through the same traffic generators in both OMNeT++ and the Linux network and*

Node in OMNet++	Node in Linux	Machine specification	Address in OMNeT++	Address in Linux
1	Hagen	Intel Pentium III, 500 MHz, 512KB L2 Cache, 128MB RAM, 3 Realtek Semiconductor Co., Ltd. RTL 8139/8139C+(rev10) network cards	1	192.168.1.1 192.168.3.1 192.168.4.1
2	Siegfried	Intel Pentium III, 500 MHz, 512KB L2 Cache, 128MB RAM, 2 Realtek Semiconductor Co., Ltd. RTL 8139/8139C+(rev10) network cards	2	192.168.1.2 192.168.2.2
3	Hildebrand	Intel Pentium III, 500 MHz, 512KB L2 Cache, 384MB RAM, 2 Realtek Semiconductor Co., Ltd. RTL 8139/8139C+(rev10) network cards	3	192.168.3.2 192.168.5.2
4	Kriemhild	Intel Pentium III, 500 MHz, 512KB L2 Cache, 384MB RAM, 2 Realtek Semiconductor Co., Ltd. RTL 8139/8139C+(rev10) network cards	4	192.168.2.1 192.168.6.1
5	Etzel	Intel Pentium III, 500 MHz, 512KB L2 Cache, 128MB RAM, 3 Realtek Semiconductor Co., Ltd. RTL 8139/8139C+(rev10) network cards	5	192.168.5.1 192.168.6.2 192.168.7.1
6	Giselher	Intel Pentium III, 500 MHz, 512KB L2 Cache, 384MB RAM, 2 Realtek Semiconductor Co., Ltd. RTL 8139/8139C+(rev10) network cards	6	192.168.7.2 192.168.8.2
7	Gernot	Intel Pentium III, 500 MHz, 512KB L2 Cache, 384MB RAM, 2 Realtek Semiconductor Co., Ltd. RTL 8139/8139C+(rev10) network cards	7	192.168.8.1 192.168.9.1
8	Brunhilde	Intel Pentium III, 500 MHz, 512KB L2 Cache, 384MB RAM, 2 Realtek Semiconductor Co., Ltd. RTL 8139/8139C+(rev10) network cards	8	192.168.4.2 192.168.9.2

Table 5.2. The mapping of Hosts to IP Addresses in SimpleNet

utilize the same performance evaluation framework again in both OMNeT++ and the Linux network, then the performance values obtained from the simulation environment should be consistent with the ones obtained from the real Linux network with minor deviations, provided our simulation environment depicts a somewhat realistic picture of a true network. The deviation in performance values might stem from the differences of the simulation environment from a true Linux network topology such as:

- The clocks in OMNeT++ are perfectly synchronized. But in a LAN of Linux machines they are synchronized using Network Time Protocol (NTP) [163], which can provide an accuracy of approximately 10 msec.
- The network stack processing time during the processing of a packet, including both control and data, is ignored in the simulation environment.
- The context-switching time from kernel space to user space and vice versa is ignored in a network simulator.
- The latency of scheduling a task, dependent on the scheduler of an operating system, is ignored in the simulation. This is of primary importance once multiple and concurrent sessions are active in a Linux router.
- The nonavailability of high resolution timers in Linux [149] might result in MPIA values different from those of OMNeT++.
- The protocol-specific characteristics of Ethernet employed at the data link layer (layer 2), its direct influence on the signals at physical layer, and the transmission medium are ignored in the simulation for the sake of simplicity.

5.5 The Motivation Behind the Design and Structure of Experiments

The basic motivation behind designing the scenarios for the experiments was to cover a broad range of operational environments and to quantify the performance values of different parameters in a systematic fashion. After extensive brainstorming, we agreed to divide our experiments into three broad categories according to the network traffic: quantum traffic engineering, real-world applications, and hybrid traffic engineering. The idea of *quantum traffic engineering* is to provide to each tested algorithm, abstract but repeatable traffic patterns in order to obtain statistically significant performance values through multiple (in our case ten) independent attempts. We utilized our Scientific Quantum Traffic Generator (SQTG), introduced in Chapter 3, for this purpose. For real-world applications traffic, we used the File Transfer Protocol (FTP) [192] to download large files. These experiments revealed quantifiable benefits, which are of pertinent interest to the industry, in employing different routing protocols. Finally we utilized the Distributed Internet Traffic Generator (D-ITG) [26, 12, 13, 14, 11] developed at the Università degli Studi di Napoli in Italy to emulate Voice over IP (VoIP) traffic [35, 4] along with UDP traffic. Moreover, utilizing a traffic generator developed by a third party would eliminate any bias that might have been introduced to our *BeeHive* algorithm by our own traffic generator. We call these experiments hybrid ones because they simulate a real-world application (VoIP) traffic but we are not using a true VoIP application . During the extensive performance evaluation and verification cycle, all performance values are an average of performance values obtained from ten independent experiments. This was done to factor out stochastic elements in the network environment or in the algorithms. The experiments were conducted for 1,000 seconds, a good enough time for factoring out transients.

5.6 Discussion of the Results from the Experiments

In this section, we discuss in detail the results obtained from our extensive experiments, both in the simulation and in real networks. We discuss the results according to their category type as introduced in the previous section.

5.6.1 Quantum Traffic Engineering

We utilized our SQTG for this series of the experiments. We conducted ten experiments by varying the values of different parameters of the traffic generator. The parameter values are shown in Table 5.3. We generated UDP [190] traffic because it supports an unbiased evaluation of a routing protocol, in contrast to TCP [191], which has a complex mechanism to adaptively control the size of the congestion window if packets are lost.

Exp	sSize	pSize	MSIA	MPIA	β_c	Source	Destination	Hot Spot	Routers Down
1	2130000	512	1.8	0.005	3000	Hagen	Gernot	none	none
2	2130000	512	1.2	0.005	3000	Hagen	Gernot	none	none
3	2130000	512	1.0	0.005	3000	Hagen	Gernot	none	none
4	2130000	512	0.9	0.005	3000	Hagen	Gernot	none	none
5	1065000	1024	1.8	0.005	1500	Hagen	Gernot	none	none
6	1065000	1024	1.2	0.005	1500	Hagen	Gernot	none	none
7	1065000	1024	1.0	0.005	1500	Hagen	Gernot	none	none
8	4055000	512	1.15	0.005	3000	Etzel	Hagen	none	none
9	1270000	512	1.15	0.005	3000	Etzel	Hagen	Hagen(100,800)	none
10	2580000	512	1.15	0.005	3000	Etzel	Hagen	none	Hildebrand(300,600)

Table 5.3. Parameters for traffic generator for experiments 1 to 10

Experiments 1, 2, 3, and 4

The purpose of this set of experiments is to study the behavior of the algorithms
under saturated network loads. The MSIA value was decreased from 2.0 sec to 0.9
sec in these experiments. The size of the session was kept constant at 2,130,000
bytes and the packets were sent from Hagen to Gernot (see Figure 5.12(b)). The
experiments with the same parameters were conducted in both OMNeT++ and the
real network topology. Figure 5.13 shows the packet delivery ratio and 90th percentile of packet delay distribution obtained both from OMNeT++ and Linux. As
expected, the packet delivery ratio of *BeeHive* scales better than that of *OSPF* in both
OMNeT++ and Linux. The packet delivery ratio of *OSPF* drops significantly from
99.9% to about 50% as the MSIA value is decreased from 2.0 sec to 0.9 sec respectively. The reason is obvious: *OSPF* always utilizes the single path Hagen-Brunhilde-
Gernot while *BeeHive* distributes packets over two paths Hagen-Brunhilde-Gernot
and Hagen-Hildebrand-Etzel-Gieselher-Gernot. Consequently, *BeeHive* achieves 17
Mbits/sec throughput (see Table 5.4), which is approximately twice the one achieved
by *OSPF*. *OSPF* saturates the queues of path Hagen-Brunhilde-Gernot; as a result,
a significant queue delay is experienced by data packets. Consequently, one can see
from Figure 5.13(b) that the 90th percentile of packet delays of *OSPF* is significantly higher (1,100 msec) than that of *BeeHive* (20–25 msec). A reader might find
an anomaly in Figure 5.13(b) because it suggests that the packet delays of the algorithms obtained from a real Linux topology are smaller (approximately 10 msec)
compared to the ones obtained from OMNeT++. Our investigation revealed that this
is the result of a clock synchronization problem. Remember that NTP, as discussed
before, can provide a resolution of 10 msec only. Consequently, the packet delay in
simulation and real networks might differ in the 10–15 msec range.

The effect of saturated network load is more prominent in the session completion ratio, as shown in Figure 5.14(a). *OSPF* is hardly able to complete any session
at an MSIA value of 1.2 sec or less. Note in Figure 5.14(b) that the session delay
of *BeeHive* is significantly less than that of *OSPF*, in both OMNeT++ and simulation. However, the difference in session delay values obtained from the simulation
and the real network is significant. We investigated the problem and found that nonavailability of high resolution timers in Linux [149] is responsible for it. Our MPIA
value is 5 msec while Linux has a scheduling uncertainty of 10 msec. As a result, a

session takes longer to finish at the source, and this explains the larger session delay for the algorithms in Linux compared with OMNeT++. With an MSIA value of 0.9 sec the resources of Linux machines were nearly consumed, and the performance values differ significantly from the simulation values mainly because of a significant difference in the operational environments of simulation and reality. We have collected the important performance values from the Linux and OMNeT++ environments in Table 5.4. Note that most of the parameter values, with few exceptions, are easily traceable between simulation and reality. One can safely comment: *the relative tendency of the performance values of the two algorithms in both environments is approximately similar.*

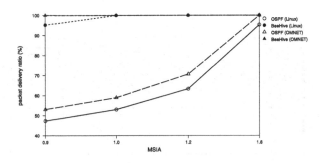

(a) Packet delivery ratio

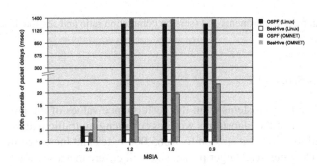

(b) 90th percentile of packet delays

Fig. 5.13. Experiments 1–4 (packet delivery ratio and packet delay)

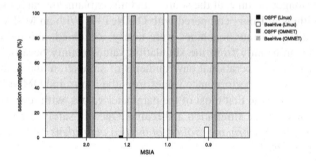

(a) Session completion ratio

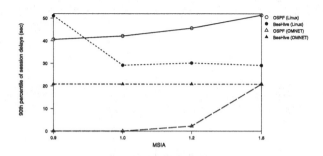

(b) 90th percentile of session delays

Fig. 5.14. Experiments 1–4 (session completion ratio and session delay)

Experiments 5, 6, and 7

The tolerable error that stems from the performance values of the previous experiments led us to test the algorithms with less challenging traffic patterns. Therefore, we decreased the size of a session to 1,065,000 bytes and increased the size of the packet to 1,024 bytes. The larger packet size will ensure that a session finishes faster than in the scenarios of the previous subsection. A smaller session delay will decrease the number of sessions that are concurrently active at the source node. Consequently, it will reduce the error stemming from the scheduling latency of the Linux scheduler. We decreased the MSIA value from 1.8 sec to 1.0 sec. We hoped to see a better relatedness of the performance values obtained from OMNeT++ to those of the Linux topology. Figure 5.15(a) shows the packet delivery ratio of *OSPF* and *BeeHive*, respectively, obtained both from simulations and the real network topology. Both algorithms were able to deliver approximately all the packets to their destination under such a small network load. However, the 90th percentile of the packet

Algorithm		BeeHive				OSPF			
env	MSIA	2.0	1.2	1.0	0.9	2.0	1.2	1.0	0.9
LINUX	P_{gen}	2040050	3402747	4082759	4497561	2056660	3382901	4059611	4511485
	P_{rec}	2040050	3402747	4082667	4274467	2056660	2144274	2155929	2157964
	P_{loop}	0	0	0	0	0	0	0	0
	S_{to}	500	830	995	1108	502	843	996	1107
	S_{co}	500	830	994	94	502	13	0	0
	S_o	2.785	7.961	9.261	9.706	0	0	0	0
	R_o	0.091	0.096	0.089	0.081	0	0	0	0
	t_d	1.8	3	3	4.4	6.9	1245	1247	1248
	S_d	24	29.7	28.7	46.6	21	42.1	39.6	38.3
	J_d	7	7.1	7	7	12.2	6.9	18.3	19.6
	J_{90d}	8	8.1	8	8	8.8	7	21.2	35
	T_{av}	8.8	13.9	16.7	17.5	8.8	8.8	8.8	8.8
	h_{av}	2.6	3	3	3	2	2	2	2
OMNET++	P_{gen}	2060837	3433429	4120063	4577176	2061032	3433665	4120134	4577404
	P_{rec}	2060824	3433399	4120007	4577110	2061026	2425433	2428200	2429529
	P_{loop}	0	0	0	0	0	0	0	0
	S_{to}	501	834	1000	1112	500	834	1001	1112
	S_{co}	490	816	980	1088	490	0	0	0
	S_o	2.9	5.7	8.6	10.6	0	0	0	0
	R_o	0.1	0.1	0.1	0.1	0	0	0	0
	t_d	6.5	7.8	12.4	15	3	1216	1218	1219
	S_d	20.8	20.8	20.8	20.8	20.8	2.2	0	0
	J_d	5	5	5	5	5	5	5	5
	J_{90d}	8.8	9	9	9	5	5	5	5
	T_{av}	8.4	14	16.9	18.8	8.4	9.9	10	10
	h_{av}	2.6	2.7	2.9	3	2	2	2	2

Table 5.4. Important performance values for Experiments 1 to 4 from

delays of *BeeHive* is significantly better than that of *OSPF* because *BeeHive* distributes the traffic load on multiple paths and this results in significantly smaller queue lengths compared with *OSPF*. Both algorithms are able to complete approximately all the sessions (see Figure 5.16(a)), and with approximately the same session delay (see Figure 5.16(b)). Note that the difference in session delays obtained from the simulations and the real network topology is about two seconds, which is reasonably acceptable and supports our thesis: *a concurrent number of sessions have to be scheduled by a Linux scheduler, and because of ambiguity in the scheduling latency, the traffic patterns could differ from the ones in the simulation.* The other important performance values are collected in Table 5.5.

Experiment 8

We designed this experiment to unveil the true performance of *BeeHive*. We selected Etzel as the source node and Hagen as the destination node. Three distinct paths exist between Etzel and Hagen: Etzel-Hildebrand-Hagen, Etzel-Kriemhild-Siegfried-Hagen, and Etzel-Gieselher-Gernot-Brunhilde-Hagen (see Figure 5.12(b)). We theoretically expect that *BeeHive* should be able to deliver approximately three times more packets to their destination than *OSPF*. The important parameters for the traffic generator are listed in Table 5.3. Figure 5.17(a) empirically substantiates our expectations: the packet delivery ratio of *BeeHive* is approximately 99.9% compared to 35% of *OSPF*. Figure 5.17(b) shows that the packet delay of *BeeHive* is 20 msec

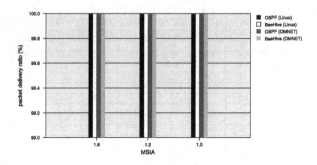

(a) Packet delivery ratio

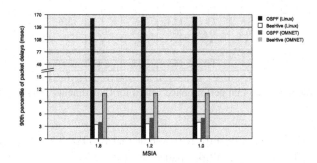

(b) 90th percentile of packet delays

Fig. 5.15. Experiments 5–7 (packet delivery ratio and packet delay)

compared to 1,431 msec of *OSPF*. The other important performance values are collected in Table 5.6. As expected, the throughput of *BeeHive* is approximately 24 Mbits/sec, which is three times higher than the 8 Mbits/sec of *OSPF*. At this load, *OSPF* is unable to complete even a single session (Table 5.6), compared to more than 90% for *BeeHive*. The session delay of *BeeHive* is significantly higher in a real network topology compared with OMNeT++ because of the reasons already discussed. During a few runs, *OSPF* was able to finish one or two sessions, and the session delay was approximately 60 seconds. Note that the jitter of *BeeHive* is significantly smaller than that of *OSPF* in the real network topology. The performance values of the algorithms are easily traceable from simulation to the real network of Linux routers.

Algorithm		BeeHive			OSPF		
env	MSIA	1.8	1.2	1.0	1.8	1.2	1.0
LINUX	P_{gen}	576914	862343	1035056	575898	861758	1035427
	P_{rec}	576914	862343	1035056	575898	861758	1035427
	P_d	100	100	100	100	100	100
	P_{loop}	0	0	0	0	0	0
	S_{to}	556	831	998	555	831	999
	S_{co}	556	831	998	555	831	999
	S_c	100	100	100	100	100	100
	S_o	0.738	2.354	3.026	0	0	0
	R_o	0.014	0.011	0.109	0.003	0.002	0.002
	t_d	2.4	2.5	2.9	160	164	164
	S_d	7.3	7.3	7.3	7.3	7.3	7.3
	J_d	6.8	7	7	7	7	7
	J_{90d}	8.9	8.9	8.9	7	7	7
	T_{av}	4.7	7	8.5	4.7	7	8.5
	h_{av}	2.281	2.6	2.642	2	2	2
OMNET++	P_{gen}	577368	865758	1038789	577343	865777	1038732
	P_{rec}	577363	865751	1038781	577341	865773	1038727
	P_d	99.999	99.999	99.999	99.999	99.999	99.999
	P_{loop}	0	0	0	0	0	0
	S_{to}	556	834	1000	556	834	1000
	S_{co}	553	829	995	553	829	995
	S_c	99.5	99.5	99.5	99.5	99.5	99.5
	S_o	1.636	2.469	2.981	0	0	0
	R_o	0.031	0.036	0.039	0	0	0
	t_d	7.5	7.6	7.6	4	4	4
	S_d	5.2	5.2	5.2	5.2	5.2	5.2
	J_d	6	6.6	6.6	5	5	5
	J_{90d}	9.6	9.5	9.5	5	5	5
	T_{av}	4.7	7.1	8.5	4.7	7.1	8.5
	h_{av}	2.6	2.6	2.6	2	2	2

Table 5.5. Important performance values for Experiments 5 to 7

Environment		Linux		OMNET++	
Algorithm		BeeHive	OSPF	BeeHive	OSPF
P_{gen}		3077866	3049156	3311173	3311178
P_{rec}		3073359	1067326	3311071	1204964
P_d		99.853	35.003	99.998	36.39
P_{loop}		0	0	0	0
S_{to}		433	433	435	435
S_{co}		432	0	401	0
S_c		99.8	0	92.82	0
S_o		13.994	0	14.243	0
R_o		0.224	0	0.151	0
t_d		7.2	1348	13.7	1203
S_d		50.4	60.3	39.6	0
S_{90d}		54.4	67.7	39.6	0
J_d		4.8	24.4	5	5
J_{90d}		12.8	77	7.3	5
h_{av}		3	2	2.9	2

Table 5.6. Important performance values for Experiment 8

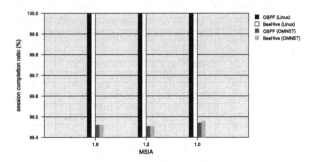

(a) Session completion ratio

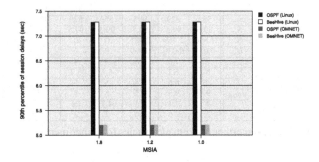

(b) 90th percentile of session delays

Fig. 5.16. Experiments 5–7 (session completion ratio and session delay)

Experiment 9: Hot spot

The purpose of this experiment is to study the behavior of the algorithms under hot spot traffic. A sessionless hot spot traffic with MPIA = 0.001 sec from Etzel to Hagen was superimposed on the normal session-oriented traffic. The parameter values for session-oriented traffic are shown in Table 5.3. The hot spot remained active from 100 seconds to 800 seconds. Figure 5.18(a) clearly shows that the hot spot traffic is successfully delivered by *BeeHive*, in both simulation and real network. In contrast, the packet delivery ratio under *OSPF* decreased by 3% and 6%, in simulation and the real network respectively. The packet delay of *BeeHive*, as shown in Figure 5.18(b), is significantly smaller than under *OSPF*. Similarly, Figure 5.19(a) shows that the hot spot has significantly effected the ability of *OSPF* to successfully complete the sessions while on *BeeHive* it has a negligible effect. The hot spot traffic has significantly degraded the session delay of *OSPF* in the real Linux network and it

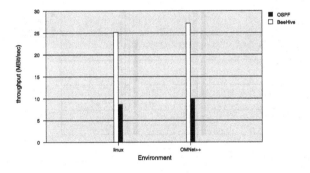

(a) Throughput

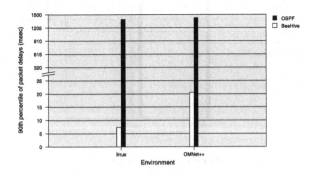

(b) 90th percentile of packet delays

Fig. 5.17. Experiment 8 (throughput and packet delay)

has negligible effect on *BeeHive*. The hot spot has helped in reducing the jitter (see Table 5.7) because now more packets flow between each (source, destination) pair, and this reduces the inter-arrival time of the packets originating at the same node at their destination node. One can easily see a significant difference in the number of packets generated during hot spot traffic between OMNeT++ and Linux in Table 5.7. The reason is again due to the nonavailability of high resolution timers in Linux, as described in the previous subsections. This explains the negligible effect of the hot spot traffic on the jitter values in the real network topology. The other important parameters are collected in Table 5.7.

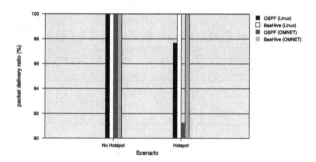

(a) Packet delivery ratio

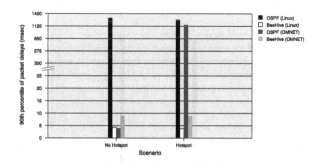

(b) 90th percentile of packet delays

Fig. 5.18. Hot spot experiments (packet delivery ratio and packet delay)

env	Linux				OMNET++			
algo	BeeHive		OSPF		BeeHive		OSPF	
scen	No Hotspot	Hotspot	No Hotspot	Hotspot	No Hotspot	Hotspot	No Hotspot	Hotspot
P_{gen}	2134388	2367310	2133851	2202463	2145269	2845271	2145479	2845395
P_{rec}	2134388	2367310	2133851	2151464	2145256	2845257	2145472	2596391
P_{loop}	0	0	0	0	0	0	0	0
S_{to}	868	869	868	868	870	870	870	870
S_{co}	868	869	868	355	859	859	859	552
S_o	2.578	4.199	0	0	3.403	4.769	0	0.797
R_o	0.116	0.111	0.001	0	0.062	0.075	0	0
t_d	4.3	3.6	1288	1246	6.575	5.8	3	409
S_d	17.2	18	17.2	28.2	12.4	12.4	12.4	12.4
J_d	7	6.6	7	11.4	5	1.9	5	1
J_{90d}	8.1	8	7	18.8	7.3	3	5	1
T_{av}	8.7	9.7	8.7	8.8	8.8	11.7	8.8	10.6
h_{av}	2.8	2.8	2	2	2.7	2.4	2	1.7

Table 5.7. Important performance values for hot spot experiments

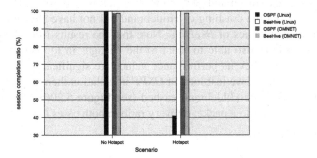

(a) Session completion ratio

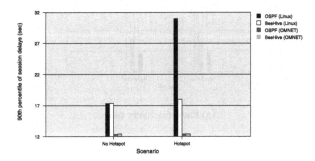

(b) 90th percentile of session delays

Fig. 5.19. Hot spot experiments (session completion ratio and session delay)

Experiment 10: Router down

The purpose of this experiment was to study the fault-tolerant behavior of *BeeHive* as compared with *OSPF*. A good routing algorithm should be able to quickly adapt its routes if a router crashes. The input parameters for this experiment are in Table 5.3. In this experiment, Hildebrand crashed from 300 seconds to 600 seconds. We also repeated the experiments with the same traffic pattern but without any router crash. Figure 5.20(a) shows that the packet delivery ratio of *BeeHive* is significantly higher than that of *OSPF* even when Hildebrand crashed. Similarly, Figure 5.20(b) shows that the packet delay of *BeeHive* is significantly smaller (15 msec) compared with that of *OSPF* (1,800 msec). The crashing of Hildebrand also has a negligible effect on both the session completion ratio (see Figure 5.21(a)) and the session delay (see Figure 5.21(b)) of *BeeHive*. The important parameters are collected in Table 5.8. *Bee-Hive* was able to quickly reroute the network traffic over two alternate paths: Etzel-

Kriemhild-Siegfried-Hagen and Etzel-Gieselher-Gernot-Brunhilde-Hagen. One can see in Table 5.8 that the crashing of Hildebrand did not have any significant effect on the performance values of *BeeHive*. Note that the results from the experiments suggest that *OSPF* is also able to react to the changes in topology but due to its single-path routing policy its performance values are significantly inferior to those of *BeeHive*. However, if we just look at *OSPF* in isolation then its packet delay significantly degraded once Hildebrand cashed (see Figure 5.20(b)). However, *OSPF* has approximately the same packet delivery ratio in both cases (see Figure 5.20(a)).

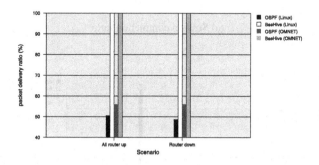

(a) Packet delivery ratio

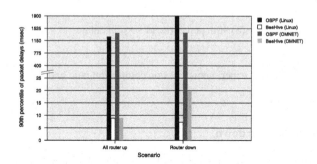

(b) 90th percentile of packet delays

Fig. 5.20. Router down (packet delivery ratio and packet delay)

5.6.2 Real-World Applications Traffic Engineering

The purpose of these experiments are twofold. Firstly, to repudiate a strong thesis held by the networking community: *a stochastic routing algorithm brings subsequent*

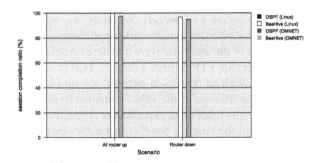

(a) Session completion ratio

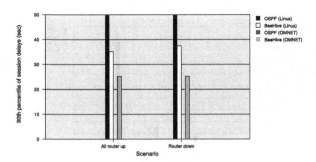

(b) 90th percentile of session delays

Fig. 5.21. Router down (session completion ratio and session delay)

env	Linux				OMNET++			
algo	BeeHive		OSPF		BeeHive		OSPF	
scen	No crash	crash	No crash	crash	No crash	crash	No crash	crash
P_{gen}	4289765	42792034	4260629	4275420	4329813	4330172	4330022	4330022
P_{rec}	4289765	42780121	2153291	2082480	4329781	4329640	2426096	2429090
P_{loop}	0	0	0	0	0	0	0	0
S_{to}	866	866	867	911	870	870	870	870
S_{co}	866	838	0	1	848	826	0	0
S_o	2.743	19.98	0	0.24	7.246	9.687	0	1.675
R_o	0.08	0.085	0	0	0.105	0.095	0	0
t_d	7	7	1249.2	1364.6	7	10.1	1217	1214
S_d	35	35.1	47	43	25.2	25.2	0	0
J_d	6.4	4.3	18.9	18.8	5	5	5	5
J_{90d}	10.1	7.4	28.1	28.1	7.3	7.3	5	5
T_{av}	8.5	17.6	8.8	8.5	17.7	17.8	9.9	10
h_{av}	2.3	3.1	2	2.3	2.7	3	2	2.3

Table 5.8. Important performance values for router down experiments

packets out of order at their destination which then might confuse the TCP protocol as it expects an in-order delivery of packets. Secondly, we wanted to demonstrate the benefits of the *BeeHive* protocol in real-world applications. We installed an FTP server at Hagen, and files of different size were transfered from Etzel to Hagen using an FTP client. Remember that FTP utilizes a reliable TCP protocol at the transport layer. We repeated the download process of each file ten times, and all the reported performance values are an average of the values obtained from the ten independent runs. We consider the results from these experiments crucial for convincing the networking community about the tangible benefits that nature-inspired algorithms like *BeeHive* might be able to provide. We conducted two sets of experiments: first, we started a download of a file of a particular size; second, we started downloading files, each of same size; after every minute, for 15 minutes.

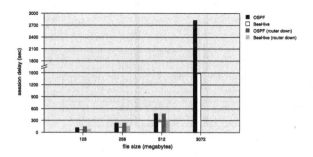

(a) Session delay of FTP transfers

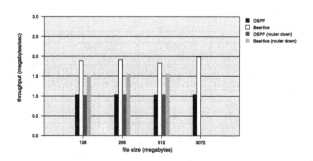

(b) Throughput of FTP transfers

Fig. 5.22. FTP experiments

FTP: Experiment 1, 2, 3, and 4

In these experiments we downloaded four files of different sizes one after the other. The sizes of the files were 128 megabytes, 256 megabytes, 512 megabytes, and 3 gigabytes. In another variation of the experiments Hildebrand crashed during the transfer. We wanted to investigate whether the algorithms could quickly react to the changes in the topology and reroute the traffic on the other paths. The files were download from Etzel to Hagen. Figure 5.22(a) shows the time it took (in seconds) to transfer these files. Two observations are obvious from Figure 5.22(a): *BeeHive* requires approximately half the download time compared with *OSPF* in all scenarios. Also, the crashing of Hildebrand during the transfer did not significantly degrade the performance of the algorithms. The reason for the significantly smaller delay is that *BeeHive* is able to maintain higher throughput than *OSPF* (see Figure 5.22(b)). The results of the experiments are instrumental because they suggest that *BeeHive* is able to successfully transport TCP traffic, and even with this protocol its performance is significantly better than that under *OSPF*. We believe that the results will help in establishing the suitability of nature-inspired routing algorithms in real-world networks.

FTP: Experiment 5

In this experiment we started downloading files of 128 megabytes each from Etzel to Hagen every minute for 15 minutes. We wanted to investigate which algorithm has a smaller turnaround time and a smaller time to complete a single download. Figure 5.23 shows that the time to download each file in *BeeHive* is significantly smaller compared with *OSPF*. The average time for each download in the case of *BeeHive* is approximately 1,000 seconds less compared with that of *OSPF*. *OSPF* took 3,642 seconds to finish the download of all 15 sessions compared to 2,042 seconds taken by *BeeHive*. This experiment further backs the results of the experiments from the previous subsection: *BeeHive* is able to seamlessly work with TCP, and its performance values are also significantly better than those of *OSPF*.

5.6.3 Hybrid Traffic Engineering

The motivation of doing experiments with hybrid traffic engineering is twofold: first, to use a traffic generator that has been developed by a third party to factor out any bias generated either in the traffic patterns or in the performance evaluation by our SQTG; second, to statistically evaluate different performance values for multimedia applications like Voice over IP (VoIP). VoIP is expected to capture a significant share of voice traffic in the telecommunication industry. D-ITG served our purpose. It can synthetically generate TCP, UDP, and VoIP traffic and then evaluate important performance values at the destination using a performance evaluation module. We generated UDP traffic to verify that the tendency observed in the experiments using our SQTG is traceable to the results obtained from the D-ITG experiments. We also did experiments with VoIP traffic only, and with VoIP traffic coupled with high UDP traffic load.

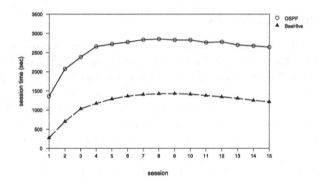

Fig. 5.23. FTP experiments with 15 downloads

UDP traffic

In D-ITG, a user can set the rate of packet generation, the size of a packet, and its source and destination nodes. We chose a packet size of 512 bytes and a packet rate of 6,000 packets/sec, and the packets are sent from Etzel to Hagen (see Figure 5.12(b)). This scenario is semantically similar to the traffic load of Experiment 8 discussed in the previous section. We would like to emphasize an important observation here: D-ITG is not an open-loop traffic generator; rather, it tries to generate only the number of packets a routing protocol can easily transport without significant packet drops. One has to carefully interpret the results obtained from D-ITG because the packet delivery ratio of the algorithms compared might be the same; but then the number of packets generated and delivered are usually different. Consequently, the throughput values would be significantly different. This behavior of D-ITG makes it difficult to systematically interpret and compare the performance values of different algorithms because the traffic patterns are adapted in response to the behavior of an algorithm.

The important performance values are collected in Table 5.10. Note that the throughput of *BeeHive* (24 Mbits/sec) is approximately three times higher than that of *OSPF* (9 Mbits/sec). The average delay of *BeeHive* is 34 msec compared to 71 msec of *OSPF*. However, we should remember that *BeeHive* is handling three times more packets than *OSPF* in this scenario. The jitter value of *OSPF*, as expected, is better than that of *BeeHive*. Finally, we have collected important performance values for *BeeHive* obtained from the current experiment and from Experiment 8 of our SQTG in Table 5.9. In Table 5.9, one can easily correlate different performance values obtained from two different traffic generators. However, we can-not make such a comparison for *OSPF* because D-ITG iteratively reduces the network load if an algorithm is unable to cope with it.

VoIP traffic

The purpose of these experiments is to verify that *BeeHive* is able to handle multi-media traffic in a similar or better fashion than *OSPF*. We started ten parallel VoIP

Traffic Generator	T_{av}	t_d	J_d	P_{gen}	P_{rec}	P_d
SQTG	25.1	7.2	4.8	3077866	3073359	99.9
ITG	24.5	3.4	3.7	2991185	2985347	99.8

Table 5.9. Performance values from SQTG and D-ITG

sessions between Etzel and Hagen which utilized G.711 [2] codec with one sample per packet. One VoIP session requires approximately 64 kbits/sec bandwidth. Ten sessions will need 640 kbits/sec, which is significantly less than the 10 Mbits/sec bandwidth available in simpleNet. The performance values are collected in Table 5.10. As expected, both algorithms are able to deliver approximately the same number of packets, and maintain the same throughput, packet delay, and jitter values. We could not simulate a large number of VoIP sessions because D-ITG simply cannot manage a large network load. We expect that *BeeHive* can outperform *OSPF* at high loads.

Scenario	Algorithm	T_{av}	t_d	J_d	P_{av}	P_{gen}	P_{rec}	P_d
UDP	OSPF	8.9	72	1.9	2170	2170526	2170526	100
	BeeHive	24.5	34.2	3.7	5981	5994045	5982370	99.8
VoIP	OSPF	0.7	0.2	0.012	995	1000000	995719	99.57
	BeeHive	0.7	0.4	0.2	995	1000000	996039	99.6
VoIP& UDP	OSPF	8.5	48.6	5	2899	2899329	2899158	99.9
	BeeHive	13	2.7	1	3991	3999980	3991418	99.79

Table 5.10. Important performance values for UDP and VoIP experiments

VoIP traffic + UDP traffic

The above-mentioned scenario for VoIP traffic generally does not exist in the Internet [83]. Most of the time voice traffic is multiplexed with data traffic. In parallel, some users are surfing, some are downloading large audio/video files, and some are involved in audio or video conferencing. The purpose of this experiment was to create an environment in which the ten VoIP sessions have to share the network with a UDP session, generating 3,000 packets/sec of 512 bytes size per packet. All sessions are between Etzel and Hagen. We have collected the relevant parameters in Table 5.10. *BeeHive* is able to deliver approximately one million more packets to their destination by maintaining a throughput of 13 Mbits/sec, compared to 8.5 Mbits/sec for *OSPF*. The packet delay of *BeeHive* is 2.6 msec, compared to 48 msec under *OSPF*. At this load, *BeeHive* is able to maintain a jitter value of 1 msec, compared to 4 msec of *OSPF*. One can easily conclude that in the case of hybrid traffic patterns, *BeeHive* is able to outperform *OSPF* considering all performance values. We intend to modify D-ITG in such a manner that it can manage a higher number of VoIP sessions with UDP and TCP traffic to further validate our assumption that the true merit of *BeeHive* is visible at large network loads. However, the results from the current experiments can still be considered as a break through for nature-inspired routing algorithms because for the first time, to our knowledge, a nature-inspired routing protocol that has been tested in a real topology of Linux routers, and it has shown throughout quantifiable benefits with real applications.

5.7 Summary

The major contribution of the work is a *Natural Engineering* approach that we followed during the development of our routing protocol *BeeHive*, and our realizing it inside the network stack of the Linux operating system. The engineering approach was instrumental in our incorporating only those features in the *BeeHive* algorithm into the OMNeT++ simulator that are also available in the network stack of Linux. We also designed a protocol verification framework in which we developed a comprehensive performance evaluation framework, and implemented it in both OM-NeT++ and a Linux network. The framework allowed us to generate similar traffic patterns, both in the simulation and in a real network of Linux machines. The basic assumption of the framework is: *if the simulation results of* BeeHive *are of any significance then its behavior should be traceable in a real network of Linux machines as well.* We designed a variety of experiments which utilized real-world applications like FTP and synthetic traffic generators like SQTG and D-ITG to demonstrate that the *behavior of BeeHive from simulation is consistent with its behavior in a real network of Linux routers*. The results obtained from simulation and a real network of Linux machines also show a strong correlation, and in our view this is breakthrough work in favor of nature-inspired routing algorithms. To our knowledge, it is the first study that has empirically refuted the suspicious notion held by the telecommunication industry that *nature-inspired algorithms are not realizable with the existing resources of software and hardware*. Of course, our findings will have to be further confirmed in large networks. Our current results are quite an intriguing stimulation for such plans.

We have deliberately used the low-end Pentium-III-based machines for the implementation of *BeeHive* to prove the fact that the algorithm is realizable with the existing infrastructure and yet its performance benefits are unequivocally noticeable in real world networks. Future work includes testing the algorithm on large-scale topologies.

6

A Formal Framework for Analyzing the Behavior of *BeeHive*

The major contribution of this chapter is a formal framework that helps in analyzing the behavior of our BeeHive *routing protocol. To the best of our knowledge, researchers in our community have paid little attention to developing such a formal framework that provides analytical insight into the behavior and performance of nature-inspired routing algorithms. Instead, most of them follow a classical well-known engineering philosophy: inspire, abstract, design, develop, and validate. As a consequence, the protocols are designed on the basis of heuristics, and then their performance is evaluated in a network simulator. In line with this practice, we verified the correctness of our model by comparing its estimated values with the results obtained from network simulations in OMNeT++. An important outcome of the work is that the estimated and measured values only differ by a small deviation. We believe that this work will significantly contribute in developing a formal theory of nature-inspired routing protocols that will ultimately help in their acceptability in the networking industry.*

6.1 Introduction

Researchers working in the domain of nature-inspired routing protocols have mostly followed the same protocol engineering philosophy: inspire, abstract, design, develop, and validate. As a result, nature-inspired protocol engineering is predominately biased towards heuristics and their empirical evaluation in a simulation environment.[1] The formal model of a routing protocol is an important element of *protocol engineering* [140]. To the best of our knowledge, no comprehensive research, except the preliminary work reported in [18], has been conducted in developing a formal framework that provides insights into the behavior of nature-inspired routing protocols.

---

[1] Most of this chapter is reproduced by permission of the publisher, IEEE, from our paper: S. Zahid, M. Shehzad, S. Usman Ali, and M. Farooq. *A comprehensive formal framework for analyzing the behavior of nature inspired routing protocols.* In Proceedings of Congress on Evolutionary Computing (CEC), pages 180–187. IEEE, Singapore, September 2007.

The lack of formal treatment of nature-inspired routing protocols is often cited as an important shortcoming of these protocols. Researchers even today have little formal understanding of the reasons behind the superior performance of such routing protocols. The obvious reason for lack of a formal treatment of a dynamic and adaptive routing algorithm is that it demands expertise in analyzing non-linear, time-varying, real-time, and dynamic optimization problems. Active research is in progress to develop analytical tools for modeling such problems. The major contribution of the work presented in this chapter is that we have developed a formal framework for analyzing the behavior of *BeeHive*. This is the first cardinal step towards development of a general framework for analyzing nature-inspired routing protocols. With the help of this model, we are able to not only understand the formal logic behind different design options adopted in *BeeHive* but also systematically study their effect on its behavior. We also used the formal model in representing relevant performance parameters like average throughput and average packet delay. We conducted our experiments on two topologies in a well-known OMNeT++ network simulator to verify the semantic and functional correctness of our formal model. The behavior of *BeeHive* and relevant performance parameters estimated by our formal model match the behavior and results obtained from OMNeT++ simulations.

6.1.1 Organization of the Chapter

In Section 6.2, we present a formal treatment of two different formulas utilized in *BeeHive* for evaluating the quality of a path. Section 6.3 presents our comprehensive formal model of the *BeeHive* protocol. We have modeled goodness, total delay, and throughput of the links in the network. In Section 6.4 we verify the correctness of our model by comparing its results with the results obtained from OMNeT++. Finally, we conclude our work with an outlook to the future.

6.2 Goodness

Recall from Chapter 3 that g_{nd} is defined as the goodness of a neighbor n of node i for reaching a destination d. In order to gain a better understanding of the formal model in terms of the graph theory concepts, we want to redefine different parameters in terms of an edge or a link (i, n) that connects current node i with neighbor n. Following this convention q_{in}, for example, will mean the queuing delay of link (i, n), and gl_{in} will mean the goodness of link (i, n). We model a given topology as an undirected graph $G(\mathcal{V}, \mathcal{E})$ in which $\mathcal{V}$ is a set of vertices in the graph and $\mathcal{E}$ represents the set of edges.

Remember that we use two formulas (refer to Equations (3.5) and (3.6) in Chapter 3) for calculating the goodness of a link. Now we provide formal reasons to show why *BeeHive* delivered the same performance even with a simple function like Equation (3.6). The important symbols used in the chapter are tabulated in Table 6.1. Assume a network in which all the nodes are connected, either directly or indirectly, with one another. Thus the network is a connected graph. The network consists of a

gl_{in}	goodness of a link (i, n)
ld_{in}	delay factor of a packet from node i to n
Ld	matrix whose entries are ld_{in}
p_{in}	propagation + transmission delay for link (i, n)
q_{in}	queuing delay for the link (i, n)
b_{in}	proportion in which the bees traverse the link (i, n)
x	iteration index
y	q_{in}/p_{in}
β_i	rate of the bees entering the network at node i
β	matrix whose entries are β_i
γ_i	flow of bees at node i
γ	matrix whose entries are γ_i
ξ_i	rate of the data packets entering the network at node i
ξ	matrix whose entries are ξ_i
η_i	flow of data at node i
η	matrix whose entries are η_i
bf_{in}	flow of bees on link (i, n)
∂f_{in}	flow of data on link (i, n)
v_{in}	total flow of traffic on link (i, n)
S_{in}	service rate of the link (i, n)
tx_{in}	transmission delay for the link (i, n)
pd_{in}	propagation delay of the link (i, n)
$cd_{n\mathcal{D}}$	cumulative delay from node n to node $\mathcal{D}$
Cd	matrix whose entries are $cd_{n\mathcal{D}}$
td_{in}	total delay from node i to $\mathcal{D}$ through n
Td	matrix whose entries are td_{in}
α	ratio of size of bee packet to data packet
T_{in}	throughput of the link (i, n)
$\mathcal{V}$	set of all nodes in netwrok
$\mathcal{N}_i$	set of neighbours of node $i \in \mathcal{V}$
$\mathcal{E}$	set of links in the network

Table 6.1. Symbols used in the chapter

set of nodes $\mathcal{V}$. Neighbors of each node $i \in \mathcal{V}$ are in the corresponding set $\mathcal{N}_i$. All links in the network belong to the set $\mathcal{E}$. In our model, we send data to a single destination $\mathcal{D}$, and we assume that two nodes are not directly connected by more than one link.

The goodness of link from node i to node n is given by:

$$gl_{in} = \frac{\frac{1}{ld_{in}}}{\sum_{k \in \mathcal{N}_i}\left(\frac{1}{ld_{ik}}\right)}, \tag{6.1}$$

where

$$ld_{in} = \left(\frac{1}{p_{in}}(e^{-\frac{q_{in}}{p_{in}}}) + \frac{1}{q_{in}}(1 - e^{-\frac{q_{in}}{p_{in}}})\right)^{-1}. \tag{6.2}$$

In Equation (6.2), p_{in} represents the sum of the propagation delay and the transmission delay for the link (i, n), q_{in} represents the queuing delay from node i to n in order to reach the destination $\mathcal{D}$, and ld_{in} is the delay factor of the link connecting i to n. One can easily conclude from Equations (6.1) and (6.2) that the probability that a packet traverses a certain path depends on the delays of the path. Queuing delays under low loads are small. As a result, the goodness of a path is determined by its propagation delay.

Mathematical functions like exponentials take a significantly large number of cycles to compute. Recall from Chapter 3 that our simple non-exponential version in Equation (3.6) has ten times smaller processing complexity as compared to Equation (3.5), but it does not degrade the performance of *BeeHive*. The goodness formula is same as Equation (6.1), where ld_{in} is given by

$$ld_{in} = p_{in} + q_{in}. \tag{6.3}$$

In our model the values of ld_{in} are maintained in a matrix Ld of dimensions $(|\mathcal{V}| - 1) \times (|\mathcal{V}| - 1)$.

Lemma 1:
The two goodness formulas in Equations (6.2) and (6.3) converge to the same goodness value under high traffic load.

Proof:
Substituting

$$y = \frac{q}{p} \tag{6.4}$$

in Equation (6.2), we get

$$\frac{1}{p}(e^{-\frac{q}{p}}) + \frac{1}{q}(1 - e^{-\frac{q}{p}}) = \frac{1}{q}\left\{1 + ye^{-y} - e^{-y}\right\}. \tag{6.5}$$

Power series of e^{-y} is given by

$$e^{-y} = 1 - y + \frac{y^2}{2!} - \frac{y^3}{3!} + \cdots . \tag{6.6}$$

Substituting Equation (6.6) in Equation (6.5) and simplifying, the right-hand side of Equation (6.5) becomes

$$= \frac{1}{q}\left\{\sum_{k=1}^{\infty}(-1)^{k+1}\left(\frac{k+1}{k!}\right)y^k\right\}. \tag{6.7}$$

Similarly, using Equation (6.3), we obtain

$$\frac{1}{p+q} = \frac{1}{q}\left\{\frac{y}{y+1}\right\}. \tag{6.8}$$

On applying binomial expansion, the left-hand side of Equation (6.8) becomes

$$= \frac{1}{q}\left\{\sum_{k=1}^{\infty}(-1)^{k+1}y^k\right\}. \tag{6.9}$$

Now we can see that Equations (6.7) and (6.9) differ only by the factor $\frac{n+1}{n!}$.

Using another approach to show the similarity between the two goodness formulas in Equations (6.5) and (6.8), let

$$f_1(y) = 1 + ye^{-y} - e^{-y}, \tag{6.10}$$

$$f_2(y) = \frac{y}{y+1}. \tag{6.11}$$

Applying limit $y \to \infty$ to Equation (6.10),

$$lim_{y\to\infty} \, f_1(y) = lim_{y\to\infty} \left(1 + ye^{-y} - e^{-y}\right)$$
$$= 1 + \, lim_{y\to\infty} \left(\frac{y}{e^y}\right) - 0. \tag{6.12}$$

Using L'Hopital's rule for the second term on the right-hand side

$$lim_{y\to\infty} \left(\frac{y}{e^y}\right) = lim_{y\to\infty} \left(\frac{\frac{d(y)}{dy}}{\frac{d(e^y)}{dy}}\right)$$
$$= lim_{y\to\infty} \frac{1}{e^y}$$
$$= 0. \tag{6.13}$$

So Equation (6.12) becomes

$$lim_{y\to\infty} \, f_1(y) = 1. \tag{6.14}$$

Again, applying limit $y \to \infty$ to Equation (6.11),

$$lim_{y\to\infty} \, f_2(y) = lim_{y\to\infty} \left(\frac{y}{y+1}\right)$$
$$= lim_{y\to\infty} \left(\frac{1}{1+\frac{1}{y}}\right)$$
$$= 1. \tag{6.15}$$

From Equations (6.14) and (6.15), one can easily deduce that the two goodness formulas give same results under heavy traffic load. This can be verified graphically as well. In Figure 6.1 we plot $f_1(y)$ and $f_2(y)$, as these functions represent the only difference between Equations (6.5) and (6.8). In Figure 6.1 one can see that the difference between the two factors decreases as y increases. This behavior is expected because an increase in y is possible only due to an increase in q (see Lemma 1) that results due to heavy network traffic load.

6.3 Analytical Model

In this section we present our formal model that can be of significant help in analyzing the behavior of the *BeeHive* protocol. We will utilize relevant concepts of deductive mathematics and queuing theory in our formal model. Markov transition matrices coupled with stochastic processes [159, 129, 198] serve as the foundation of our framework. We model the network traffic by calculating three parameters:

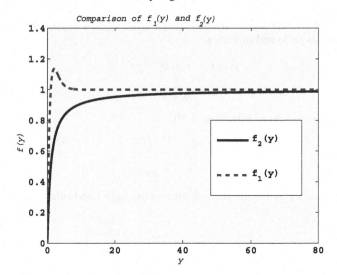

Fig. 6.1. Comparison between Equations (6.10) and (6.11)

- goodness of a link from node i to node n, gl_{in};
- total delay experienced by a packet (propagation + queuing + transmission) to traverse a certain path in the network;
- throughput of the links in the network.

The packet-routing probabilities (goodness) can be collected using the Markov Transition Matrix, in which each entry represents the routing probability between the corresponding two nodes. This is a $|\mathcal{V}| \times |\mathcal{V}|$ matrix that follows the properties of a Markov Transition Matrix, which are:

- $\sum_n gl_{in} = 1$.
- It is a square matrix.
- Its entries are nonnegative.

We also use a discrete iteration index $x \in \mathbb{N}$. Furthermore, we maintain a matrix B of dimensions $(|\mathcal{V}| - 1) \times (|\mathcal{V}| - 1)$. The entries of this matrix actually determine the proportion in which a node forwards the *bee agents* towards its neighbors. This matrix is defined as:

$$B(x) = \mathfrak{b}_{in}(x) \qquad i, n = 1, 2, \cdots, |\mathcal{V}| \wedge i \neq \mathcal{D}, \tag{6.16}$$

where

$$\mathfrak{b}_{in}(x) = \begin{cases} \frac{1}{n(\mathcal{N}_i)} & (i, n) \in \mathcal{E} \\ 0 & i = n \vee (i, n) \notin \mathcal{E}. \end{cases} \tag{6.17}$$

Similarly, we maintain another matrix $\mathcal{G}$ of dimensions $(|\mathcal{V}| - 1) \times (|\mathcal{V}| - 1)$. The entries of this matrix determine the goodness of the corresponding links. This matrix is given as:

$$\mathcal{G}(x) = gl_{in}(x) \qquad i, n = 1, 2, \cdots, |\mathcal{V}| \wedge i \neq \mathcal{D}, \tag{6.18}$$

where

$$gl_{in}(x) = \begin{cases} \dfrac{\frac{1}{td_{in}(x-1)}}{\sum_{k \in \mathcal{N}_i} \left(\frac{1}{td_{ik}(x-1)}\right)} & (i, n) \in \mathcal{E} \\ 0 & i = n \vee (i, n) \notin \mathcal{E}, \end{cases} \tag{6.19}$$

where td_{in} is the total delay from node i to $\mathcal{D}$ through neighbor n. The values of td_{in} are maintained in matrix Td of dimensions $(|\mathcal{V}| - 1) \times (|\mathcal{V}| - 1)$. The values for $b_{i\mathcal{D}}(x)$ and $g_{i\mathcal{D}}(x)$ cannot be directly calculated from Equations (6.16) and (6.18) respectively. They can be calculated using the following formulas:

$$b_{i\mathcal{D}}(x) = 1 - \sum_{n \in (\mathcal{N}_i \setminus \mathcal{D})} b_{in}(x), \tag{6.20}$$

$$g_{i\mathcal{D}}(x) = 1 - \sum_{n \in (\mathcal{N}_i \setminus \mathcal{D})} gl_{in}(x). \tag{6.21}$$

6.3.1 Node Traffic

The flow rate of a node or a link is the amount of the traffic (*bee-agents* and data packets) passing through the node or the link respectively per unit time.

Bee traffic

In *BeeHive*, every node generates *bee agents* that traverse all possible paths available to a node in the network and update routing tables at every visited node. The size of these packets is small compared to data packets. As a result, they use less than 1% of the bandwidth. We introduce a constant α which is the ratio of the size of a bee packet to the data packet. We multiply the bee generation rate of every node with this constant to specify the degree to which *bee agents* load the network.

Let β_i be the rate of bee traffic entering the network at node i. The values for all $\beta_i, i = 1, 2, \cdots, |\mathcal{V}| \wedge i \neq \mathcal{D}$ are collected in a $(|\mathcal{V}| - 1)$-dimensional vector represented by β. The bee generation rate is assumed to be constant during an iteration x. *Bee agents* explore the complete network instead of just sampling the best paths. The flow of *bee agents* at node i, $\gamma_i(x)$, is given by:

$$\gamma_i(x) = \beta_i + \sum_{m \in \mathcal{N}_i} \gamma_m(x) b_{mi}(x) \qquad i = 1, 2, \cdots, |\mathcal{V}| \wedge i \neq \mathcal{D}. \tag{6.22}$$

The solution of the recursive or iterative functions is complex because a large number of simultaneous equations need to be reduced and simplified by algebraic techniques. Therefore, we use matrices which are simpler to handle. Solution to Equation (6.22) is given by:

$$\gamma(x) = \beta \left(I - B^t(x)\right)^{-1}, \tag{6.23}$$

where $B^t(x)$ represents the transpose of $B(x)$ and $\gamma(x)$ is a $(|\mathcal{V}| - 1)$-dimensional vector that contains the flow rates of *bee agents* of all nodes except $\mathcal{D}$. The proof of Equation (6.23) is given at the end of the chapter.

Data traffic

The nodes in the network also generate data packets that are destined for $\mathcal{D}$. Different packet generation rates can be assigned to each node in the network. Let ξ_i be the rate at which data packets are sent into the network by node i. The values for all $\xi_i, i = 1, 2, \cdots, |\mathcal{V}| \wedge i \neq \mathcal{D}$, are collected in a $(|\mathcal{V}| - 1)$-dimensional vector represented by $\boldsymbol{\xi}$. The data generation rate is considered constant during different iterations (x).

Note that every node i acts as a forwarding node for the data packets generated by the neighbors of i. However, it will act only as a sink node if data packets are destined for it. As a result, we derive flows at nodes as probabilistic recursive functions. These functions model the goodness of the links of a node. As a consequence, the rate of packet arrival and departure is dependent on the goodness of incoming and outgoing links respectively. The flow of data at node i, $\eta_i(x)$, is given by:

$$\eta_i(x) = \xi_i + \sum_{m \in \mathcal{N}_i} \eta_m(x) g_{mi}(x) \qquad i = 1, 2, \cdots, |\mathcal{V}| \wedge i \neq \mathcal{D}. \tag{6.24}$$

The solution to Equation (6.24) is given by:

$$\boldsymbol{\eta}(x) = \boldsymbol{\xi}\left(I - \mathcal{G}^t(x)\right)^{-1}, \tag{6.25}$$

where $\boldsymbol{\eta}(x)$ is also a $(|\mathcal{V}| - 1)$-dimensional vector that contains the flow rates of data packets of all nodes except $\mathcal{D}$.

6.3.2 Link Flows

Once the node flow rate (number of data packets being forwarded by a node in unit time) of a node is known, its corresponding link flow rate (amount of data flowing on the link in unit time) can be calculated as a function of its neighbors and their goodness. These are represented as:

$$\mathfrak{bf}_{in}(x) = \gamma_i(x)\mathfrak{b}_{in}(x), \tag{6.26}$$

$$\mathfrak{df}_{in}(x) = \eta_i(x)gl_{in}(x), \tag{6.27}$$

where $\mathfrak{bf}_{in}(x)$ and $\mathfrak{df}_{in\mathcal{D}}(x)$ represent the bee link flow rate and data link flow rate respectively from node i to node n. The total link flow rate, v_{in}, can be calculated as:

$$v_{in}(x) = \mathfrak{bf}_{in}(x) + \mathfrak{df}_{in}(x). \tag{6.28}$$

6.3.3 Calculation of Delays

In computer networks the arrival of packets is assumed to follow a Poisson distribution, and the queues that are built are as a consequence M/M/1 queues [245, 251, 136]. Moreover, the service or departure rate can be preassigned because they are independent of the network conditions. We can calculate the queuing delay associated with every traversed link by utilizing the M/M/1 queuing theory if we know the arrival and departure rates. Representing the service rate of link (i, n) by S_{in}, we have the result [245]:

$$q_{in}(x) = \begin{cases} \frac{1}{S_{in} - v_{in}(x)} & v_{in} < S_{in} \\ \infty & v_{in} \geq S_{in}. \end{cases} \tag{6.29}$$

The total link delay $ld_{in}(x)$ experienced by the packets to traverse a link (i, n) can be found by using the approach given in either Equation (6.2) or (6.3).

Using the approach given in Equation (6.2) we get:

$$ld_{in}(x) = \left(\frac{1}{p_{in}} (e^{-\frac{q_{in}(x)}{p_{in}}}) + \frac{1}{q_{in}(x)} (1 - e^{-\frac{q_{in}(x)}{p_{in}}}) \right)^{-1}. \tag{6.30}$$

Using the approach given in Equation (6.3) we get:

$$ld_{in}(x) = p_{in} + q_{in}(x), \tag{6.31}$$

where

$$p_{in} = tx_{in} + pd_{in}.$$

tx_{in} and pd_{in} are the transmission and propagation delays respectively of the link between node i and node n.

Once the packet has reached an intermediate node n, it might again have multiple paths to reach the destination node through its neighbors, and we need to calculate the delays for all possible routes from n to the destination $\mathcal{D}$. By again using the concept of recursive probabilistic functions we get:

$$td_{in}(x) = ld_{in}(x) + cd_{n\mathcal{D}}(x), \tag{6.32}$$

where $cd_{n\mathcal{D}}(x)$ is the cumulative delay experienced by a packet in order to reach $\mathcal{D}$ from n through any neighbors of n. It is given by:

$$cd_{n\mathcal{D}}(x) = \sum_{k \in \mathcal{N}_n} \left(\alpha b_{nk}(x) + (1-\alpha) g_{nk}(x) \right) \left(ld_{nk}(x) + cd_{k\mathcal{D}}(x) \right)$$
$$n = 1, 2, \cdots, |\mathcal{V}| \wedge n \neq \mathcal{D}. \tag{6.33}$$

The solution to Equation (6.33) is given by:

$$cd(x) = \left(I - \alpha B(x) - (1-\alpha) G(x) \right)^{-1} \delta(x). \tag{6.34}$$

$\delta(x)$ is given by:

$$\delta_n(x) = \sum_{k \in \mathcal{N}_n} \left(\alpha b_{nk}(x) + (1-\alpha) g_{nk}(x) \right) ld_{nk}(x)$$
$$n = 1, 2, \cdots, n \ (\mathcal{V}) \wedge n \neq \mathcal{D}, \tag{6.35}$$

where $\delta(x)$ and $cd(x)$ are vectors of dimensions $(|\mathcal{V}| - 1)$ each. The solution similar to Equation (6.34) is derived at the end of the chapter.

The values of $cd(x)$ can be obtained from Equation (6.34), the values of $td_{in}(x)$ can be obtained from Equation (6.32), and we get a new value of goodness by using Equation (6.1). Here, we give it as:

$$gl_{in}(x+1) = \begin{cases} \dfrac{\frac{1}{td_{in}(x)}}{\sum_{k \in \mathcal{N}_i} (\frac{1}{td_{ik}(x)})} & (i,n) \in \mathcal{E} \\ 0 & i = n \vee (i,n) \notin \mathcal{E}. \end{cases} \qquad (6.36)$$

So, starting from a goodness value in Equation (6.19), we reached Equation (6.36), which is the goodness value for the next iteration.

In the case when link flow rate becomes greater than the service rate, the value of $q_{in} \to \infty$, and consequently $td_{in} \to \infty$. Thus the goodness factors given by Equations (6.19) and (6.36) cannot be calculated from these equations. In order to keep the model flexible under such circumstances, we take $gl_{in} = b_{in}$ for this particular link, (i,n), until the link flow rate again becomes smaller than the service rate.

6.3.4 Throughput

Finally, we present our model for calculating the throughput of the algorithm:

$$T_{in} = \frac{\text{effective data transferred on } (i,n)}{ld_{in}(x)}.$$

The expression on the right-hand side is actually the data flow rate of the link (i,n) given by Equation (6.27); so throughput is given by:

$$T_{in} = \partial f_{in}(x). \qquad (6.37)$$

6.4 Empirical Verification of the Formal Model

We now verify our model on two topologies. The first topology is a four-node square topology shown in Figure 6.2, while the other one is simpleNet (see Figure 3.5). In the first topology our focus is to study the impact of data packet rates and service rates of nodes and links on the goodness of the links. In the second example we compare the results obtained from our formal model with the ones obtained from extensive simulations of the *BeeHive* protocol in OMNeT++. The values obtained in every iteration are averaged using the exponential moving average method. This is done to dampen the oscillations in the graph. This exponential moving average is taken in the simulation of the formal model as well as in the results obtained from OMNeT++.

6.4.1 Example 1

The topology of Figure 6.2 is simple. For this particular example we assume that Node 4 is the destination node and the remaining nodes can send data to it. First we use the following parameters for our simulation:

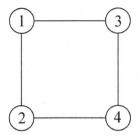

Fig. 6.2. Topology 1

- transmission + propagation delay, $p_{in} = 0.1$
- ratio of size of bee packet to data packet, $\alpha = 0.01$
- bee generation rates of nodes, $\beta = [2\ 2\ 2\ 0]$
- data generation rates of nodes, $\xi = [5\ 8\ 0\ 0]$
- service rate of links, $S_{in} = 10\ \forall\ (i,\ n) \in \mathcal{E}$
- starting value of total delay, $td_{in}(0) = 1$

We now show a single iteration of the model that will provide a better insight to the reader about the model. For the given topology, using Equations (6.16) and (6.17) gives:

$$B(1) = \begin{bmatrix} 0 & 0.5 & 0.5 \\ 0.5 & 0 & 0 \\ 0.5 & 0 & 0 \end{bmatrix}$$

Substituting $td_{in}(0) = 1$ in Equation (6.19) gives $gl_{in}(1)$. When $gl_{in}(1)$ is substituted in Equation (6.18), the resulting matrix $\mathcal{G}(1)$ is:

$$\mathcal{G}(1) = \begin{bmatrix} 0 & 0.5 & 0.5 \\ 0.5 & 0 & 0 \\ 0.5 & 0 & 0 \end{bmatrix}$$

Equations (6.22) to (6.25) can be utilized to find the vectors γ and η:

$$\gamma(1) = \begin{bmatrix} 8 & 6 & 6 \end{bmatrix}$$

$$\eta(1) = \begin{bmatrix} 18 & 17 & 9 \end{bmatrix}$$

Evaluation of Equation (6.26) to (6.28) and (6.31) gives:

$$Ld(1) = \begin{bmatrix} 0 & 1.1417 & 1.1417 \\ 0.7803 & 0 & 0 \\ 0.2828 & 0 & 0 \end{bmatrix}$$

Td is estimated by using Equation (6.32):

$$Td(1) = \begin{bmatrix} 0 & 1.1298 & 1.1049 \\ 1.1563 & 0 & 0 \\ 1.1315 & 0 & 0 \end{bmatrix}$$

Substituting Td(1) in Equation (6.36) gives the value of goodness factor for the next iteration.

$$\mathcal{G}(2) = \begin{bmatrix} 0 & 0.4944 & 0.5056 \\ 0.4610 & 0 & 0 \\ 0.4601 & 0 & 0 \end{bmatrix}$$

One can keep on iterating the above-mentioned steps to arrive at the steady state

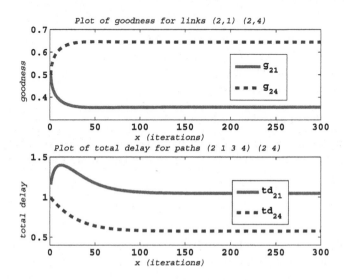

Fig. 6.3. Plots of goodness and delay

values of the goodness,

$$\mathcal{G}(\infty) = \begin{bmatrix} 0 & 0.4657 & 0.5343 \\ 0.3558 & 0 & 0 \\ 0.1861 & 0 & 0 \end{bmatrix}.$$

From Figure 6.3 we see that the goodness for the link $(2, 1)$ is less than that for the link $(2, 4)$ as the total delay for the packets to reach the destination from 2 by traversing the link $(2, 4)$ is significantly less than following the path $(2 \rightarrow 1 \rightarrow 3 \rightarrow 4)$. This phenomenon is discussed in more detail in the next example, where we show that the probability that a packet follows loops is small.

If Node 1 has some data to send, then it can utilize two paths to route its data packets, i.e., one path is $(1 \rightarrow 2 \rightarrow 4)$ and the second one is $(1 \rightarrow 3 \rightarrow 4)$. From the pre-assigned data generation rates of Nodes 2 and 3, the goodness of the link $(1, 3)$ is expected to be more than that of $(1, 2)$.

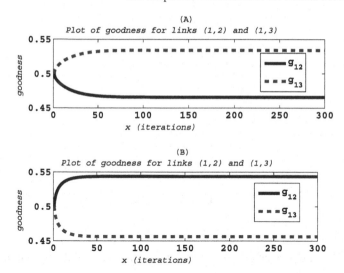

Fig. 6.4. Goodness for links (1, 2) and (1, 3)

The results of our formal model in Figure 6.4(A) verify the above-mentioned intuitive logic. The goodness for link $(1, 3)$ stabilizes at approximately 0.534 while that of link $(1, 2)$ stabilizes at 0.466. So more data is sent on the link $(1, 3)$ in order to achieve better performance.

Now we take new parameters for another simulation.

- transmission + propagation delay, $p_{in} = 0.1$
- ratio of size of bee packet to data packet, $\alpha = 0.01$
- bee generation rates of nodes, $\beta = [2\ 2\ 2\ 0]$
- data generation rates of nodes, $\xi = [5\ 2\ 2\ 0]$
- service rate of link $(1,3)$, $S_{13} = 8$
- service rate of links, $S_{in} = 15\ \forall\ (i, n) \in \mathcal{E} \setminus (1, 3)$
- starting value of total delay, $td_{in}(0) = 1$

For these parameters we can see that service rate is less for the queues of the link $(1, 3)$ than of the link $(1, 2)$. As both Nodes 2 and 3 have the same data traffic rates, the delay experienced via path $1 \to 3 \to 4$ should be more than that via $1 \to 2 \to 4$ because the service rate of Node 3 is smaller than the service rate of other nodes. As a result, the goodness of the link $(1, 3)$ is less than that of the link $(1, 2)$. One can clearly see from Figure 6.4(B) that the goodness of both paths, estimated by our formal model, is as expected.

6.4.2 Example 2

Now we verify our model on simpleNet, shown in Figure 6.5. We reproduce the figure here for convenient tracking of the experiments. We compare the results obtained from our model with the results obtained from extensive simulations in OMNeT++.

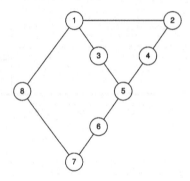

Fig. 6.5. simpleNet

The following parameters are used to obtain the results of the formal model:

- *transmission + propagation delay,* p_{in} = 0.1
- *ratio of size of bee packet to data packet,* α = 0.01
- *bee generation rates of nodes,* β = [3 2 2 2 3 2 0 2]
- *data generation rates of nodes,* ξ = [2 2 2 2 2 2 0 2]
- *service rate of links,* S_{in} = 34 $\forall$ $(i,n) \in \mathcal{E}$
- *starting value of total delay,* $td_{in}(0) = 1$
- *destination node,* $\mathcal{D} = 7$

Figures 6.6 and 6.7 compare the delays of the paths from Node 1 to Node 7 via Node 3 and from Node 1 to Node 7 via Node 8 respectively. One can see that the values of the delay estimated by our formal model are approximately the same as those obtained from the network simulator.

In Figure 6.8 the goodness of the link $(1,3)$ is graphically depicted. Again, the results obtained from our formal model are validated by the OMNeT++ simulations. In Figure 6.9 we compare the goodness of link $(1,8)$. The steady-state value of goodness in both cases is 0.41. The goodness of link $(1,2)$ can be found by subtracting the sum of the goodnesses of links $(1,3)$ and $(1,8)$ from 1, i.e., $(1 - (0.41 + 0.31)) = 0.28$.

Finally, we show that the probability that a packet will enter into loops is significantly small in *BeeHive*. In this topology we consider the loop $(2 \rightarrow 1 \rightarrow 3 \rightarrow 5 \rightarrow 4 \rightarrow 2)$ and find out the probability that a packet originating at Node 2 and traveling towards the destination Node 7 will follow this cyclic path. From Figure 6.10 we can find the steady-state goodness of all these links. Multiplying the steady-state values of goodness gives the probability that a packet will move in this loop. The resulting probability comes out to be 0.01063, which is significantly small and shows that only 1% of the packets can follow cyclic paths.

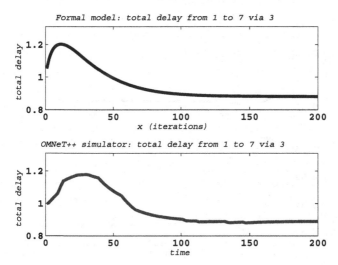

Fig. 6.6. Total delay experienced by a packet to reach Node 7 from Node 1 via Node 3

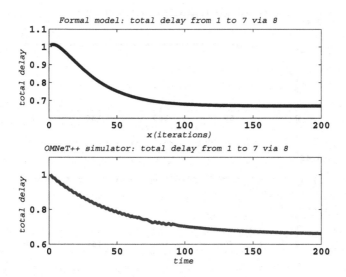

Fig. 6.7. Total delay experienced by a packet to reach Node 7 from Node 1 via Node 8

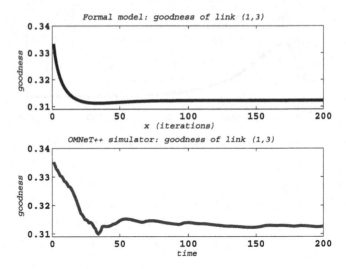

Fig. 6.8. Goodness of link $(1, 3)$

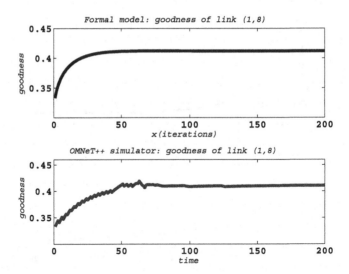

Fig. 6.9. Goodness of link $(1, 8)$

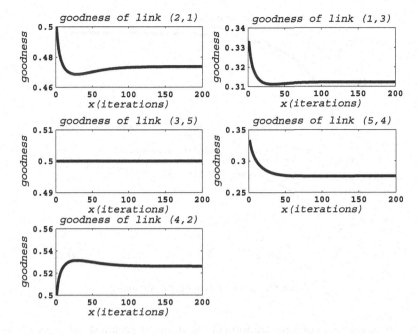

Fig. 6.10. Goodness of different links in simpleNet

Finally we compare the cumulative throughput of the system obtained by the formal mathematical model and the OMNeT++ simulation. In Figure 6.11 one can again see that the results from the model match with the results of the simulation.

6.5 Summary

In this chapter we have introduced a formal framework for analyzing our *Bee-Hive* protocol. We have used probabilistic recursive functions for analyzing on-line stochastic packet-switching behavior of the algorithm. The queuing delays experienced due to the congestion have been analyzed using the formal concepts of (M/M/1) queuing theory. The formal model is simple yet general enough to capture the relevant features of the *BeeHive* protocol.

We have also developed an empirical verification framework in OMNeT++ to validate the correctness of our formal model. We validated our formal model on two topologies and compared its results with the results obtained from the extensive simulations. The estimated performance values of the model have only a small deviation from the real values measured in the network simulator.

In the future we want to extend the framework to model the formation of *foraging regions* and *foraging zones*. *BeeHive* utilized these relevant concepts for scalability.

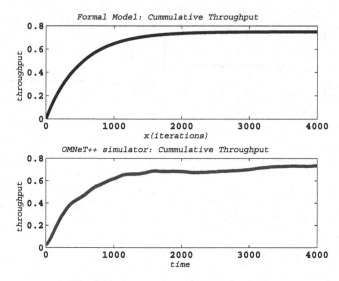

Fig. 6.11. Average throughput in simpleNet

Once these relevant concepts are formally modeled, we can test our framework on large topologies like NTTNet. This is an ongoing research effort.

In our current work we have treated computer networks as "Stochastic Networks" [218] in which the behavior of a network is modeled in terms of the equilibrium probability distribution of packets on different routes. The equilibrium distribution is used to define objective functions or constraints to model performance values such as throughput and packet delay. Such probabilistic methods are also utilized to model the Internet and the Web [16]. However, the "Stochastic Networks" are unable to model the self-organizing and emerging behavior of nature-inspired routing algorithms.

Recently, physicists have started applying the principle of "Statistical Physics" to come up with an empirical and theoretical research framework for analyzing the behavior of such self-organizing network systems [184, 80]. We already mentioned in Chapter 3 that Sumpter introduced the Weighted Synchronous Calculus of Communicating Systems (WSCCS) to formally model an agent-based system for a honeybee colony. Using this model, he studied the foraging behavior of a honeybee colony [242], which of course is a self-organizing system. We did some preliminary work in [272, 150], in which we modeled the behavior of *bee agents* in our *BeeAdHoc* [269, 270] and *BeeHive* algorithms by utilizing WSCCS. The initial results from the theoretical models are encouraging. We believe that combining our current stochastic framework with WSCCS will help in developing a generic formal framework for analyzing self-organizing network systems.

Proof

Theorem 1:

The solution of the equation

$$\gamma_i = \beta_i + \sum_{m \in \mathcal{N}_i} \gamma_m \mathfrak{b}_{mi} \qquad i = 1, 2, \cdots, |\mathcal{V}| \wedge i \neq \mathcal{D} \tag{6.38}$$

is given by

$$\gamma = (I - B^t)^{-1} \beta.$$

Proof:

$$\gamma_i = \beta_i + \sum_{m \in \mathcal{N}_i} \gamma_m \mathfrak{b}_{mi} \qquad i = 1, 2, \cdots, |\mathcal{V}| \wedge i \neq mathcalD. \tag{6.39}$$

Substituting $i = 1, 2, \cdots, |\mathcal{V}| \wedge i \neq \mathcal{D}$ in Equation (6.38) and using from Equation (6.17) that $\mathfrak{b}_{in} = 0$ when $i = n \vee (i, n) \notin \mathcal{E}$,

$$
\begin{aligned}
\gamma_1 &= \beta_1 + \gamma_1 \times 0 + \gamma_2 \mathfrak{b}_{21} + \cdots + \gamma_{|\mathcal{V}|} \mathfrak{b}_{|\mathcal{V}|1}, \\
\gamma_2 &= \beta_2 + \gamma_1 \mathfrak{b}_{12} + \gamma_2 \times 0 + \cdots + \gamma_{|\mathcal{V}|} \mathfrak{b}_{|\mathcal{V}|2}, \\
&\vdots \qquad \vdots \qquad \vdots \qquad \vdots \qquad \vdots \qquad \vdots \\
&\vdots \qquad \vdots \qquad \vdots \qquad \vdots \qquad \vdots \qquad \vdots \\
\gamma_{|\mathcal{V}|} &= \beta_{|\mathcal{V}|} + \gamma_1 \mathfrak{b}_{1|\mathcal{V}|} + \gamma_2 \mathfrak{b}_{2|\mathcal{V}|} + \cdots + \gamma_{|\mathcal{V}|} \times 0.
\end{aligned}
$$

On the right-hand side of all the equations given above there is always a term that becomes zero. Moreover, $\gamma_\mathcal{D}$ is also 0. So there are $(|\mathcal{V}| - 1)$ equations, each having $(|\mathcal{V}| - 1)$ terms involving γ_{in}.

Rearranging the above equations we get:

$$
\begin{aligned}
\beta_1 &= \gamma_1 - \gamma_2 \mathfrak{b}_{21} - \gamma_3 \mathfrak{b}_{31} - \cdots - \gamma_{|\mathcal{V}|} \mathfrak{b}_{|\mathcal{V}|1}, \\
\beta_2 &= -\gamma_1 \mathfrak{b}_{12} + \gamma_2 - \gamma_3 \mathfrak{b}_{32} - \cdots - \gamma_{|\mathcal{V}|} \mathfrak{b}_{|\mathcal{V}|2}, \\
&\vdots \qquad \vdots \qquad \vdots \qquad \vdots \qquad \vdots \qquad \vdots \\
&\vdots \qquad \vdots \qquad \vdots \qquad \vdots \qquad \vdots \qquad \vdots \\
\beta_{|\mathcal{V}|} &= -\gamma_1 \mathfrak{b}_{1|\mathcal{V}|} - \gamma_2 \mathfrak{b}_{2|\mathcal{V}|} - \gamma_3 \mathfrak{b}_{3|\mathcal{V}|} - \cdots + \gamma_{|\mathcal{V}|}.
\end{aligned}
$$

Representing the above set of equations in matrix form, we obtain:

$$
\begin{bmatrix} \beta_1 \\ \beta_2 \\ \vdots \\ \vdots \\ \beta_{|\mathcal{V}|} \end{bmatrix} = \begin{bmatrix} 1 & -\mathfrak{b}_{21} & -\mathfrak{b}_{31} & \cdots & -\mathfrak{b}_{|\mathcal{V}|1} \\ -\mathfrak{b}_{12} & 1 & -\mathfrak{b}_{32} & \cdots & -\mathfrak{b}_{|\mathcal{V}|2} \\ \vdots & \vdots & \vdots & \cdots & \vdots \\ \vdots & \vdots & \vdots & \cdots & \vdots \\ -\mathfrak{b}_{1|\mathcal{V}|} & -\mathfrak{b}_{2|\mathcal{V}|} & -\mathfrak{b}_{3|\mathcal{V}|} & \cdots & 1 \end{bmatrix} \begin{bmatrix} \gamma_1 \\ \gamma_2 \\ \vdots \\ \vdots \\ \gamma_{|\mathcal{V}|} \end{bmatrix}.
$$

Using the symbols of these matrices defined in Section 6.3 we obtain:

$$\beta = (I - B^t)\gamma. \tag{6.40}$$

Premultiplying both sides of (6.40) by $(I - B^t)^{-1}$, we finally obtain our required solution:

$$\gamma = (I - B^t)^{-1} \beta.$$

7

An Efficient Nature-Inspired Security Framework for *BeeHive*

The major contribution of the work presented in this chapter is a novel security framework, BeeHiveAIS, inspired by the principles of Artificial Immune Systems (AISs), for nature-inspired routing protocols in general and for BeeHive in particular. We designed this security framework after conducting a comprehensive review of the security threats that could be launched in a network utilizing BeeHive and systematically analyzing their impact on the behavior of BeeHive. We enhanced the performance evaluation framework, introduced in Chapter 3, to empirically demonstrate that the new framework provides the same security level as BeeHiveGuard, a digital signature-based security framework. However, the processing and control overheads of BeeHiveAIS are significantly smaller than those of BeeHiveGuard.

7.1 Introduction

We recall from Chapter 3 that nature-inspired routing protocols like *BeeHive* launch agents that gather routing information about the state of a network in a local region and then update the routing tables on the basis of this information. This feature enables the routers to make routing decisions without requiring access to the complete network topology. However, we must emphasize here that we, in our protocol *Bee-Hive*, and the authors of all other protocols discussed in Chapter 2, always implicitly trusted the identity of the agents and their routing information. This assumption, however, remains no more valid for real-world networks, where malicious intruders or compromised nodes can wreak havoc. To our knowledge, little attention has been paid to analyzing the security threats of nature-inspired routing protocols and efficiently countering them. We believe that the router vendors will not be willing to deploy any routing protocol in real-world routers if its security threats are not properly investigated. In addition to this, a scalable security framework for such protocols, which has acceptable processing and control overheads, is the logical solution to convince router vendors to deploy protocols like *BeeHive* in their routers.

The lack of any comprehensive work in this important domain provided us the motivation to first conduct the vulnerability analysis of our protocol *BeeHive* and then propose two security frameworks for it: *BeeHiveGuard* and BeeHiveAIS. *BeeHiveGuard* is a digital signature-based authentication and integrity-checking framework for *bee agents* while *BeeHiveAIS* is an immune-inspired security framework that utilizes the principles of AIS to counter the security threats of malicious nodes in a *BeeHive*-running network. In this chapter, we show that processing and control overheads of *BeeHiveAIS* are significantly smaller than those of *BeeHiveGuard*, but it provides the same security level as that achieved by *BeeHiveGuard*.

7.1.1 Organization of the Chapter

In the next section, 7.2, we provide a brief review of four classes of robustness to which a security algorithm can belong. We then provide, in Section 7.2.1, a list of different types of attacks that a malicious router is able to launch in a network running *BeeHive* protocol. We then introduce our signature-based security framework, *BeeHiveGuard*, in Section 7.3. Its high processing and control overheads motivate the need for a simple, efficient, and scalable security framework, *BeeHiveAIS*, which is discussed in Section 7.4. We then report in Section 7.5 the results of our experiments that conclude: (1) Malicious nodes can significantly alter the behavior of *BeeHive*; (2) *BeeHiveGuard* and *BeeHiveAIS* are able to successfully counter the attacks of malicious nodes; (3) *BeeHiveAIS* shows a scalable behavior on NTTNet and Node150 topologies. Finally, we conclude the chapter by stating that the performance of *BeeHiveAIS* under normal operating conditions is also comparable to that of *BeeHive*.

7.2 Robustness and Security Analysis of a Routing Protocol

In [188], Perlman has first introduced the classical *Byzantine Generals Problem* [144]. This is a theoretical problem in which generals defending a besieged city have to take the same decision (attack or retreat) by a majority vote. But if a few generals are traitors, then they can influence other generals to take a wrong decision. This problem, if understood, could help a system's designer to calculate the number of malicious components that a system can tolerate for its proper functioning. We know that routing algorithms work asynchronously in a distributed fashion; therefore, the *Byzantine Generals Problem* is not directly relevant [188] except in that the algorithm should be resilient to the presence of malicious nodes in the network. Perlman, argues that two types of failures are mainly responsible for anomalous behavior of a routing protocol: Simple and Byzantine. *Simple failures* occur once a node crashes or a link goes down while *Byzantine failures* occur due to the malicious nodes that launch malicious agents into the network. Such agents can significantly alter the routing behavior of a routing protocol. Perlman introduces a taxonomy of four important classes of robustness for a routing algorithm:

1. *Simple robustness.* The algorithms in this class are robust against *simple failures.* This means that the behavior of an algorithm returns to its normal state once the crashed node or link is up again.
2. *Self-stabilization.* The algorithms in this class can adapt their behavior if a *simple failure* occurs. Similarly, if a *Byzantine failure* occurs, the behavior of an algorithm also returns to normal once the malicious node is removed from the network [68].
3. *Byzantine detection.* The algorithms in this class cannot function in a normal fashion in the presence of *Byzantine failures.* Nevertheless, the algorithm is able to locate the malicious nodes in the network.
4. *Byzantine robustness.* The algorithms in this class are able to function in a normal fashion even in the presence of *Byzantine failures* in the network.

In this chapter, we want to aim at *BeeHive* achieving *Byzantine robustness* because we have already shown in Chapter 3 that *BeeHive* has *simple robustness* and is *self-stabilizing* if *simple failures* occur.

7.2.1 Security Threats to Nature-Inspired Routing Protocols

In [117, 119, 118], the authors have provided a comprehensive description of relevant attacks that malicious nodes can launch in Mobile Ad Hoc Networks (MANETs). We believe that many of these attacks also pertain to nature-inspired routing protocols for fixed networks. We now provide a list of attacks that malicious nodes can launch on a network running *BeeHive* [275].

- *Fabrication attacks* are launched by a router to change the normal route of a data packet. This is accomplished by retransmitting old agents, modifying the information of agents, or launching bogus agents. A *fabrication attack* can be further classified as:
  - An *update storm or malicious flooding.* In this attack a malicious router injects a large number of agents in a short interval of time into the network. As a result, the information (mostly bogus) carried by its agents spreads faster in the network than the true information of other routers. Consequently, the malicious router can divert data packets towards itself.
  - A *replay attack.* In this attack a router retransmits old agents that carry outdated information in the network.
  - A *rushing attack.* This attack in only possible in those routing protocols in which the agents are identified with a unique sequence number. An attacker launches the agents whose source address is of some other node. Moreover, it assigns them a significantly high sequence number. In this way it forces other routers to accept its bogus agents and drop the real ones.
- *Dropping attacks* are powerful because they can divide a network into several partitions. They are of two types: blackhole attack and network partition.
  - *Blackhole attack.* In this attack, an attacker diverts data packets towards itself and simply drops them.

- *Network partition.* An attacker tries to separate a network into k $(k \geq 2)$ partitions. As a result, the nodes in one partition cannot communicate with the nodes in the other partitions.
- *Tampering attack.* In this attack, a malicious node simply modifies the routing information carried by an agent for its own benefit.
  - *Identity impersonating or spoofing.* In this attack, a router impersonates another router by launching bogus agents. As a result, the malicious router can force data packets not to follow a path over another router or it can divert them towards itself.
  - *Detour attack.* An attacker forces its neighbors to route all their network traffic over it.

In our current work, we focus on only fabrication and tampering attacks because they are the most pertinent to routing protocols. Consequently, we ignore the misbehavior detection of malicious nodes due to the dropping attacks because they are not related to the routing protocols.

7.2.2 Existing Works on Security of Routing Protocols

In [174, 228, 256, 118, 106], the authors have developed techniques to counter some of the above-mentioned security threats in classical routing algorithms. They utilized standard cryptography techniques, i.e., digital signatures or Hashed Message Authentication Codes (HMACs) to avert fabrication and tampering attacks. In these techniques, a router verifies that the originator of a control message is the node that is indicated in the header. In [228], the authors have secured distance vector routing protocols by incorporating the information about a node and its predecessor node in the control packet. Sequence numbers are used to identify an old or an obsolete control packet. However, none of these approaches try to analyze and counter the security threats related to the specific features of nature-inspired routing algorithms, with the exception of the preliminary work of Zhong and Evans [301]. They studied the anomalous behavior of *AntNet* [62] under three types of attacks: fabrication, dropping, and tampering. Their experiments clearly demonstrate that the malicious nodes can disrupt the normal routing behavior of *AntNet* by launching these attacks. However, the issue of providing a security framework for *AntNet* was never properly addressed in their work.

7.3 BeeHiveGuard: A Digital Signature-Based Security Framework

We now introduce our signature-based security framework, *BeeHiveGuard*. In [219] and [228], the authors have provided a number of guidelines that, if followed, can enable a routing protocol to counter a number of the above-mentioned threats. The most important principles are: (1) To ensure that the control packets (in our case *bee agents*) are authenticated; and (2) to ensure the integrity of routing information carried by control packets. By authentication, it is meant that a node is able to check whether the received packet is actually sent by the sender mentioned in the header or some other node has spoofed its address. Similarly, the integrity of routing information demands that no intermediate node in between the sender and the receiver be able to alter the routing information in an illegitimate fashion.

The above-mentioned two objectives can be easily achieved if a node employs classical public-key algorithms like RSA [195] for signing its *bee agents* with its digital signature in order to encrypt their information. The algorithm consists of three steps: key generation, encryption, and decryption. In order to sign a message, the sender node encrypts a message with its private key (known only to a node) and other nodes decrypt the message with the public key of the sender (known to everybody in the network). The algorithms like RSA assume that a secure Public Key Infrastructure (PKI) is available that reliably distributes the public keys of nodes to all other nodes in the network. We now introduce two relevant features of our *BeeHiveGuard* framework: agent integrity and routing information integrity.

7.3.1 Agent Integrity

A *bee agent* carries two types of information fields: management and routing. The management information fields do not change during the flooding of an agent and the most relevant management fields are an agent's identifier and the identifier of its replicas, its source address, its time-to-live (TTL) timer, and the address of its *foraging region*. These fields must be protected so that their values cannot be modified or impersonated by an intermediate router. The source node signs these fields with its private key and puts the corresponding signature *sig1* in the *bee agent*. If a traitorous router tries to change these fields or impersonate someone else, then other nodes can easily detect and discard the corresponding bogus *bee agents*. This is possible because the malicious router cannot generate a valid signature *sig1* of the sender (it requires the private key of the sender to do that).

7.3.2 Routing Information Integrity

The purpose of this extension is to secure the routing information, i.e., the propagation delay or the queuing delay of a *bee agent*. The delays are used to estimate the quality of a visited path. The routers calculate the delay values and then modify them accordingly in the *bee agents*; therefore, it becomes a challenging task to differentiate a valid modification from a fake one. We can do it if we assume that *no two subsequent routers on a route fake their routing information*. The basic idea is that a *bee agent* carries the signed routing information of a node and its predecessor node. *sig2* is the signature obtained by signing the queuing and propagation delays of a visited node and *sig3* is the signature for its predecessor node. Figure 7.1 shows how digital signatures are used to secure the routing information.

Node 1 launches two replicas of its *bee agent* towards Nodes 0 and 2 respectively. Node 1 has no predecessor; therefore, *sig2* represents the signed delays of the Node

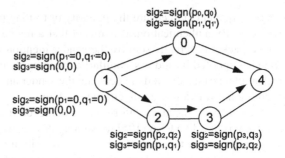

Fig. 7.1. Securing routing information

1, and *sig3* is obtained by signing a 0 value for both delays. Once the replica arrives at Node 2, *sig3* is set to *sig2* and the delays of this node are signed in *sign2*. This process continues until the replica reaches Node 4. Then Node 4 estimates its delays to Node 1 by adding its delay values with the ones of Nodes 3 and 2. As a result, Node 3 can only manipulate its own delays but not the cumulative delays from Nodes 1 to 3. Moreover, a node also compares the delay values in *sig2* with the ones in *sig3*. If the delay values in *sig2* are less than or equal to the ones in *sig3*, then the predecessor node has provided fake delay values. As a result, *sig2* value at this node is calculated with the help of the delay values in *sig3*, and the *bee agent* continues its exploration. Since a node utilizes the information of its predecessor node, and the predecessor node of its predecessor node, a predecessor node cannot significantly influence the routing behavior by faking only its own routing information.

7.3.3 Architecture of *BeeHiveGuard*

The top-level architecture of our framework, *BeeHiveGuard*, is shown in Figure 7.2. We have decoupled the operations in our security framework from the core opera-

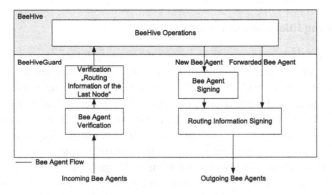

Fig. 7.2. Top-level *BeeHiveGuard*

tions of *BeeHive*. Consequently, we can incorporate a new security framework for the *BeeHive* algorithm without any modifications to the core routing algorithm. The incoming *bee agents* are validated before being handed to *BeeHive*, and similarly outgoing *bee agents* are also signed before further retransmission. Recall that the core functions of *BeeHive* in Algorithm 1 of Chapter 3 are *launchBeeAgents* and *processBeeAgents*. For brevity, we skip the architectural details and show in Algorithm 3 the extended version of these two functions to provide the reader with a better insight as to how the above-mentioned two features are realized in a secure *BeeHive*. A potential shortcoming of *BeeHiveGuard* is that it puts in each *bee agent* three signatures to protect it against any illegitimate enroute modification by any malicious node. If we utilize a recommended key length of 1,024 bytes only, then the size of a single signature is 128 bytes, which results in a size of (128 * 3 + 48 = 432 bytes) for *bee agents* in *BeeHiveGuard*. Recall that *bee agents* in *BeeHive* are just 48 bytes, so it is obvious that the security framework has increased their size approximately ten times. In addition to this, the cost of processing a *bee agent* at intermediate routers is also expected to increase manifold due to complex computation-intensive functions that authenticate an agent and check the integrity of its routing information. This intuition-based analysis will be shortly confirmed in the results of our experiments. Therefore, we need to design a security framework that has low processing and control overheads so that it can be deployed in real time on real-world routers. We now introduce an efficient, scalable, and robust AIS-based security framework to achieve this objective.

7.4 BeeHiveAIS: an Immune-Inspired Security Framework for *BeeHive*

We will first briefly introduce the principles of Artificial Immune Systems (AISs) and then introduce our *BeeHiveAIS* framework [274].

7.4.1 Artificial Immune Systems (AISs)

AISs [54, 55] are inspired by the principles of the human immune system. The features of AIS that are particularly relevant to providing security in routing protocols are: self-identity, anomaly detection, and learning. The self-identity enables an AIS to understand normal behavior of the agents in a routing protocol and to generate corresponding self-antigens. The AIS then generates a repository of antibodies, which can detect an anomalous behavior due to malicious agents (non-self-antigens). The antigens and antibodies must be in a shape space format [185, 215] to facilitate the definition of affinity between them, which is often a mathematical distance function. The negative selection algorithm [92] randomly generates a number of antibodies and adds only those to the repository whose affinity with the self-antigens is not above a certain threshold value. This generation process for the antibodies is known as *thymus model* [54, 55]; it enables an AIS to detect anomalies through self and non-self differentiation.

Algorithm 3 Extending bee launching and bee processing with authentication and integrity

---

procedure launchBeeAgents(s,n_i,n)

  if t % Δt = 0 or n_i % Packet_Limit = 0 **then**

    decide, if the bee is a short distance or a long distance bee agent

    $digitalsignature\_2 = sig(0,0)$

    $digitalsignature\_3 = sig(0,0)$

    for $x \leftarrow 1$ to N **do**

      create a replica b_s^{xv} of bee agent b_s^v

      find address of neighbor at index x

      $digitalsignature\_1 = sig(s,v,x,h_s)$

      launch replica b_s^{xv} to neighbor at index x

      x++

    end for

  end if

end procedure

procedure processBeeAgents(b_s^{xv},i)

  if b_s^{xv} already visited i or its hop limit reached **or** $verify(digitalsignature\_1) == true$ **then**

    kill b_s^{xv} {avoid loops }

  else if *queuing delay and propagation delay protected by digitalsignature_3 < queuing delay and propagation delay protected by digitalsignature_2* **then**

    $digitalsignature\_3 = digitalsignature\_2$

    $digitalsignature\_2 = sig(q_i^n + q_{i-1}^n, p_i^n + p_{i-1}^n)$ {q_{i-1}^n and p_{i-1}^n are protected by old digitalsignature_2}

    if b_s^{xv} is inside FR_s **then**

      Update local propagation and queuing delay (IFZ) and the delays of the bee agent

    else

      Update local propagation and queuing delay (IFR) and the delays of the bee agent

    end if

  else

    $digitalsignature\_2 = sig(q_i^n + q_{i-1}^n, p_i^n + p_{i-1}^n)$ {q_{i-1}^n and p_{i-1}^n are protected by digitalsignature_3}

    if b_s^{xv} is inside FR_s **then**

      Update local propagation and queuing delay (IFZ) and the delays of the bee agent

    else

      Update local propagation and queuing delay (IFR) and the delays of the bee agent

    end if

  end if

  if b_s^{jv} already reached i {$\forall j \neq x$} **then**

    kill b_s^{xv}

  else

    use priority queues to forward b_s^{xv} to all neighbors of i except p

  end if

end procedure

The security framework based on AIS provides a number of benefits: small processing overhead due to a simple anomaly detection algorithm; no significant increase in control overhead because the agents need not carry any signatures; and a reasonably small size database to store antibodies. These benefits of AIS make it perfectly suitable for securing agent-based adaptive routing protocols in an efficient manner in real time.

7.4.2 Behavioral Analysis of *BeeHive* for Designing an AIS

As a first step, we analyzed the normal behavior of the *BeeHive* algorithm to design our AIS-based security framework for it. We now provide important conclusions of our empirical analysis aimed at understanding the normal behavior of *BeeHive*.

Propagation delay is greater than queuing delay

An important outcome of our analysis is that *the propagation delay in most of the cases was significantly greater than the queuing delay*. As a result, the goodness of a neighbor for reaching a certain destination varied most of the time within a small window from its average goodness value. If a router can estimate this window, then it can counter tampering of routing information (propagation and queuing delays). This conclusion is based on our experiments in which we analyzed the histograms of *bee agents* for delay stability (d_{stab}) values, where $d_{stab} = \frac{qd}{pd}$ (qd and pd are cumulative queuing and propagation delays respectively that a *bee agent* carries in its header for a certain destination). We show representative histograms in Figures 7.3, 7.4, and 7.5

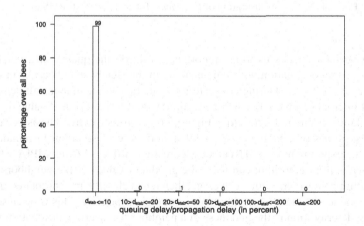

Fig. 7.3. NTTNet – histogram of *bee agents* for $d_{stab} = \frac{qd}{pd}$ for MSIA = 5

for MSIA = 5, MSIA = 2.6, and MSIA = 1 respectively while MPIA = 0.005 in all scenarios. The rest of the parameters remain the same as reported in Chapter 3. The motivation behind these MSIA values is to study (d_{stab}) under low (MSIA = 5), high (MSIA = 2.6), and very high traffic load (MSIA = 1) scenarios. We can conclude from these figures that in the case of MISA = 5 and MSIA = 2.6, approximately more than 98% of *bee agents* report a queuing delay that is significantly smaller than propagation delay ($d_{stab} \leq 0.1$). Figure 7.5 for MSIA = 1.6 is, however, an exception where only 57% of *bee agents* report $d_{stab} \leq 0.1$. Recall from Chapter 3 that this scenario is only reported as a benchmark and is highly unlikely to be encountered in real-world networks. Even in this scenario, 90% *bee agents* never report a queuing delay that is five times greater than the propagation delay.

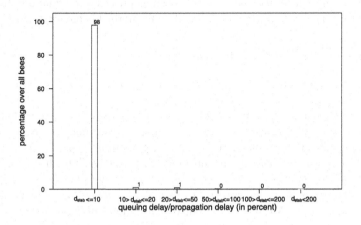

Fig. 7.4. NTTNet – histogram of *bee agents* for $d_{stab} = \frac{qd}{pd}$ for MSIA = 2.6

This behavior leads us to an important corollary: the quality of a neighbor to reach a certain destination will remain stable in the steady state because in most of the cases $d_{stab} \leq 0.1$. During congestion the quality of a neighbor will degrade for a small interval of time but it will be again restored to the original value due to the load-balancing feature of *BeeHive*. Figure 7.6 confirms our hypothesis because the goodness values of neighbors 5, 7, and 9 at Node 6 for destination 11 remain stable during the experiment in NTTNet (see Figure 3.6 (MSIA = 2.6 and MPIA = 0.005)). This means if an algorithm can learn the goodness values of the neighbors, which vary within a threshold from the average, then it can classify any abrupt deterioration in the goodness value of a neighbor due to malicious activity. This is because of the fact that deterioration in the goodness of a neighbor for a certain destination without a deterioration in the goodness of other neighbors (or vice versa) is a strong hint of an attack by a malicious node (under high traffic load the goodness of all neighbors for reaching a certain destination are expected to deteriorate).

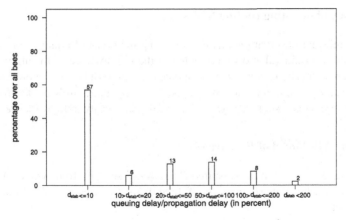

Fig. 7.5. NTTNet – histogram of *bee agents* for $d_{stab} = \frac{qd}{pd}$ for MSIA = 1

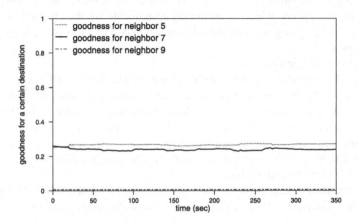

Fig. 7.6. NTTNet – goodness trend

Stable flooding pattern of *bee agents*

We observed that the flooding pattern that defines the outreach radius of *bee agents* to different nodes from their source nodes remains mostly stable, i.e., a particular replica of a *bee agent* always arrives at a node through the same neighbor. The only exception is that under high traffic loads a particular replica of a *bee agent* might have to wait in a priority queue due to the fact that the transmission of a data packet is in progress at its intended output interface. Our observation is that this does not significantly alter the flooding pattern. Remember from Chapter 3 that we did not enforce a fixed flooding pattern for the sake of simplicity in our *BeeHive* protocol.

Frequency of updating routing tables

Another relevant observation was that reducing the update frequency of routing tables below a certain value does not enhance the performance of the algorithm. The reason behind this behavior is that the update interval must be large enough to capture a significant change in the queueing delay. This systematic study of *BeeHive* proved useful in designing a simple, efficient, and scalable security framework, *BeeHiveAIS*.

7.4.3 The AIS Model of *BeeHiveAIS*

We now discuss our immune-inspired *BeeHiveAIS* security framework for our *BeeHive* protocol. *BeeHiveAIS* consists of three distinct phases: initialization, learning, and protection.

1. *BeeHiveAIS* initialization phase
   During the initialization phase, the AIS model learns the normal flooding pattern of the *bee agents*, which is eventually responsible for the creation of *foraging zones* and *foraging regions*. Recall from Chapter 3 that this phase lasts about 30 seconds.
2. *BeeHiveAIS* learning phase
   After 30 seconds starts the learning phase, in which data traffic is injected into the network; our framework assumes that no malicious nodes are present in the network. The major objective is to learn the normal behavior of the *BeeHive* algorithm in order to build a repository of the self-antigens from which a database of the antibodies to counter different types of threats is generated. This phase lasts for 50 seconds after the data traffic has been injected into the network.
3. *BeeHiveAIS* protection phase
   Finally, the protection phase starts in which the AIS detects the security threats of the malicious nodes and then counters them through the antibodies.

Modeling of self-antigens in *BeeHiveAIS*

The important AIS concepts utilized in *BeeHiveAIS* and their mappings to *BeeHiveAIS* are cataloged in Table 7.1. The framework uses two types of antigens: type 1 and type 2.

Type 1 antigens

The purpose of type 1 self-antigens is to ensure that a malicious node cannot tamper with the agent-specific data. This is achieved through self-antigens that consist of tuples like <*sourceaddress, neighbor, hops*>. The antibodies are randomly generated, and if their affinity with the self-antigens is above a threshold value, they are discarded (negative selection). Currently, if an antigen is matched to an antibody during the protection phase, then the *bee agent* that triggered the match is dropped.

AIS		*BeeHiveAIS*
Self Cells		Well-behaving nodes
Nonself Cells		Misbehaving nodes
Self-Antigens	type 1	Correct incoming direction and region of a bee agent
	type 2	Correct propagation and queuing delays tendencies
Antigens	type 1	Incorrect incoming direction and region of a bee agent
	type 2	Incorrect propagation and queuing delays tendencies
Antibody	type 1	Pattern that can detect antigens of type 1
	type 2	Pattern that can detect antigens of type 2

Table 7.1. Mapping of concepts from AIS to *BeeHiveAIS*

Type 2 antigens

The type 2 self-antigens ensure that a malicious node cannot manipulate the propagation and queuing delays. These self-antigens consist of average goodness values, which are gathered for a certain destination through a certain neighbor over a sliding window of the delays of five subsequent *bee agents*. In addition, the self-antigens also contain a flag which determines whether the change in goodness value is due to the presence of a malicious node or due to the change in traffic patterns. As a result, upper and lower threshold goodness values are assigned to a neighbor for reaching a destination. The creation process for the antibodies is similar to the one explained for type 1. A bee agent is dropped if an antibody matches an antigen: the reported goodness value of a neighbor is either above or below the upper or lower learned threshold values respectively. In this case the goodness value of the neighbor is set to the average goodness value.

Update frequency

As discussed before, we added in our AIS an auxiliary feature which limits the update frequency of the routing tables to a certain value. If a *bee agent* arrives before the expected time, then it is dropped to counter the Denial of Service (DoS) attack, which basically results due to significant reduction in the bee-launching interval.

Size of detector's database

As mentioned before, a type 1 self-antigen consists of a tuple of the form <*source address, neighbor, hops*>. We can store each value in four bytes; therefore, one tuple takes 12 bytes. We need to store same number of tuples as the number of entries in the routing table. Recall from Chapter 3 that this number for NTTNet is 78. Consequently we need just $78 \times 12 = 936$ bytes for storing type 1 self-antigens. For a type 2 self-antigen we can store the average goodness value in a floating-point number (eight bytes) and one more byte to store the flag. Therefore, a single type 2 self-antigen can be stored in nine bytes and the total memory required for a complete window of size x is $9 \times x$. If we use a window size of 5 ($x = 5$), then for NTTNet, we need just $78 \times 5 \times 9 = 3,510$ bytes. Finally, the size of the detector's database for NTTNet is just 4,446 bytes, which is an acceptable memory overhead. Similarly, the number of routing table entries for Node150 (refer to Chapter 3) is 193. It means that for type 1 self-antigens, we will require $193 \times 12 = 2,316$ bytes, and for type 2 self-antigens will require $193 \times 5 \times 9 = 8,685$ bytes. Therefore, we will require 11,001 bytes for the detector's database in the Node150 topology, which is again a reasonable increase compared to the 4,446 bytes of NTTNet.

7.4.4 Top-Level *BeeHiveAIS*

The block-level diagrams of the important functions in *learning* and *protection* phases are shown in Figures 7.7 and 7.8 respectively. Here, however, the operations of *BeeHiveAIS* are not completely independent of the operations of *BeeHive* because matching of type 2 antigens requires an access to the routing table of *BeeHive*, as shown in Figure 7.8

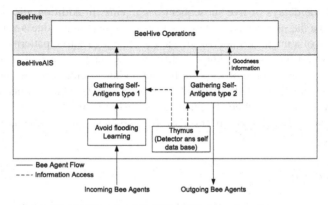

Fig. 7.7. Top-level *BeeHiveAIS* – Learning phase

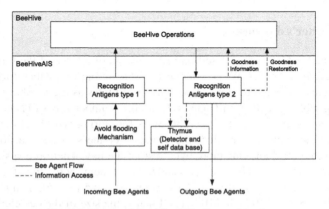

Fig. 7.8. Top-level *BeeHiveAIS* – Protection phase

Algorithm 4 Extending bee processing with an AIS

---

procedure processBeeAgents(b_s^{xv},i) \\in the learning phase

  if b_s^{xv} already visited i or its hop limit reached **then**

    kill b_s^{xv} {avoid loops}

  else

    insert self antigen $< s, p, hop\_limit >$ into the thymus

    create n antibodies of type-1

    run a negative selection algorithm with the antibodies of type-1 as input

    add valid antibodies of type-1 to the thymus

    if b_s^{xv} is inside FR_s then

      Update local propagation and queuing delay (IFZ) and delays of a bee agent

    else

      Update local propagation and queuing delay (IFR) and delays of a bee agent

    end if

  end if

  insert propagation delay and queuing delay to the thymus

  create x antibodies of type-2

  run a negative selection algorithm with the antibodies of type-2 as input

  add valid antibodies of type 2 to the thymus

  if b_s^{jv} already reached i $\{\forall j \neq x\}$ then

    kill b_s^{xv}

  else

    use priority queues to forward b_s^{xv} to all neighbors of i except p

  end if

end procedure

procedure processBeeAgents(b_s^{xv},i) \\in the monitoring phase

  if *time to the last bee agent b_s^{xv} is below the threshold a* **then**

    kill b_s^{xv}

  else if *an antibody of type 1 has a high affinity to $< s, p, hop\_limit >$* **then**

    kill b_s^{xv}

  else if b_s^{xv} already visited i or its hop limit reached **then**

    kill b_s^{xv} {avoid loops}

  else if b_s^{xv} is inside FR_s **then**

    Update local propagation and queuing delay (IFZ) and delays of a bee

  else

    Update local propagation and queuing delay (IFR) and delays of a bee

  end if

  if *an antibody of type 2 has a high affinity to the sequeuce of the last queuing and propagation delays and no bee agent reports a high queuing delay* **then**

    kill b_s^{xv}

    Restore the goodness of this source and neighbor

  end if

  if b_s^{jv} already reached i $\{\forall j \neq x\}$ then

    kill b_s^{xv}

  else

    use priority queues to forward b_s^{xv} to all neighbors of i except p

  end if

end procedure

It is pertinent to mention that *BeeHiveAIS* is a passive framework that is deployed at each router to monitor the incoming *bee agents* and then classify them as normal or malicious. Again ignoring the architectural details, we show in Algorithm 4 the modified "processBeeAgents" function. In order to highlight the distinct actions that are performed during the learning and protection phases, we show them separately in the processBeeAgents function by presenting the function twice.

7.5 Simulation Models of Our Security Frameworks

We realized *BeeHiveGuard* and *BeeHiveAIS* in the OMNeT++ simulator because *BeeHive* had been originally realized in the OMNeT++ simulator. We first integrated a standard cryptography library, OpenSSL [53], in OMNeT++, and then utilized its functions that implement relevant cryptography techniques like digital signatures, symmetric and asymmetric cryptography, and hash functions in order to realize our *BeeHiveGuard* framework.

We designed an attacker framework and embedded it into OMNeT++. The framework gives a designer an opportunity to designate any node a malicious node. The malicious node can launch a number of types of attacks as listed in Section 7.2.1. The designer can also give a starting and an ending time of the attack. We used the performance evaluation framework introduced in Chapter 3 to measure relevant performance parameters. We extended it to log two additional auxiliary parameters: the quality of a neighbor for reaching a destination and the total number of packets switched by an intermediate node. These two auxiliary parameters provide useful insight to study the impact of an attack on a routing algorithm.

We need to empirically validate the functional behavior of both of our security frameworks because it is very difficult to contemplate the emerging behavior of *BeeHive* with the security enhancements. Moreover, *BeeHiveAIS* is a stochastic classifier and it must be tested under a variety of operational scenarios to inspire the confidence that *BeeHiveAIS* scales all of them. Therefore, our validation principle is: *BeeHiveAIS must provide the same security level as BeeHiveGuard does, but with significantly smaller processing and control overheads. Moreover, the relevant performance values of the secure algorithm must be within an acceptable range of those of BeeHive (without any attack).* Our reported values are an average of the values obtained from ten independent runs and the confidence level is 95%.

7.5.1 Attack Scenarios on Simple Topologies

We now show on simple topologies that how a malicious node can alter the routing behavior of nature-inspired algorithms like *BeeHive*.

Tampering control messages

We simulated in topology Net1 a malicious node, which alters the queuing delay and propagation delay fields of the *bee agents* passing through it. In Net1, a traffic session is started between Nodes 4 and 1. Node 3 modifies the queuing and propagation delays of the *bee agents* launched by Node 1. As a result, it artificially increases the quality of the path 4-3-2-1 as compared to 4-0-1.

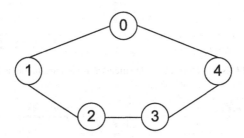

Fig. 7.9. Net1: Node 3 is tampering with the queuing and the propagation delay

Impersonating, detour and flooding attacks

In topology Net2, we created a scenario which can simulate impersonating, detour, and flooding attacks. A traffic session is started between Nodes 3 and 0. In a normal mode, data packets take the path 3-2-1-0. However, Node 4 launches the three attacks by injecting a large number of *bee agents*, which have Node 0 as their source node instead of Node 4. In this way, data packets also take the path 3-2-4-2-1-0.

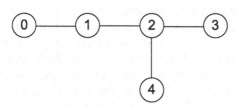

Fig. 7.10. Net2: Impersonating, detour, and flooding attacks launched by Node 4

7.5.2 Analysis of Attacks and Effectiveness of Security Frameworks

We now show that both of our security frameworks provide *Byzantine robustness* to the *BeeHive* protocol for above-mentioned attacks.

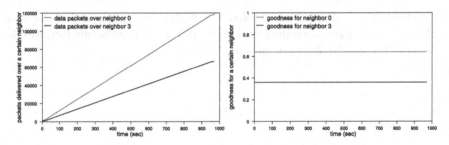

Fig. 7.11. Net1: Normal mode (measurements are made at Node 4)

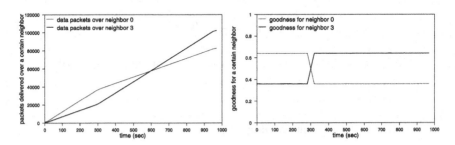

Fig. 7.12. Net1: Under attack (measurements are made at Node 4)

Tampering control messages

In Figure 7.11, one can see that in a normal mode the path 4-0-1 is rated higher than the path 4-3-2-1 because of its smaller delays. As a result, more data packets are routed on the path 4-0-1 as compared to the path 4-3-2-1. However, the situation is drastically changed once Node 3 launches its attack at 300 seconds by tampering with the information in the *bee agents* (see Figure 7.12). The impact of the attack is significantly reduced in *BeeHiveGuard* (see Figure 7.13) because Node 3 can now just manipulate its own queuing and propagation delays. Remember that in *BeeHive* it can manipulate the delays of the complete path 3-2-1. Similarly, we can see in Figure 7.14 that *BeeHiveAIS* also successfully counters the attack because the AIS framework classified the sudden change in the quality values as a non-self-antigen, and subsequently the corresponding malicious *bee agents* were not allowed to update the routing tables. If we compare Figure 7.13 with Figure 7.14, then it is obvious that the learning-based countermeasure strategy of *BeeHiveAIS* is even more effective than the signature-based strategy of *BeeHiveGuard*.

Impersonating, detour, and flooding attacks

One can see in Figure 7.15 that in a normal mode all data packets are routed on the path 3-2-1-0. Node 4 launches its attacks by transmitting bogus *bee agents* at 300 seconds. As a result, it has successfully detoured a significantly large number of data

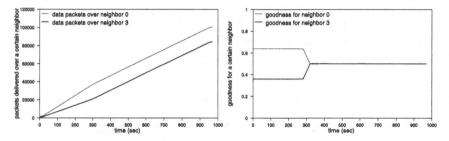

Fig. 7.13. Net1: BeeHiveGuard (measurements are made at Node 4)

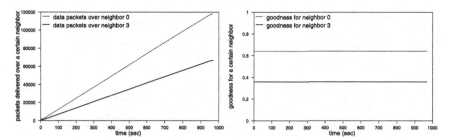

Fig. 7.14. Net1: BeeHiveAIS (measurements are made at Node 4)

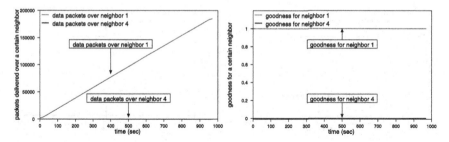

Fig. 7.15. Net2: Normal mode (measurements are made at Node 2)

packets towards itself (see Figure 7.16). Remember that the left subfigure of Figure 7.16 shows the number of packets that followed either the path 3-2-1 or the path 3-2-4. Consequently, the number of data packets that followed the path 3-2-4-2-1 is not counted for neighbor 1. Figures 7.17 and 7.18 show that in both *BeeHiveGuard* and *BeeHiveAIS*, Node 4 is not able to influence the routing decisions by propagating its bogus *bee agents*.

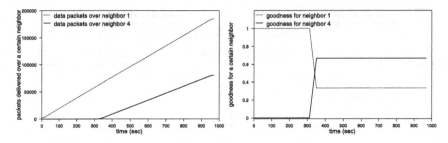

Fig. 7.16. Net2: Under attack (measurements are made at Node 2)

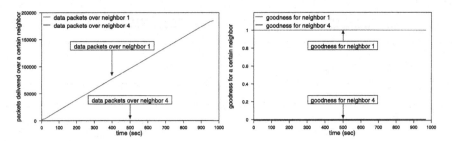

Fig. 7.17. Net2: BeeHiveGuard (measurements are made at Node 2)

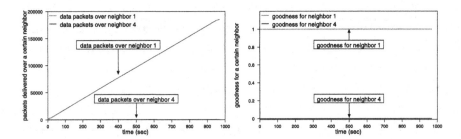

Fig. 7.18. Net2: BeeHiveAIS (measurements are made at Node 2)

Cost analysis

We have collected relevant cost and performance values in Table 7.2. Recall from Chapter 3 that A_a is the average number of processor cycles required to process a *bee agent*, A_t is the total number of cycles that a node spends in processing the *bee agents*, R_o models the bandwidth consumed by the *bee agents*, S_o models the additional bandwidth consumed by data packets if they do not follow the shortest hop path, T_{av} is the average throughput (Mbits/sec), and t_d is the average delay (in milliseconds). It is easy to conclude from Table 7.2, that both *BeeHiveAIS* and *BeeHive-Guard* are able to successfully counter the threats and provide the same throughput and packet delay (the two important performance values) as *BeeHive* under normal

Net	Algorithm	A_a	R_o	S_o	T_{av}	t_d
Net1	BeeHive	16699	0.022	0.473	0.79	0.004
Net1	BeeHive under Attack (1)	16547	0.022	0.73	0.79	0.004
Net1	BeeHiveGuard (2)	11579849	0.121	0.601	0.791	0.004
Net1	BeeHiveAIS (3)	33955	0.022	0.474	0.79	0.004
Difference of (2) & (1) (%)		69800	450	17.6	0.1	0
Difference of (3) & (1) (%)		105.2	0	0	0	0
Net2	BeeHive	13924	0.018	0	0.79	0.004
Net2	BeeHive under Attack (4)	14703	0.057	1.443	0.79	0.008
Net2	BeeHiveGuard (5)	23603742	0.133	0	0.79	0.004
Net2	BeeHiveAIS (6)	25287	0.044	0	0.79	0.004
Difference of (5) & (4) (%)		160400	133.3	100	0	50
Difference of (6) & (4) (%)		71.9	22.8	100	0	50

Table 7.2. Performance and costs of BeeHiveGuard and BeeHiveAIS in Net1 and Net2

conditions. But the processing complexity of *bee agents* (in Net1) in *BeeHiveGuard* is $69,800\%$ greater than that of *bee agents* of *BeeHive*. In comparison, the *bee agents* in *BeeHiveAIS* take just 105% more cycles compared to the *bee agents* of *BeeHive*. This shows that the processing complexity of *bee agents* in *BeeHiveGuard* is 650 and $2,220$ times more than that of in *BeeHiveAIS* in Net1 and Net2 respectively. Similarly, the control overhead of *BeeHiveGuard* is 450 and six times greater than that of *BeeHiveAIS* in Net1 and Net2 respectively. We now need to evaluate the scalability of *BeeHiveAIS* on larger networks under a number of operational scenarios.

7.5.3 NTTNet

Recall from Chapter 3 that NTTNet (see Figure 3.6) is a well-known topology, and here we will discuss some of the representative attack scenarios on it. We reproduce the figure here for convenience. We abbreviate *BeeHive*, *BeeHiveGuard*, and *BeeHiveAIS* with BHive, BHG and BHAIS respectively to save space in the subsequent tables. The network traffic is session-oriented with MSIA = 2.6 and MPIA = 0.005

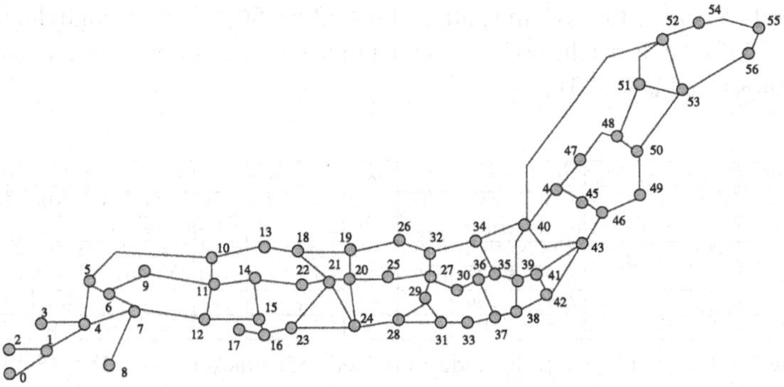

Fig. 7.19. NTTNet

and session size = 2,130,000 bits. The reported results are an average of the values obtained from ten independent runs, each lasting 1,000 seconds.

The number of switched packets by a node did not provide useful insight for large networks due to the scaling issues. Therefore, we defined a new parameter, *routing affinity*, for large networks. The *routing affinity* is defined as the ratio of the packets routed through a node to the total number of packets generated in the network. We equipped each node with a traffic scope that measures its *routing affinity*. Moreover, it can generate a traffic chart, showing the routing affinity of a node for a selected algorithm (see Figure 7.20). Due to cyclic paths, this ratio can exceed a value of 1. If a malicious node cannot significantly increase its *routing affinity* by launching different attacks, then we can safely conclude that the security framework did achieve its objective. Since NTTNet is a large network, we will now use our *routing affinity* measure to show the impact of different attacks.

We now report four relevant types of security threats that malicious nodes can launch in this network: impersonating, tampering of routing information, a combined super attack, and Denial of Service (DoS) attack. We will use symbol "x-y-z" to represent an attack scenario, where x, y, and z are the node ids of sender, receiver, and attacker respectively. If x and y contain value "all," then all nodes in the network send and receive data packets, and in this case, z will represent the number of malicious nodes.

Impersonating

In this scenario, Node 35 transferred data to Node 55. Node 25, which does not lie on a desirable route from Node 35 to Node 55, impersonated the Node 55 (Node 25 launched *bee agents* by faking the source address of Node 55). As a result, Node 25 is expected to significantly disrupt the routing behavior of *BeeHive*. Figure 7.20 and Table 7.3 confirm this hypothesis. The attacker node, Node 25, is able to significantly enhance its routing affinity by attracting a large numbers of packets towards itself even though it does not lie on the route (note that the packets from Node 35 can reach Node 55 either through the path "34-40-52-53-56-55" or through the path "34-40-52-54-55"). As a result, 60% of the data packets are dropped due to looping in cyclic paths (see Table 7.3).

Algorithm	p_d	P_{loop}	S_c	R_o	A_a	t_d	t_{90d}	S_d	S_{90d}	J_d	T_{av}	h_{ex}	S_o	T_o	R_{ent}
(1)BHive	99.9	0	99.8	0.167	30683	0.027	0.028	2.62	2.77	0.004	0.844	0.443	0.038	0.206	65.9
(2)BHive(a)	41.4	29.9	9.51	0.232	25009	2.29	5.49	2.62	2.77	0.904	0.336	26.5	0.918	1.149	69.4
(3)BHG(a)	99.9	2.11	99.7	4.15	8801801	0.031	0.036	2.63	2.77	0.006	0.82	2.52	0.213	4.36	73.3
(4)BHAIS(a)	99.9	0	99.7	0.197	52433	0.027	0.028	2.62	2.77	0.004	0.812	0.443	0.037	0.234	65.8
(3)-(1) in %	-	-	-	2385	28586	-	-	-	-	-	-	-	461	-	
(4)-(1) in %	-	-	-	18	71	-	-	-	-	-	-	-	2.6	-	

Table 7.3. NTTNet – data from Node 35 to Node 55; attacker: Node 25 ("35-55-25")

One can easily conclude from Figure 7.20 and Table 7.3 that *BeeHiveAIS* and *BeeHiveGuard* are able to successfully counter the attack of Node 25. The additional

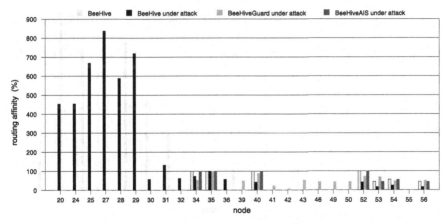

Fig. 7.20. NTTNet – data from Node 35 to Node 55; attacker: Node 25 ("35-55-25")

processing and control overheads of *BeeHiveGuard* are 28,586% and 2,385% respectively compared to *BeeHive*. It is important to note that the increase in processing and control overheads of *BeeHiveAIS* are only 71% and 18% respectively compared to *BeeHive*. This is certainly attributable to the AIS utilized by *BeeHiveAIS*, which requires no additional information to be transmitted in the *bee agents* for correct functionality of its simple anomaly detection model. In *BeeHiveGuard*, about 2% of the data packets enter cyclic paths. But this is due to the fact that extremely large size *bee agents* carrying digital signatures in *BeeHiveGuard* altered their reachability pattern. As a result, data packets can reach Node 55 also through Node 43. Note that the performance values of *BeeHiveAIS* are comparable to *BeeHive* (without any attack).

Tampering of routing information

In this attack, a malicious node tries to manipulate the routing information carried by the *bee agents*. As a result, a node can artificially alter the goodness of different nodes for its benefit. In this scenario, Node 55 transfered data to Node 48, and Node 54 manipulated the routing information to enhance its routing affinity.

Algorithm	p_d	P_{loop}	S_c	R_o	A_a	t_d	t_{90d}	S_d	S_{90d}	J_d	T_{av}	h_{ex}	S_o	T_o	R_{ent}
(1)BHive	99.9	0	99.8	0.167	30487	0.018	0.024	2.61	2.76	0.005	0.799	0.533	0.043	0.212	66.1
(2)BHive(a)	99.9	0.001	99.8	0.167	29933	0.019	0.027	2.62	2.76	0.006	0.803	0.763	0.063	0.231	66.1
(3)BHG(a)	99.9	0.006	99.6	3.5	8850628	0.018	0.025	2.62	2.76	0.005	0.826	0.573	0.048	3.55	70.2
(4)BHAIS(a)	99.9	0	99.6	0.167	53144	0.018	0.024	2.61	2.76	0.005	0.801	0.534	0.043	0.211	65.6
(3)-(1) in (%)	-	-	-	1996	28931	-	-	-	-	-	-	-	12	-	
(4)-(1) in (%)	-	-	-	0	74	-	-	-	-	-	-	-	0	-	

Table 7.4. NTTNet – data from Node 55 to Node 48; attacker: Node 54 ("55-48-54")

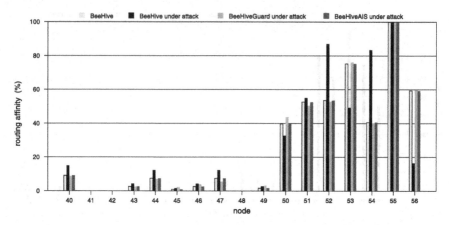

Fig. 7.21. NTTNet – data from Node 55 to Node 48; attacker: Node 54 ("55-48-54")

It is evident from Figure 7.21 that in *BeeHive* Node 54 is successful in its attack because it is able to enhance its routing affinity by 40%, as a consequence decreasing the routing affinity of Node 56 from 60% to less than 20%. One can easily conclude from Figure 7.21 and Table 7.4 that *BeeHiveAIS* and *BeeHiveGuard* are able to successfully neutralize the impact of this attack. However, the processing and control overheads of *BeeHiveAIS* are significantly smaller than those of *BeeHiveGuard*. The reasons for this substantial difference are already described in the impersonating attack. The performance values of *BeeHiveAIS* are approximately the same as those of *BeeHive* (without any attack).

A combined super attack

The motivation behind this experiment was to study the impact of a combined super attack collectively launched by a number of nodes (Nodes 21, 24, 40, and 43), each node selecting a different type of the attack, in the network. Node 40 and 43 simply set the delay value of each *bee agent* passing through them to 0. In this attack all nodes acted as a source or destination node of a traffic session.

Algorithm	p_d	P_{loop}	S_c	R_o	A_a	t_d	t_{90d}	S_d	S_{90d}	J_d	T_{av}	h_{ex}	S_o	T_o	R_{ent}
(1)BHive	99.6	3.35	93.2	0.293	31972	0.121	0.426	2.74	3.1	0.026	46.5	1.83	8.77	9.06	78.4
(2)BHive(a)	97.8	7.15	80.2	0.472	36152	0.238	0.831	2.84	3.43	0.06	45.7	2.58	12.1	12.6	84
(3)BHG(a)	98.7	5.7	83.7	7.87	9118316	0.204	0.731	2.81	3.38	0.041	46.2	2.03	9.66	17.5	80.2
(4)BHAIS(a)	99.2	4.45	88.5	0.333	65302	0.162	0.551	2.77	3.18	0.034	46.5	1.91	9.18	9.51	77.3
(3)-(1) in (%)	-	-	-	2586	28420	-	-	-	-	-	-	-	-	10	-
(4)-(1) in (%)	-	-	-	14	104	-	-	-	-	-	-	-	-	4.68	-

Table 7.5. NTTNet – normal traffic; four nodes attack ("all-all-4")

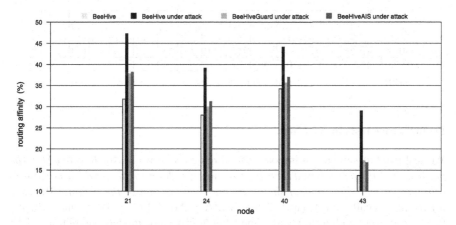

Fig. 7.22. NTTNet – normal traffic; four nodes attack ("all-all-4")

One can see in Figure 7.22 that in *BeeHive* the nodes that launched the attack, are able to significantly enhance their routing affinity. The impact of the attacks of Nodes 43 and 21 is more than that of the other two nodes. It is also important to note that *BeeHiveAIS* and *BeeHiveGuard* are able to successfully counter the attacks in Nodes 24, 40, and 43. However, their countermeasures are not completely successful in Node 21. It is worth mentioning that Node 21 occupies a pivotal position in NT-TNet; hence it is not easy to counter all side effects caused by its tampering with the routing information. Nevertheless, *BeeHiveAIS* provides the same security level as that provided by *BeeHiveGuard* but at significantly smaller processing and control overheads (see Table 7.5). Again the performance values of *BeeHiveAIS* are closest to *BeeHive* (without any attack) and are significantly better than those of *BeeHiveG-uard* (under attack).

DoS attack

This attack can be launched by substantially reducing the bee launching interval of the *bee agents* in order to saturate the priority queues. As a result, a router is always busy processing the *bee agents* (due to their higher priority) and never gets the chance to do its actual task of packet switching. This attack was launched by 19 nodes in the network. In this scenario all nodes acted as a source or destination node of a traffic session.

One can conclude from Table 7.6 that this attack has the strongest impact on the performance values of *BeeHive*. However, *BeeHiveAIS* is able to successfully counter the attack and its performance values are approximately same as those of *BeeHive* (without any attack). The difference in control overhead stems from the fact that a neighbor node of the malicious node in *BeeHiveAIS* has to still process a significantly higher number of *bee agents* before dropping them. *BeeHiveGuard* simply crashed in this scenario.

Algorithm	p_d	P_{loop}	S_c	R_o	A_a	q_{av}	t_d	t_{90d}	S_d	S_{90d}	J_d	T_{av}	h_{ex}	S_o	T_o
(1)BHive	99.6	3.35	93.2	0.293	31972	0.012	0.121	0.426	2.74	3.1	0.026	46.5	1.83	8.77	9.06
(2)BHive(a)	67.5	1.73	39.8	34.6	22379	0.296	1.56	4.84	2.88	3.85	0.183	31.3	1.55	5.01	39.6
(3)BHAIS(a)	99.2	2.87	87.6	6.69	46570	0.021	0.193	0.629	2.8	3.24	0.04	46	1.74	8.25	14.9
(3)-(1) in (%)	-	-	-	2183	46	-	-	-	-	-	-	-	-	-	5.93

Table 7.6. NTTNet – normal traffic; 19 nodes launch DoS attack (”all-all-19”)

Performance tests in NTTNet

We still need to validate an important requirement that we set for *BeeHiveAIS*: *The performance of an AIS-based security framework under normal operating conditions (absence of malicious nodes) should be the same as that of the original* BeeHive *algorithm*. Recall from Chapter 3 that MSIA = 2.6, MPIA = 0.005, and MPIA = 0.0005 represent high traffic loads. The latter MPIA value models bursty traffic. The important performance parameters for the two algorithms are tabulated in Table 7.7. Note that the most important performance parameters like throughput (T_{av}), delay (both t_d and t_{90d}), and jitter (J_d) of the two algorithms are approximately the same. This validates the above-mentioned requirement.

Algorithm	p_d	P_{loop}	S_c	R_o	A_a	t_d	t_{90d}	S_d	S_{90d}	J_d	T_{av}	h_{ex}	S_o	T_o	R_{ent}
MSIA = 2.6, MPIA = 0.005															
BHive	99.6	3.79	93.7	0.293	29733	0.125	0.432	2.75	3.14	0.027	46.4	1.9	9.11	9.41	78.7
BHAIS	99.5	2.91	91.7	0.277	65108	0.135	0.464	2.75	3.1	0.028	46.7	1.68	8.11	8.39	77.2
MSIA = 2.6, MPIA = 0.0005															
BHive	98	4.18	83.2	0.294	31041	0.474	1.113	0.989	1.71	0.086	45.9	1.96	9.28	9.57	78.4
BHAIS	97.6	3.88	82.5	0.252	71610	0.458	1.101	0.938	1.63	0.08	45.7	1.78	8.42	8.68	76.9

Table 7.7. Performance comparison of *BeeHive* and *BeeHiveAIS* in NTTNet

7.5.4 Node150

The results of the experiments on NTTNet clearly indicate that *BeeHiveAIS* is able to provide the same security level as signature-based *BeeHiveGuard* but at significantly smaller processing and control overheads. Moreover, it provides an AIS-based lightweight security framework that provides the same performance as the original *BeeHive* under no attack. Therefore, we now want to evaluate its scalability in a large topology, Node150 (see Figure 3.7 in Chapter 3). We have dropped *BeeHiveGuard* because of its high processing and control overheads (one experiment of *BeeHiveGuard* on this topology took more than four days to complete).

A super attack of 17 nodes

Our attacker frame assigned malicious nodes randomly with the only restriction that two adjacent nodes not be malicious in a single experiment. It is evident in Figure 7.23 that the impact of malicious nodes in Node150 is relatively small, i.e., the routing affinity of a malicious node even in the best-case scenario increases by just 2%. Moreover, *BeeHiveAIS* is able to successfully counter the attacks of 17 nodes in the network. The performance parameters tabulated in Table 7.8 further confirm the *Byzantine robustness* of *BeeHiveAIS*. But the performance of *BeeHive* under attack degrades because the number of packets in loops increases to 3%.

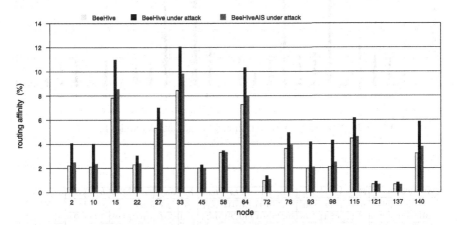

Fig. 7.23. Node150 – normal traffic; 17 nodes attack

Algorithm	p_d	P_{loop}	S_c	R_o	A_a	t_d	t_{90d}	S_d	S_{90d}	J_d	T_{av}	h_{ex}	S_o	T_o	R_{ent}
(1)BHive	99.9	0.561	98.8	0.806	34175	0.034	0.11	2.63	2.8	0.007	123	0.571	2.31	3.11	192
(2)BHive(a)	99.7	3.43	97	1.435	34627	0.049	0.194	2.66	2.89	0.011	122	1.25	5.03	6.47	204
(3)BHAIS(a)	99.9	0.542	98.1	1.117	72350	0.037	0.127	2.64	2.81	0.007	122	0.594	2.39	3.51	189
(2)-(1) in %	-	-	-	78.0	1.3	-	-	-	-	-	-	-	-	108	-
(3)-(1) in %	-	-	-	38.6	111.7	-	-	-	-	-	-	-	-	12.9	-

Table 7.8. Node150 – normal traffic; 17 nodes attack

A super attack of 54 nodes

Finally, we created a serious threat scenario in which 54 nodes (about 33%) acted as malicious nodes. Even under these stressful conditions one can see in Figure 7.24 that *BeeHiveAIS* is not only able to successfully counter such a large attack storm, but is also able to deliver the same performance as that of *BeeHive* under normal conditions (see Table 7.9).

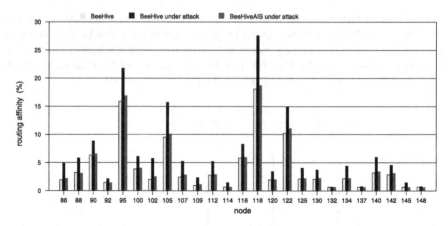

Fig. 7.24. Node150 – normal traffic; 54 nodes attack (Nodes 86 to 149)

Algorithm	p_d	P_{loop}	S_c	R_o	A_a	q_{av}	t_d	t_{90d}	S_d	S_{90d}	J_d	T_{av}	h_{ex}	S_o	T_o	R_{ent}
BHive	99.9	0.561	98.8	0.806	34175	0.002	0.034	0.11	2.63	2.8	0.007	123	0.571	2.31	3.11	192
BHive(a)	96.8	8.42	65.1	2.46	36246	0.026	0.201	0.756	2.74	3.16	0.068	119	2.72	10.6	13.1	228
BHAIS(a)	99.6	1.23	93	1.76	72013	0.007	0.061	0.228	2.67	2.92	0.014	122	0.728	2.93	4.7	188
(2)-(1) in %	-	-	-	205.2	6.1	-	-	-	-	-	-	-	-	-	321.2	-
(3)-(1) in %	-	-	-	118.3	110.7	-	-	-	-	-	-	-	-	-	51.1	

Table 7.9. Node150 – normal traffic; 54 nodes attack

Performance tests in Node150

We now report in Table 7.10 that the performance of *BeeHiveAIS* under normal conditions is comparable to that of *BeeHive*. The parameters for the traffic generator are the same as those reported in the case of the NTTNet scenario in the previous subsection. Again note that the most important performance parameters like throughput (T_{av}), delay (both t_d and t_{90d}), and jitter (J_d) of the two algorithms are approximately the same.

Algorithm	p_d	P_{loop}	S_c	R_o	A_a	t_d	t_{90d}	S_d	S_{90d}	J_d	T_{av}	h_{ex}	S_o	T_o	R_{ent}
					MSIA = 2.6, MPIA = 0.005										
BHive	99.9	0.561	98.8	0.806	34175	0.034	0.11	2.63	2.8	0.007	123	0.571	2.31	3.11	192
BHAIS	99.9	0.53	98.5	0.795	72936	0.035	0.117	2.63	2.8	0.007	123	0.547	2.21	3.01	189

Table 7.10. Performance comparison of *BeeHive* and *BeeHiveAIS* in Node150

7.6 Summary

In this chapter, we analyzed the security threats of our *BeeHive* routing protocol. However, the analysis can be easily extended to any nature-inspired routing protocol. The results of our extensive experiments reveal that the algorithm is susceptible to a number of Byzantine attacks like tampering, impersonating, detouring, flooding, rushing, and dropping that a malicious node can launch in the network. We then proposed two security frameworks to counter these attacks: (1) *BeeHiveGuard*, which is developed using signature-based cryptographic techniques; and (2) *BeeHiveAIS*, which utilizes the principles of AIS. The results of our experiments indicate that *BeeHiveAIS* provides the same level of security in countering the attacks, launched by malicious nodes, as that of *BeeHiveGuard*. But its processing and control overheads are approximately 200 and 20 times respectively smaller than *BeeHiveGuard*. *BeeHiveAIS* needs only 4 Kbytes and 12 Kbytes of memory in NTTNet and Node150 respectively to store a repository of antibodies. We believe that *BeeHiveAIS* is a simple, efficient, and scalable security framework for providing *Byzantine robustness* to routing protocols in general and nature-inspired routing protocols in particular. The framework is lightweight, so it will not cause any performance bottleneck in the normal packet-switching task of a real-world router.

8

Bee-Inspired Routing Protocols for Mobile Ad Hoc and Sensor Networks

In this chapter we shift our focus from fixed telecommunication networks to Mobile Ad Hoc Networks (MANETs) and Wireless Sensor Networks (WSNs). These emerging networks are receiving a significant amount of attention from researchers. We first present BeeAdHoc, a new routing algorithm for energy-efficient routing in MANETs, and then BeeSensor, a power-aware routing protocol for WSNs. Both algorithms mainly utilize two types of agents, scouts and foragers, for routing. Both algorithms are reactive source routing algorithms and they consume less energy compared to existing state-of-the-art routing algorithms because they utilize fewer control packets for routing. The results of our experiments indicate that both algorithms consume significantly less energy compared to state-of-the-art routing algorithms, without making any compromise on traditional performance metrics (packet delivery ratio and delay). We also discuss implementation of BeeAdHoc inside the Linux kernel on mobile laptops and its testing and evaluation in a real-world MANET consisting of eight mobile nodes.

8.1 Introduction

We now shift our focus from fixed communication networks to MANETs, which is becoming an active area of research [203]. These types of networks do not require any network infrastructure for their deployment; therefore, they are ideally suited to be used in war theaters and for disaster management. All nodes in such networks act not only as producers or consumers of data streams, but also as routers and switch packets destined for other nodes. The most important challenges in MANETs are: mobility and limited battery capacity of the nodes. Mobility of nodes results in continuously evolving new topologies, and consequently the routing algorithms have to discover or update the routes in real time but with small control overhead. The limited battery capacity requires that the packets, if possible, be distributed on multiple paths, which would result in the depletion of the batteries of different nodes at an equal rate, and hence as a result, the lifetime of the network would be increased [226, 129]. The metrics for energy-efficient routing are also introduced in [129], and it is obvious that energy-aware routing algorithms will degrade traditional performance metrics like throughput and packet delay (see Chapter 3). Therefore, an important challenge in MANETs is to design a routing algorithm that is not only energy-efficient but also delivers performance same as or better than existing state-of-the-art routing algorithms.

The routing algorithms for MANETs can be broadly classified as proactive algorithms or reactive algorithms. Proactive algorithms periodically launch control packets which collect the new network state and update the routing tables accordingly. On the other hand, reactive algorithms find routes on demand only. Reactive algorithms look more promising from the perspective of energy consumption in MANETs. Each category of the above-mentioned algorithms is further classified as source routing (host-intelligent) or next-hop-routing based (router-intelligent) in our taxonomy in Chapter 2. *DSR* (Dynamic Source Routing) is a reactive source routing algorithm [128] while *AODV* (Ad Hoc On-demand Distance Vector Routing) is a reactive next hop routing algorithm [187]. *DSDV* (Dynamic Destination-Sequenced Distance-Vector) is a proactive next hop routing algorithm [186]. *AODV* and *DSR* are considered to be state-of-the-art routing algorithms developed by the networking community for MANETs. However, these algorithms are not designed for energy-efficient routing. Feeney reported in [87] the energy consumption behavior of *DSR* and *AODV* and concluded that the algorithms are not optimized for energy consumption. Even most of the energy aware algorithms introduced in [226, 129] utilize either *DSR* or *AODV* for route discovery and maintenance but then use energy as a cost metric for routing. In this chapter, we report our bee-inspired MANET routing algorithm, *BeeAdHoc*, which delivers performance same as or better than that of *DSR*, *AODV*, and*DSDV* but consumes less energy compared to them. The algorithm achieves these objectives by transmitting fewer control packets and by distributing data packets on multiple paths.

8.1.1 Existing Works on Nature-Inspired MANET Routing Protocols

The first algorithm which presents a detailed scheme for MANET routing based on ant colony principles is *ARA* [103]. The algorithm has its roots in *ABC* [214] and *AntNet* [62] routing algorithms for fixed networks, which as mentioned in Chapter 2 are inspired by the pheromone-laying behavior of ant colonies. The algorithm floods ants to the destinations while establishing reverse links to the source nodes of the ants. Nodes launch ant agents in a reactive manner in order to limit control overhead. *AntHocNet*, which is a hybrid algorithm having both reactive and proactive components, has been proposed in [82, 66]. The algorithm tries to keep most of the features of the original *AntNet* and shows promising results in simulations compared to *AODV*. *Termite* is another MANET routing algorithm inspired by termite behavior [201, 200, 202]. In this algorithm, no special agents are needed for updating the routing tables; rather, data packets are delegated this task. Each data packet follows the pheromone for its destination and leaves the pheromone for its source. Pheromone is a quality metric representing the goodness of a link. The data packets are biased toward the paths that have higher pheromone values. An exponential pheromone decay is introduced as a means of negative feedback to prevent old routes from remaining in the routing tables.

8.1.2 Organization of the Chapter

The rest of the chapter is organized as follows. In Section 8.2 we will introduce our bee agent model and on its basis we will describe our routing algorithm, *BeeAdHoc*, in Section 8.3. We will first explain the complete experimental framework in Section 8.4 and then discuss the results obtained from the extensive simulations. We will then discuss realization of *BeeAdHoc* in real-world MANETs in Section 8.5. We report our testing framework for real world MANETs and then discuss results obtained from them in Section 8.6. Subsequently, we conclude our discussion on *BeeAdHoc* by introducing a number of security frameworks for it in Section 8.7. We then shift our focus to routing challenges in wireless sensor networks in Section 8.8. We introduce our *BeeSensor* protocol in Section 8.9. Later, in Section 8.10, we discuss our performance evaluation framework for WSNs and then report results obtained from experiments in Section 8.11. Finally, we conclude the chapter by highlighting the potential of nature-inspired routing protocols in future networks.

8.2 Bee Agent Model

Our *Bee Agent Model* is inspired by the foraging principles of a honeybee colony. Our agent model consists of four types of agents: *packers*, *scouts*, *foragers*, and *beeswarms*. In the rest of the chapter we use the term scout for scout agent, forager for forager agent, and so on.

8.2.1 Packers

Packers mimic the task of a food-storer bee. Packers reside inside a network node, and receive data packets from and store them in the upper transport layer (see Fig. 8.1). Their main task is to find a forager for the data packet at hand. Once the forager is found and the packet is handed over, the packer agent is removed from the system.

8.2.2 Scouts

The task of scouts is to discover new routes from their launching node to their destination node. A scout is broadcast to all neighbors in range using an *expanding time-to-live (TTL) timer*. At the start of the route search, a scout is generated, its TTL is set to a small value (e.g., 3), and it is broadcast. If, after a certain amount of time, the scout is not back with a route, the strategy consists of the generation of a new scout and the assignment of a TTL higher than in the previous attempt. In this way, the search radius of the generated scouts is incrementally enlarged, increasing the probability of a scout's reaching the searched destination. When a scout reaches the destination, it starts a backward journey on the same route that it has followed while moving forward toward the destination. Once the scout is back at its source node, it recruits foragers for its route by utilizing a mechanism derived from the waggle dance of scout bees in nature. A dance is abstracted into the number of clones that could be made of the same scout, which is encoded in their *dance number* (corresponding to recruiting forager bees in nature).

8.2.3 Foragers

Foragers are the main workers in the *BeeAdHoc* algorithm. They are bound to the "bee hive" of a node. They receive data packets from packers and deliver them to their destination in a source-routed modality. To "attract" data packets foragers use the same metaphor of a waggle dance as scouts do. Foragers are of two types: delay and lifetime. From the nodes they visit, *delay foragers* gather end-to-end delay information, while *lifetime foragers* gather information about the remaining battery power. Delay foragers try to route packets along a minimum-delay path, while lifetime foragers try to route packets so that the lifetime of the network is maximized. A forager is transmitted from node to node using a unicast, or point-to-point modality. Once a forager reaches the searched destination and delivers the data packets, it waits there until it can be piggybacked on a packet bounded for its original source node. In particular, since TCP acknowledges received packets, *BeeAdHoc* piggybacks the returning foragers in the TCP acknowledgments. This reduces the overhead generated by control packets, saving energy at the same time.

8.2.4 Beeswarm

Beeswarms are the agents used to explicitly transport foragers back to their source node when the applications are using an unreliable transport protocol like UDP, such that no acknowledgments are sent for the received data packets. To optimize forager transport, one beeswarm agent can carry multiple foragers: one forager is put in the header of the beeswarm while the others are put in the agent's payload. The beeswarm is launched once the difference between the incoming foragers from a certain node i and the outgoing foragers to the same node i exceeds a threshold value at a destination node j. Once the beeswarm arrives at the node i, the foragers are extracted from the payload and stored on the dance floor.

8.3 Architecture of BeeAdHoc

In *BeeAdHoc*, each MANET node contains at the network layer a software module called *hive*, which consists of three parts: the *packing floor*, the *entrance*, and the *dance floor*. The structure of the hive is shown in Fig. 8.1. The entrance floor is an interface to the lower MAC layer, while the packing floor is an interface to the upper transport layer. The dance floor contains the foragers and the routing information to route locally generated data packets. The functional characteristics of each floor composing the hive are explained in the following.

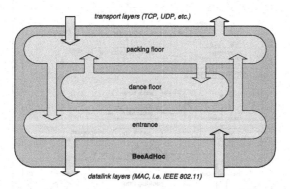

Fig. 8.1. Overview of the BeeAdHoc architecture

8.3.1 Packing Floor

The packing floor is an interface to the upper transport layer (e.g., TCP or UDP). Once a data packet arrives from the transport layer, a matching forager for it is looked up on the dance floor. If a forager is found then the data packet is encapsulated in its payload. Otherwise, the data packet is temporary buffered waiting for a returning forager. If no forager comes back within a certain predefined time, a scout is launched which is responsible for discovering new routes to the packet's destination. Figure 8.2 explains the series of actions performed at a packing floor.

8.3.2 Entrance

The functions performed in the entrance are shown in Figure 8.3. The entrance is an interface to the lower-level MAC layer. The entrance handles all incoming and outgoing packets. Actions on the dance floor depend on the type of packet that entered the floor from the MAC layer. If the packet is a forager and the current node is its destination, then the forager is forwarded to the packing floor; otherwise, it is directly routed to the MAC interface of the next hop node. If the packet is a scout, it is broadcast to the neighbor nodes if its TTL timer has not expired yet or if the current node is not its destination. The information about the ID of the scout and its source node is stored in a local list. If a replica of a previously received scout arrives at the entrance floor, it is removed from the system. If a forager with the same destination as the scout already exists on the dance floor, then the forager's route to the destination is given to the scout by appending it to the route held so far by the scout.

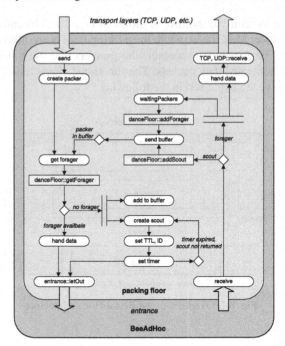

Fig. 8.2. The packing floor

8.3.3 Dance Floor

The dance floor is the heart of the hive because it maintains the *routing information* in the form of *foragers*. The dance floor is populated with routing information by means of a mechanism reminiscent of the waggle dance recruitment in natural bee hives: once a forager returns after its journey, it recruits new foragers by "dancing" according to the quality of the path it traversed. A lifetime forager evaluates the quality of its route based on the average remaining battery capacity of the nodes along its route. Mimicking forager bees in nature, it dances enthusiastically when it finds a route worth exploiting, recruiting in this way a number of foragers; a lifetime forager can be cloned many times in two distinct cases. In the first case, the nodes on the discovered route have a good amount of spare battery capacity, which means that this is a good route that can be well exploited. In the second case, a large number of data packets are waiting for the forager, so that the route needs to be exploited even though it might have nodes with little battery capacity. On the other hand, if no data packets are waiting to be transported, then a forager with a very good route might even abstain from dancing because the other foragers are fully satisfying traffic requests. This concept is directly borrowed from the behavior of scout/forager bees in nature, and it helps automatically regulate the number of foragers for each route.

The central activity of the dance floor module consists of sending a matching forager to the packing floor in response to a request from a packer. The foragers whose lifetime has expired are not considered for matching. If multiple foragers can

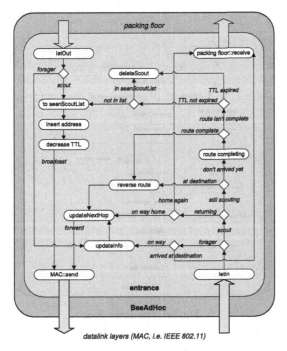

Fig. 8.3. The entrance

be identified for matching, then a forager is selected in a random way. This helps in distributing the packets over *multiple paths*, which in turn serves two purposes: avoiding congestion under high loads and depleting batteries of different nodes at a comparable rate. A clone of the selected forager is sent to the packing floor and the original forager is stored on the dance floor after reducing its dance number, that is, the number of permitted clones. If the dance number is 0, then the original forager is sent to the packing floor, removing it in this way from the dance floor. This strategy aims at favoring young over old foragers, since the former represent fresher routes, which are expected to remain valid in the near future with higher chances than the older ones because they represent recent state of the network.

If the last forager for a destination leaves a hive, then the hive does not have any more a route to the destination. Nevertheless, if a route to the destination still exists, then soon a forager will be returning to the hive; if no forager comes back within a reasonable amount of time, then the node has probably lost its connection to the destination node. This mechanism eliminates the need for explicitly monitoring the validity of the routes by using special Hello packets and informing other nodes through route error messages, as is done in several state-of-the-art algorithms such as AODV, as well as in several ACO implementations for MANETs. In this way, fewer control packets are transmitted, resulting in less energy expenditure. Figure 8.4 explains in detail the actions taken on a dance floor.

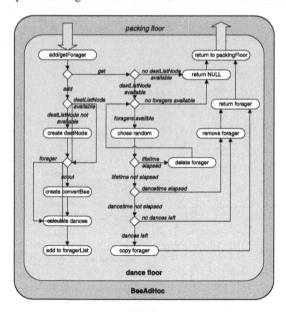

Fig. 8.4. The dance floor

8.4 Simulation Framework

We evaluated the performance of our algorithm *BeeAdHoc* using mobility enhancements made to the NS-2 simulator by the authors of [30]. The authors also evaluated the performance of different state-of-the-art algorithms like *AODV*, *DSR*, *DSDV*, and *TORA* in their work. Our test scenarios are derived from the base scenario used in [30]. We use the same implementations of *DSR*, *AODV*, and *DSDV*, which are distributed with the NS-2 simulator to factor out any implementation-related errors in the algorithms.

The scenario consists of 50 nodes moving in a rectangular area of $2,400 \times 800$ m^2. The rectangular area ensures that longer paths exist between the nodes as compared to those in a square region, provided the node density (nodes per unit area) remains the same. The nodes move according to the "random waypoint" model [128]: each node randomly selects a destination point and then moves to that point at a certain randomly selected speed. Once the node arrives at the destination point, it stops there for a certain pause time and then again randomly selects a new destination point and moves toward it at a new speed. The speed is selected from a uniform distribution between a minimum speed of 1 m/sec (walking speed) and the maximum speed of 20 m/sec (car speed within cities). All nodes generate a constant bit rate (CBR) peer-to-peer data traffic with x packets/sec. The size of a data packet is kept constant at 512 bits. We use the same models of physical and MAC layers as the authors of [30] did. The reported results are an average over five independent runs to factor out any stochastic elements in the environment or in the algorithms. The simulation time for the algorithms is set to 1000 seconds.

8.4.1 Metrics

We now define the metrics which we used in the comparison of the algorithms.

- **Energy per user data.** *The total energy consumed, including the energy consumed by the control packets, to transport one kilobyte of data to its destination.* This metric is minimum when the same number of bytes can be delivered at the destinations in few hops and with a small number of control packets. We used the model presented in [88, 87] to estimate the send/receive energy of broadcasting or point-to-point modes of transmitting packets. This metric is also referred to as energy expenditure in the rest of the chapter.
- **Success rate.** *The ratio of the number of packets successfully received by the application layer of a destination node to the number of packets originated at the application layer of each node for that destination.* This parameter is also referred to as packet delivery ratio in the rest of the chapter.
- **Delay.** *The difference between the time once the packet is received by the application layer of a destination node and the time the packet was originated at the application layer of a source node.* This definition takes care of the time that a packet has to wait at the source node while the route to its destination is to be found (reactive wait time). We report the 99th percentile of the delays distribution because it provides an insight into the deviation of the delays, an important criterion for Quality of Service (QoS) applications, in which all packets should arrive at the destination within an acceptable variance from the mean.
- **Throughput.** *If* y *number of bits are delivered within* t *time at a node then the throughput at the node could be defined as* $\frac{y}{t}$. This definition assigns a higher throughput value to an algorithm that delivers the same number of y bits in a less time. This definition of throughput implicitly strikes a good balance between the number of packets delivered at a node and their delays.
- **Network life.** *The average remaining battery capacity of the nodes.* A higher value means less depletion of the batteries and hence is a desirable property of any routing algorithm.

8.4.2 Node Mobility Behavior

The purpose of the experiments was to study the behavior of the algorithms by varying the speed of the nodes. Higher speeds reduce the stability of a network topology; as a result, an algorithm has to adapt itself to the changes in topologies. In these experiments the packet rate was 10 packets/sec (CBR source) and the pause time was 60 seconds. Figures 8.5 and 8.6 show the effect of mobility on different metrics. The packet delivery ratio (see Figure 8.5(a)) reduces with the increasing speed but *BeeAdHoc* is able to deliver approximately the same number of packets as *DSR*, the best performing algorithm. However, *BeeAdHoc* has a significantly smaller delay (see Figure 8.5(b)). Consequently, *BeeAdHoc* is able to maintain higher throughput (see Figure 8.6(a)) compared to all other algorithms. Remember that our definition of throughput favors one algorithm over the others if it is able to deliver the same number of packets but with smaller delay.

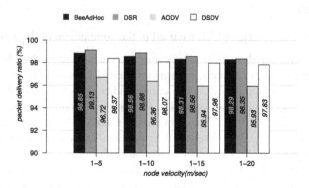

(a) Packet delivery ratio

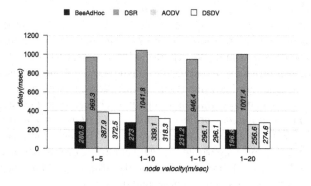

(b) Packet delay

Fig. 8.5. Effect of varying the speed of the nodes

We investigated the problem of higher packet delays of *DSR* by looking at 80th, 90th, 95th, and 99th percentile of the delay distribution (see Table 8.1). It is evident

	BeeAdHoc	DSR	AODV	DSDV
80th percentile	105.64	167.73	156.85	117.89
90th percentile	153.84	278.58	220.52	176.01
95th percentile	191.36	396.29	269.31	223.76
99th percentile	280.97	969.39	387.96	372.55

Table 8.1. Different percentiles of the delay distribution for nodes' speed 1–5 m/sec

from Table 8.1 that *BeeAdHoc* is able to deliver a majority of the packets within an acceptable deviation from the mean, while *DSR* delivers about 5% of the packets with quite large delays, and hence the 99th percentile of its delays distribution is significantly larger than that of the other algorithms. We further looked into this problem and it appeared that the most important contributor to higher delay is the packet salvaging technique used in *DSR*. Once a node finds out that the next hop in the header is down, it looks at its routing table, and if it finds a route to the destination in it, it replaces the remaining part of the header with this route. However, by the time, the packet arrives at the next node, the route again needs to be repaired. Consequently, a packet keeps on taking hops until it arrives at the destination. The basic behavior of MANETs could then be summarized as follows: *if a node finds that the next hop to a destination is no more available, then it should not try to repair the route with the old information in its routing table because there is a high probability that this old route would be no more valid as well.* Therefore, *BeeAdHoc* simply deletes the packet if it finds that the next hop is down. It is clear from Figure 8.5(a) that this simple approach results, at the maximum, in a loss of about 0.3% in the packet delivery ratio.

The simplicity of *BeeAdHoc*, which results from its simpler architecture and its using a smaller number of control packets (see Table 8.2; note that all the 50 nodes are transmitting packets), pays off once we look at its energy expenditure (see Figure 8.6(b)) in transporting the packets from their source to their destination. *BeeAdHoc* employs a simple bee behavior to monitor the validity of the routes by controlling the number of foragers, their dance, and their age parameter rather than explicitly using Hello/RERR messages. This results in the smallest amount of energy expenditure for *BeeAdHoc* (see Figure 8.6(b)).

Node Mobility	BeeAdHoc	DSR	AODV	DSDV
1-5 m/sec	83095	122313	592454	165454
1-10 m/sec	93895	224335	716235	211220
1-15 m/sec	99240	310119	836058	240234
1-20 m/sec	103119	396885	731279	253000

Table 8.2. Total number of control packets sent

Finally, Table 8.3 shows the average remaining battery capacity (%) of the nodes in the network at the end of the above-mentioned simulations. *BeeAdHoc* has higher remaining battery capacity under all conditions. The battery level of *BeeAdHoc* is better because it tries to spread the data packets over different routes rather than always sending them on the best routes. Different routes could be established to the destination nodes at higher node speeds; as a result, data packets are routed through different nodes, and this explains the increasing network life behavior of *BeeAdHoc*, with an increase in the speed of the nodes. *AODV* and *DSR* utilize a significantly larger number of control packets at higher nodes speed (see Table 8.2); therefore, the

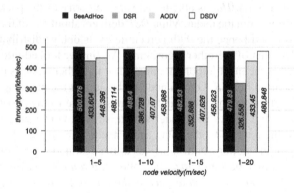

(a) Throughput

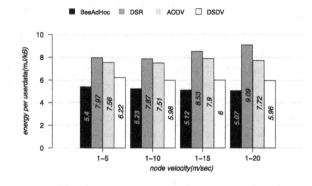

(b) Energy Expenditure

Fig. 8.6. Effect of varying the speed of the nodes

batteries of the nodes are almost completely depleted.[1]. We skip other experiments for brevity; however, an interested reader is encouraged to consult [269, 270, 271, 272] for detailed discussions of the remaining experiments.

---

[1] The description of *BeeAdHoc* is reproduced by permission of the publisher, ACM, from our paper: H. F. Wedde, M. Farooq, T. Pannenbaecker, B. Vogel, C. Mueller, J. Meth, and R. Jeruschkat. *BeeAdHoc: an energy efficient routing algorithm for mobile ad-hoc networks inspired by bee behavior.* In Proceedings of ACM GECCO, pages 153–161, Washington, June 2005.

Node Mobility	BeeAdHoc	DSR	AODV	DSDV
1-5 m/sec	6.2	1.4	1.8	2.4
1-10 m/sec	5.5	1.3	2.4	2.6
1-15 m/sec	6.1	1.0	1.6	2.6
1-20 m/sec	11.4	1.1	1.4	3.3

Table 8.3. Effect of varying speed on network life

8.5 BeeAdHoc in Real-World MANETs

We now discuss the evaluation of *BeeAdHoc* in real-world MANETs. Recall from Section 5.2 in Chapter 5 that a routing protocol can be implemented inside the network stack of Linux by using three implementation strategies: *a monolithic routing framework in the kernel space, a monolithic routing framework in user space, or a hybrid implementation in the kernel space and user space.* Each paradigm has its benefits and shortcomings. Remember that *BeeHive* was implemented as a monolithic routing framework in the kernel space. Since this approach results in the best performance, we decided to also realize *BeeAdHoc* as a monolithic framework inside the kernel space of Linux 2.6.7. We skip its details for the sake of brevity because we want to focus in this chapter on the results of empirical evaluation. However, an interested reader is encouraged to refer to Chapters 12 and 13 of [272] to get an in-depth understanding of the implementation of *BeeAdHoc* on mobile laptops running Linux. We were able to find a Linux *AODV* implementation done at Uppsala University, Sweden at *http://core.it.uu.se/adhoc/aodvuuimpl*. Similarly we decided to also compare our algorithm with a well-known proactive link-state routing algorithm *OLSR* because its implementation by INRIA was available at *http://www.olsr.org*. Unfortunately, we could not find a working implementation of *DSR* for the Linux 2.6.7 version at the time we did our experiments.

8.5.1 A Performance Evaluation Framework for Real MANETs in Linux

We now introduce our performance evaluation framework for real MANETs running on the Linux operating system. We developed a novel three-step evaluation methodology with the intent of gradually moving towards a real MANET [271, 272]: (1) simulated reality; (2) quasi-reality; and (3) reality. In simulated reality, we tested the algorithms in a virtual network of five User Mode Linux (UML) virtual machines connected through a software switch. We developed a script-based topology generator that used to periodically change one topology to a new one (selected from 15 pre-designed topologies as shown in Figures 8.7 and 8.8). As a result, we simulate mobility by periodically changing topologies. We call it simulated reality because the characteristics of wireless channel are totally ignored in this scenario. Nevertheless, the scenario provides an insight into how the algorithms, implemented on Linux virtual machines, react to topological changes.

In quasi-reality, the communication between 12 laptops was established through 802.11 wireless network cards by placing laptops in communication range of one another. A master laptop used to periodically generate a new topology from ten pre-designed topologies as shown in Figures 8.9 and 8.10. Then it used to broadcast topology to all laptops to inform them about their current neighbors. As a result, the packets from non-neighbor nodes were discarded through packet filtering at the data link layer. This scenario is one step closer to reality but mobility is still simulated. Finally, in reality, we conducted a MANET experiment on a eight laptop network in which the nodes were moving at walking speed on the north campus of Technical University of Dortmund, Germany (see Figure 8.12). Our team members gathered at Point F and synchronized their clocks and then they moved to their respective starting positions within 30 seconds. In this scenario, Goethe and Lessing exchanged data packets. Goethe, depending upon the mobility pattern, can establish a link directly (0 hops) to Lessing, or over one or more hops through Heine, Kafka, or Storm. The walking speed of nodes varied between 0 and 5 km/h.

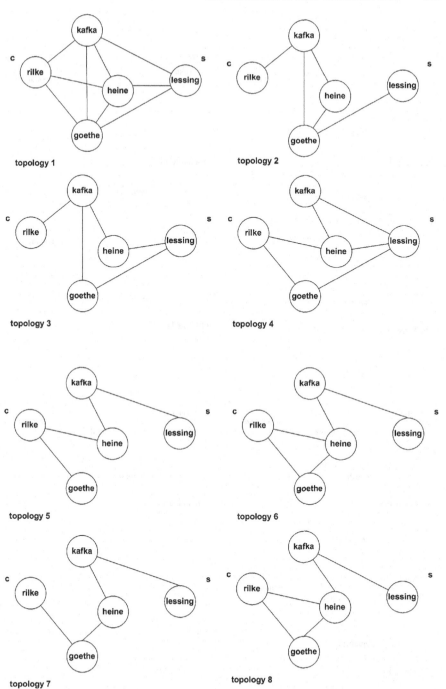

Fig. 8.7. UML predesigned topologies (1)

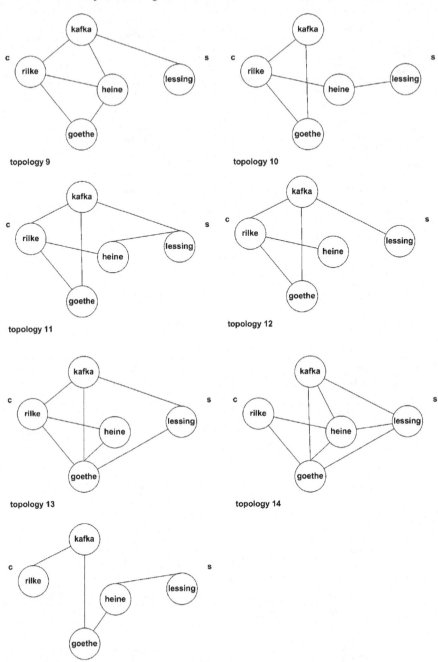

Fig. 8.8. UML predesigned topologies (2)

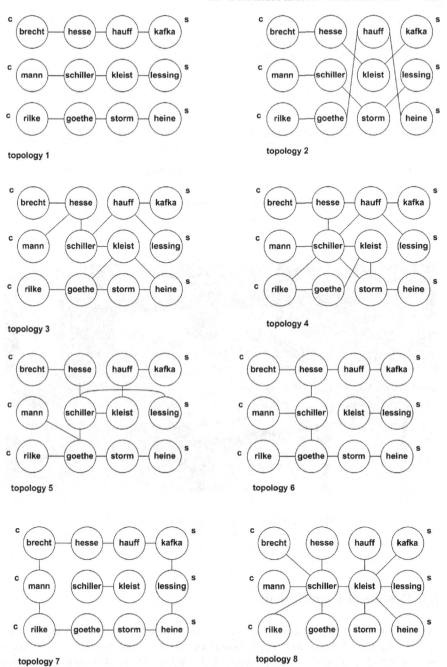

Fig. 8.9. Quasi-reality predesigned topologies (1)

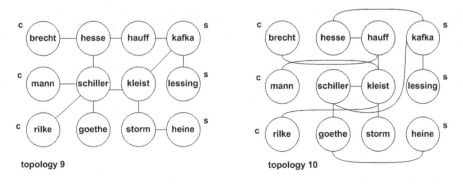

Fig. 8.10. Quasi-reality predesigned topologies (2)

Fig. 8.11. Quasi-reality setup in our networks lab

8.6 Results of Experiments

We now provide a brief overview of the results obtained from different test scenarios. We used two off-the-shelf traffic generators, Thrulay and Netperf, in our experiments. We also did some experiments with the SQTG and D-ITG traffic generators introduced in Chapter 5.

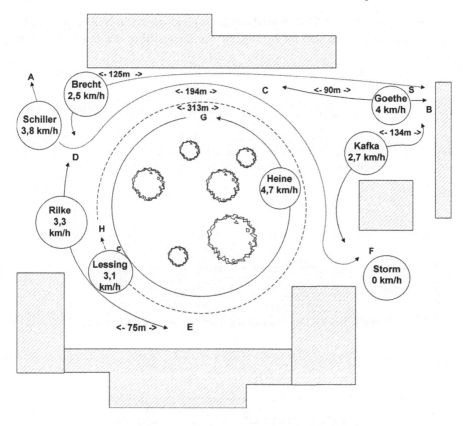

Fig. 8.12. Reality test scenario on the north campus

Simulated reality

We first report the results of the Thrulay experiments, which are tabulated in Tables 8.4, 8.5, and 8.6. The reported values are a geometric mean of five independent experiments, where each experiment lasted for 900 seconds. We can see in Table 8.4 that *BeeAdHoc* consistently delivers superior throughput because of its multi-path feature at approximately the same delay[2] as that of *AODV* (see Table 8.5). It was our observation that *OSLR* not only crashed at a higher frequency but also had problems finding new routes after a quick topology change. As expected, *BeeAdHoc* delivered this superior performance at a significantly smaller number of control packets (see Table 8.6) compared to other algorithms. It is interesting to note that under high mobility (topology change every five seconds), *AODV* has virtually become a proactive

---

[2] We report Round-Trip-Time (RTT) because it can be easily measured at a node by taking the difference of arrival time of an acknowledgment and transmission time of its corresponding packet. Due to clock synchronization problems, we cannot measure one-way delay in a real MANET.

routing protocol.[3] *BeeAdHoc* showed similar superior trends in NetPerf and SQTG experiments which we skip for brevity, but the interested reader can find them in [272].

	AODV	BeeAdHoc	OLSR
topology change every 5 sec	2.472	11.572	1.167
topology change every 10 sec	3.005	13.467	3.347
topology change every 30 sec	3.347	17.076	9.870
topology change every 60 sec	2.909	18.553	10.198
static topology	2.314	13.882	6.016

Table 8.4. Thrulay in simulated reality — throughput(Mbits/sec)

	AODV	BeeAdHoc	OLSR
topology change every 5sec	343,85	248,55	3350,22
topology change every 10sec	244,58	342,25	4097,48
topology change every 30sec	206,48	257,80	632,06
topology change every 60sec	226,08	262,32	270,20
static topology	276,04	904,81	2081,25

Table 8.5. Thrulay in simulated reality — average RTT (msec)

	AODV	BeeAdHoc	OLSR
topology change every 5sec	2997	465	2258
topology change every 10sec	2892	279	2072
topology change every 30sec	2860	155	1809
topology change every 60sec	2322	121	1689
static topology	1802	34	1199

Table 8.6. Thrulay in simulated reality — control packets

Quasi-reality

In this scenario, we report results for Thrulay experiments that lasted for 120 seconds. We observed that in quasi-reality the stability of a connection was more important under high mobility than throughput. *OLSR* used to crash frequently and its error-free operation never lasted more than 120 seconds; therefore, we selected this time for our experiments. The tests in quasi-reality took more than four weeks that included the setup of nodes in our lab and the development of test-related scripts. We invested more than 130 hours of testing and analysis time to obtain the results.

---

[3] Recall that Hello packets, periodically launched by *AODV*, are also counted as control packets

Here we report results for Thrulay only, but the interested reader can find other experiments in [272]. In these experiments, we also analyzed the impact of distance (in hops) between a sender and a receiver. At one extreme, a destination was a neighbor node (0 hops), and then we gradually increased the distance to four hops. Even at this distance, throughput became lower than 1 Mbits/sec even though the wireless network cards supported 11 Mbits/sec bandwidth. We tabulate throughput, average RTT, and number of control packets in Tables 8.7, 8.8, and 8.9 respectively. We can see in Table 8.7 that *BeeAdHoc* provides the same throughput as the other protocols. This appears to be in contradiction with the results obtained in simulated reality, where *BeeAdHoc* achieved significantly higher throughput than the other protocols. Remember that simulated reality presents an ideal MANET environment with no bottlenecks in the MAC layer. We can conclude from Table 8.7 that network layer protocols can only provide enhanced performance if the underlying 802.11 link layer protocol is able to provide stable connections with high bandwidth. Note that if the destination is four hops away, then the effective bandwidth is reduced by a factor of 10. We can see in Table 8.8 that the delays of *AODV* and *BeeAdHoc* are approximately the same and are significantly smaller than that of *OLSR*. We further emphasize that *BeeAdHoc*, as expected, achieves this comparable performance by sending significantly fewer control packets (see Table 8.9). This pattern of performance is consistent with the one observed in simulated reality.

	AODV	BeeAdHoc	OLSR	static
0 Hops	4,973	4,814	4,971	4,838
1 Hops	2,498	2,471	2,305	
2 Hops	1,695	1,653	1,082	
3 Hops	1,304	1,267	1,101	
4 Hops	1,036	1,013	0,757	

Table 8.7. Thrulay in simulated reality — throughput (Mbits/sec)

	AODV	BeeAdHoc	OLSR	static
0 Hops	366	357	366	358
1 Hops	692	683	720	
2 Hops	975	972	1365	
3 Hops	1170	1174	1439	
4 Hops	1479	1492	2002	

Table 8.8. Thrulay in simulated reality — average RTT (msec)

We also analyzed the energy consumption behavior of three algorithms by logging the remaining battery capacity with

```
cat
/proc/net/acpi/battery/BAT0/state
```

	AODV	BeeAdHoc	OLSR
0 hops	236	1	134
1 hops	235	1	161
2 hops	239	1	228
3 hops	271	1	237
4 hops	275	2	217

Table 8.9. Thrulay in simulated reality — control packets

which returns the mW-h remaining in the battery. The difference between the start and end levels gives us the energy consumed during network operations. We plotted the remaining battery levels as the time progressed in one of our experiments. Figure 8.13 shows that after 15 minutes of network activity, laptops have the highest remaining energy levels once *BeeAdHoc* has been deployed. *OLSR*, as expected, has the lowest remaining energy level because it is a proactive routing protocol.

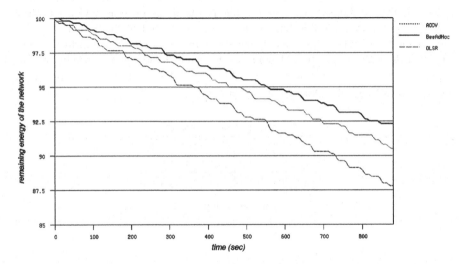

Fig. 8.13. Energy consumption behavior of routing protocols

Reality – mobility tests

We now report results for our reality scenario. We report our results for both the static and the mobile scenarios. The results for mobile and static scenarios are tabulated in Tables 8.10 and 8.11 respectively. The experiments lasted for 720 seconds. It is interesting to note that throughput in the mobile scenario is higher compared to the static one. The obvious reason is that Goethe and Lessing were always communicating over Kafka-Storm (see Figure 8.12) in the case of the static scenario, while in the case of the mobile scenario they communicated directly over a single hop through Kafka, Heine, or Storm, or over two hops through Kafka-Storm, Kafka-Heine, and so on. *BeeAdHoc* has throughput and delays comparable to those of the best performing algorithms, but it achieved this performance by transmitting a significantly smaller number of control packets. The benefit of reactive protocols is evident in Table 8.11 because in the case of the static scenario they sent a significantly smaller number of packets.

	AODV	BeeAdHoc	OLSR
Throughput	0,730 MBits/sec	0,627 MBits/sec	0,882 MBits/sec
avg. packet delay	8203 msec	2252 msec	2309 msec
control packets	1025	153	3818

Table 8.10. Thrulay in reality in the mobile scenario

	AODV	BeeAdHoc	OLSR
Throughput	0,416 Mbits/sec	0,330 MBits/sec	0,305 MBits/sec
avg. packet delay	3986 msec	4117 msec	5078 msec
control packets	91	67	2537

Table 8.11. Thrulay in reality in the static scenario

8.7 Security Threats in BeeAdHoc

We followed the same research methodology to analyze the security threats of *BeeAdHoc* as was used in Chapter 7 for *BeeHive*. We now provide a summary of our three security frameworks: (1) *BeeSec*; (2) *BeeAIS*; and (3) *BeeAIS-DC*. BeeSec [153], like *BeeHiveGuard*, utilizes a digital-signature-based security framework. Similarly, *BeeAIS* [154], like *BeeHiveAIS*, utilizes the principles of AIS to provide security. However, in MANETs, providing an AIS-based security is more challenging because of the mobility of the nodes. As a result, it was difficult to identify whether the change in the path of an agent is due to the malicious activity of a node or due to its mobility. This translates into the idea of a "self" which is changing. In order to incorporate the notion of changing "self," we have recently proposed our *BeeAIS-IDC* framework [155], which takes inspiration from danger theory [6, 7] and models the behavior and functions of dendritic cells (DCs) to provide the dynamic update of detector sets. *BeeAIS-DC* follows the dendritic cell differentiation pathways to incorporate the new or changed self after detecting the presence or absence of danger signals, and as a result, detect the non-self, thus providing protection in mobile networks against attacks from malicious nodes. According to the reported results: (1) *BeeAIS* provides the same security level as *BeeSec* but at significantly smaller processing and control overheads, which results in significant amount of energy and power savings compared to *BeeSec*; (2) *BeeAIS-IDC* under mobility provides better security than *BeeAIS*; (3) the performance of *BeeAIS-IDC* and *BeeAIS*, even with the overhead of providing security, is significantly better than that of *AODV* and *DSR*, and is approximately similar to that of the original *BeeAdHoc* algorithm. However, *BeeAIS-IDC* scales significantly better under mobility scenarios compared to *BeeAIS* due to its ability to model the changing "self." The results seem to indicate the efficacy of adopting a danger-theory-inspired AIS security framework for MANETs in power-aware systems because of its low processing and control overhead.

We now shift our focus to another important emerging type of network, wireless sensor network, and discuss challenges for routing protocols in such networks. Subsequently, we introduce our *BeeSensor* protocol.

8.8 Challenges for Routing Protocols in Ad Hoc Sensor Networks

We now focus on our research efforts to design a power-aware routing protocol for wireless sensor networks (WSNs), which are also becoming an active area of research due to their expected key role in diverse real-world civil and military applications [194]. The spectrum of applications includes target field imaging, intrusion detection, weather monitoring, security and tactical surveillance, and disaster management [8, 194]. WSNs are created in an ad hoc fashion through the wireless communication interfaces of sensor nodes when few hundreds or thousands of them are scattered in a geographical area. But each node itself has limited hardware resources [8, 194]. Such resource-constrained sensor nodes put a challenge on the routing infrastructure: power-aware routing protocols with low processing and control overheads have to be designed to deliver the sensed data to a sink node. Moreover, such routing protocols have to be self-organizing without a central controller due to the unpredictable environment in which they are expected to be deployed. Last but not least, they have to be scalable, robust, and performance-efficient [8] with an ability to keep the network operational for an extended period of time.

Now, approaching the end of this chapter, we present a simple, scalable, self-organizing, power-aware and performance optimized bee-inspired routing protocol, *BeeSensor*, for WSNs. The algorithm has been carefully engineered by taking inspiration from relevant features of *BeeAdHoc* and *BeeHive* protocols, which were based on foraging principles of honeybees. Recall that *BeeHive* delivers better performance with a simple agent model in fixed networks compared to existing algorithms, while *BeeAdHoc* delivers similar performance compared to its counterparts but at a significantly smaller energy expenditure. In our *BeeSensor* protocol [206], we want to combine benefits of two algorithms so that foragers do not need to carry complete route information in the header, and instead this information is stored at intermediate sensor nodes. As a result, foragers will be of fixed size similar to that of *bee agents* in *BeeHive*. Consequently, it improves the scalability of the protocol.

8.8.1 Existing Works on Routing Protocols for Wireless Sensor Networks

Design and development of nature-inspired routing protocols for WSNs have received little attention. Some preliminary efforts reported in [300] are limited to adapting *AntNet* for WSNs. An ant-based routing protocol based on the Ant Colony Optimization (ACO) meta-heuristic is presented in [34] and the authors have shown that it achieves better average remaining energy with a small standard deviation compared to a few of its other variants. We will shortly provide a brief overview of these protocols. The majority of routing protocols for WSNs, apart from these few nature-inspired routing protocols, are designed by the classical networking community. Sensor Protocol for Information via Negotiation (SPIN) [110] is a data-centric protocol in which energy-efficient routing is done through the negotiation of high-level metadata descriptors. Directed diffusion [122] is a popular routing paradigm which introduces the idea of aggregating the data coming from different sources to eliminate data redundancy. A number of energy-efficient variants of directed diffusion are surveyed in [8]. Power-Efficient Gathering in Sensor Information Systems (PEGASIS) [148] is a variant of the Low Energy Adaptive Clustering Hierarchy (LEACH) [109] protocol in which sensor nodes are organized into clusters in an energy-efficient manner. Geographic and Energy-Aware Routing (GEAR) [296] uses geographical information to route events toward the sink node.

A brief review of selected nature-inspired routing protocols

In this section we provide a brief review of the nature-inspired routing protocols with which we compared our *BeeSensor* routing protocol.

Flooded piggybacked ant routing (FP-Ant)

The flooding mechanism is significantly helpful in wireless networks, especially in sensor networks, where the probability of a packet loss is substantially higher compared to that of fixed networks. *FP-Ant* is a variation of the *AntNet* proposed in [300] and is based on the flooding mechanism. In *FP-Ant*, *forward ants* are combined with *data ants* which carry the list of nodes to be visited during a forward journey. Flooding of the ants is controlled in two ways. First, a node broadcasts a *forward ant* only if it is closer to the destination compared to the node from which it received the *forward ant*. Secondly, a random delay is added before further retransmission of the *forward ant*. As a result, if the node receives the same *forward ant* from other neighbors later on, then it simply abstains from rebroadcasting it. When the *forward ants* reach the destination, they not only pass the data to the destination but also remember the paths they followed. *Backward ants* subsequently visit this path in a reverse order and reinforce the link probabilities that represent the goodness of the links. The probabilities of the links are used to restrict the flooding of future *forward ants* towards the sink node. This algorithm, as expected, has significantly better packet delivery ratio, and as a result, it has large energy consumption as well. We have included *FP-Ant* in our evaluation study because it can act as a benchmark for the maximum packet delivery ratio and other protocols can be compared with it. Due to the flooding mechanism, *FP-Ant* is expected to consume more energy, but as a consequence, it depletes the energy resources of different nodes at an equal rate.

Energy-efficient ant-based routing (EEABR)

EEABR algorithm proposed in [34] is based on the Ant Colony Optimization (ACO) metaheuristic. Each node in the network launches a *forward ant* at regular intervals with the objective to find a path to a specific destination. Unlike in *AntNet*, each visited node is not saved in the memory of the *forward ant*. Rather, the ant carries the addresses of the last two visited nodes only. Intermediate nodes now require additional tables in which the record of ants received and forwarded is saved. Each record stored in the table is a quadruple of the form: <Previous node, Forward node, Ant identification, Timeout value>. When a node receives a *forward ant*, it searches its local table for detecting any possible loop. If no loop exists, the node saves the information of the ant in the table, restarts a timer, and sends the ant to the next hop. When the *forward ant* reaches the destination node, it is converted into a *backward ant* which updates the pheromone trail of the path followed by the *forward ant* to reach the destination node. The amount of pheromone trail deposited by a corresponding *backward ant* is calculated using the following formula:

$$\Delta T_k = \frac{1}{C - \left(\frac{Emin_k - Fd_k}{Eavg_k - Fd_k}\right)}, \tag{8.1}$$

where $Eavg_k$ is the average energy of the nodes stored in the memory of the *forward ant*, $Emin_k$ is the minimum energy of a node visited so far, and Fd_k, is the number of nodes visited by the *forward ant*. When a node r receives a *backward ant* coming from a neighboring node s, it updates its routing table using the following equation:

$$T_k(r, s) = (1 - \sigma)T_k(r, s) + \frac{\Delta T_k}{\phi Bd_k}, \tag{8.2}$$

where ϕ is a coefficient and Bd_k is the number of nodes visited by the *backward ant* to the node r. When the *backward ant* reaches the source node, it is discarded. EEABR is a typical proactive routing protocol that is not only targeted for energy efficiency but also for low standard deviation of remaining energy levels. We now introduce the architecture, agent model, and working of our routing protocol.

8.9 BeeSensor: Architecture and Working

8.9.1 BeeSensor Agent's Model

BeeSensor mostly works with three types of agents: packers, scouts, and foragers. Beeswarms most of the time are not needed because lower-layer link protocols in WSNs frequently exchange acknowledgments. Consequently, foragers are piggy-backed in these lower-layer acknowledgements.

Packers. Packers receive data packets from an application layer and locate an appropriate forager for them at the source node. At the sink node, they recover data from the payload of foragers and provide it to the application layer.

Scouts. Scouts in *BeeSensor* are classified as *forward scouts* and *backward scouts* depending upon the direction in which they travel. A source node that detects an event launches a *forward scout* when it does not have a route to the sink node. A *forward scout* is propagated using the broadcasting principle to all neighbors of a node. Each *forward scout* has a unique id and it also carries the detected event in its payload. In *BeeSensor*, scouts do not construct a source header in which the complete sequence of nodes to the destination is saved. As a result, scouts have a fixed size independent of the path length (number of hops) between the source and a sink node. Moreover, the forward and return paths of a scout may not necessarily be the same. Once a *forward scout* reaches the sink node, it starts its return journey as a *backward scout*. When this *backward scout* arrives at the source node, a dance number is calculated using the minimum remaining energy of the path, which indicates the number of foragers to be cloned from this scout.

Foragers. In *BeeSensor*, again foragers are the main workers that transport data packets from a source node to the sink node. They receive data packets from packers at a source node and deliver them to the packers at the sink node. The basic motivation in *BeeSensor* is to discover multiple paths to a sink node; therefore, each forager in addition to its forager id gets a unique path ID (PID) as well. A forager follows the point-to-point mode of transmission until it reaches the sink node. A forager is piggybacked to the source node in the ACKs generated by a support layer between the MAC and network layers. This support layer is part of the repository of the Routing Modeling Application Simulation Environment (RMASE) [298], a simulator we used for analysis. We modified the layer for this purpose to avoid the overhead if foragers were to be sent back using explicit network layer transmissions. Once foragers arrive at the source node, a dance number is calculated in a similar way as that in the case of a *backward scout*. Consequently, more packets are routed through better quality paths. Remember that foragers in *BeeSensor* do not carry the complete source route in their header. Rather, small forwarding tables at intermediate nodes are used to simulate the source routing behavior (see Figure 8.14). The reason for this modification is that source-routing-based protocols do not scale for large networks because the size of the foragers is directly dependent on the length of the path between a source and a sink node.

8.9.2 Protocol Description

BeeSensor works in four distinct phases: Scouting, foraging, multiple path discovery, and swarming. We discuss each of them.

Scouting. When a source node detects an event and is unable to locate any forager to carry the event to its sink node, the scouting process is initiated. The node generates a *forward scout*, puts the event in the payload of this scout, and broadcasts it to all its neighbors. When an intermediate node within a two-hop radius of the source node receives the *forward scout*, it updates its scout cache, increments the hop count, and rebroadcasts it. Intermediate nodes more than two hops away stochastically decide whether to further rebroadcast a scout or not (shown by dashed lines in Figure 8.14). We used a MATLAB function for stochastic decision making and tested it over much larger topologies (400 nodes) to ensure that the sink node always receives scouts from the farthest source node. If another replica of an already broadcasted scout is received via another path at an intermediate node, its information is stored in the scouts' cache and then it is dropped. It is worth mentioning here that the scouts launched by a source node do not search for a particular destination: rather, they carry information about the generated event. Any sink node interested in the events can respond by sending a *backward scout* to the source node. If a source node does not receive a *backward scout* after scouting a certain number of times, the scouting process is stopped, indicating that no node is interested in the events.

When a sink node receives the *forward scout*, it first updates its scout cache, and then transforms the *forward scout* into a *backward scout* and assigns it a unique path ID (PID). The sink node can get different replicas of the same scout through different neighbors and it will return a *backward scout* for each replica. This is an indication that a sink node is reachable from the source node via different neighbors of the sink node. We also associate a cost C_{ns} with each of the neighbors from which the *forward scout* is received, and it is defined as:

$$C_{ns} = E_r^n - \frac{H_{ns}}{K}, \tag{8.3}$$

where C_{ns} is the cost of selecting neighbor n for reaching the source node s, E_r^n is the remaining energy of node n, H_{ns} is the numbers of hops to reach the source node through n, and K is a weighting factor. Higher values of K lower the significance of the path length (H_{ns}). The sink node forwards the *backward scout* to the neighbor available in the scout cache for which this cost is maximum.

When a subsequent node receives this *backward scout*, it again selects the best neighbor in a similar way to that of the sink node. Then it updates its forwarding table, which contains three fields (see Figure 8.14): ID of the node from which it got the *backward scout*, NextHop; ID of the node to which the *backward scout* is to be forwarded, PrevHop; and PID, the path ID. Then it updates the minimum remaining energy field of the *backward scout* and forwards it to the selected neighbor. This process is repeated at all intermediate nodes until the *backward scout* is back at the source node. As already mentioned, the source evaluates the quality of the path by calculating the value of the dance number which depends on the minimum remaining energy of the path reported by the *backward scout*. Finally it updates the routing table. A routing table entry consists of six fields: ID of the source node and a unique message ID (these two fields uniquely identify a source node) generated at the application layer of the source node, destination ID, path ID, next hop, and number of recruited foragers.

Foraging. Source nodes also maintain a small event cache in which the events generated during the route discovery process are stored. Once the route is discovered, routing of events to the sink nodes is started. The source node selects an appropriate forager from the routing table, adds the event to its payload, and forwards it to the next hop. The source node determines the complete path of a forager and puts the corresponding path ID (PID) in the header of the forager. An intermediate node uses the path ID information of the forager to select the next hop. For example, when

Node 2, shown in Figure 8.14, receives a forager with PID = 1, it sends it to the Node 3. All subsequent nodes keep switching this forager until it reaches the sink node.

Multi-path discovery. *BeeSensor* does not establish multiple paths during the initial

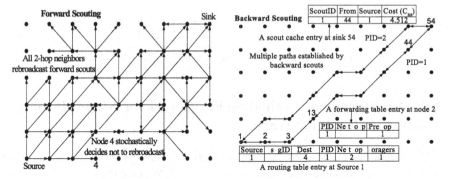

Fig. 8.14. Forward and backward scouting in *BeeSensor*

route discovery phase. Rather, a more conservative approach is adopted in which the sink node initially generates a *backward scout* and then waits for the events to start flowing towards it. After receiving a certain number of events, the sink node launches another *backward scout* with a unique PID, using the information in its scout cache, which is transported back to the source node in a similar way as initially. This *backward scout* is most likely to follow a different path, such as the path with PID = 2 shown in Figure 8.14. In this way *BeeSensor* avoids the overhead of maintaining multiple paths for applications that only generate data occasionally. However, for applications generating continuous traffic, multiple paths are established and the traffic load is distributed accordingly. But maintenance of multiple paths costs additional processing and bandwidth.

Swarming. *BeeSensor* does not use explicit swarm bees to return the foragers back to the source nodes. Rather, it piggybacks the foragers in the ACKs generated by a support layer between the MAC and network layers. Swarming is helpful in verifying that the paths to the sink node are still valid. If a source does not get back the foragers from a particular path, it is assumed that the path to the sink node is lost and the corresponding entry from the routing table is deleted. If all paths are lost, the scouting process is restarted on the arrival of next event. Similarly, if forwarding table entries at intermediate nodes or routing entries at the source node are not used for a certain period of time, they are also invalidated. As a result, no explicit special messages are needed to check the validity of links and to inform other nodes if they have become invalid. *BeeSensor* also supports the sending of a beeswarm, and PrevHop entries in the forwarding tables may be used for this purpose. In addition, sink nodes must also maintain the forwarding table to route the beeswarm back to the source node using the PrevHop field. The classification of protocols for WSNs on the basis of taxonomy in Chapter 2 is given in Table 8.12.

Characteristics	Routing protocols			
	FP-Ant	EEABR	BeeSensor	AODV
Single-path (SP) vs. Multi-path (M)	M	M	M	SP
Loop free	No	No	Yes	Yes
Reactive (R) vs. Proactive (P) vs. Hybrid (H)	H	P	R	R
Static (S) vs. Dynamic (D)	D	D	D	D
Best effort (B) vs. QoS (Q)	B	B	B	B
Load balancing	Yes	Yes	Yes	No
Fault-Tolerant	Yes	Yes	Yes	Yes
Power-aware	No	Yes	Yes	No
Nexthop (N) vs. Source routing (SR)	N	N	SR	N

Table 8.12. Classification of routing protocols for WSNs

8.10 A Performance Evaluation Framework for Nature-Inspired Routing Protocols for WSNs

We now describe our performance and evaluation framework [207] that we used for an unbiased analysis of above-mentioned protocols.[4]. We used Prowler [222], a probabilistic wireless sensor network simulator, for comprehensive empirical evaluation of *FP-Ant*, *EEABR*, *BeeSensor*, and *AODV*. Prowler offers realistic radio and Media Access Control (MAC) models for the Berkeley mote platform. It also supports an event-based structure similar to that of TinyOS/NesC which facilitates the deployment of routing algorithms on real sensor nodes. Routing Modeling Application Simulation Environment (RMASE) [298] is a framework implemented as an application in Prowler. RMASE consists of a network topology model, an application model, and a performance model. It also supports a layered architecture in which the routing algorithms can be incorporated for evaluation purposes. The performance model provided with RMASE only monitors the primary parameters: delay, packet delivery ratio, and energy consumption, as reported in [300]. We extended the default performance model distributed with RMASE to record additional relevant parameters like algorithmic complexity, control efficiency, control complexity, average remaining energy, and standard deviation that provide valuable insight into the behavior of a routing protocol for WSNs (these metrics will be shortly defined).

We evaluated all protocols in two different types of operational scenarios: (1) a static convergecast scenario in which we selected a circular region and allowed a certain percentage of the nodes within the region to act as the source nodes; (2) a well-known target tracking application. The pattern of results for both scenarios is approximately the same; therefore, we report in this chapter the results for a relatively challenging target tracking scenario. An interested reader, however, can consult [207] for results of the convergecast scenario. In the target tracking application scenario, the node generating the events kept changing dynamically during the complete run of the experiment. While tracking a moving target, a sensor node in the vicinity of a moving target generated some arbitrary sequence of events depending upon the

---

[4] The section contains, by permission of the publisher IEEE, extracts of our paper: M. Saleem and M. Farooq. *A framework for empirical evaluation of nature inspired routing protocols for wireless sensor networks*. In Proceedings of Congress on Evolutionary Computing (CEC), pages 751-758. IEEE, Singapore, September 2007.

location of the target. It stopped generating events the moment that target went out of its range. This scenario was expected to challenge the reactive protocols because every time a source is changed, the algorithm has to initiate a new route discovery process. The scenario became even more complex due to the fact that the number of events generated by a source node was not fixed. Last but not least, we ensured that the locations of the source nodes were at the farthest possible distance from the sink node.

We tested our application on a network of 49, 64, 81, 100, 121, and 144 sensor nodes that were placed randomly in the sensing field. Each experiment was performed for a duration of 300 seconds. We assumed symmetric links between the nodes. All reported values are an average obtained from five independent runs. The choice of five runs is based on an initial observation that the results obtained from five runs were approximately the same as those obtained from ten runs. The initial energy level of the nodes was 50 Joules.

8.10.1 Metrics

We now define relevant performance parameters used in our pilot study:

- **Delay.** It is defined as the difference in time between when an event is generated at a source and when it is delivered at the sink node. We do not include lost packets in calculating delay. We report the average delay.
- **Packet delivery ratio.** It is the ratio of the total number of events received at a sink node to the total number of events generated by the nodes in the sensor network (an event is dispatched in a packet). The loss ratio is:

$$Loss ratio = 1 - Packet delivery ratio. \qquad (8.4)$$

- **Energy consumption.** We report total energy consumed by the sensor nodes in a network during each experiment.
- **Energy efficiency.** It is defined as the total energy consumed per 1,000 bits delivered at the destination (J/kbits).
- **Lifetime.** We recorded both the average remaining energy of the nodes and its standard deviation. We are not reporting these two parameters independently. Rather, we report the lifetime of the network, which is defined as the difference of the average remaining energy levels of the sensor nodes and their standard deviation. The basic motivation behind this definition is that an algorithm should try to maximize the average remaining energy levels of nodes with a small standard deviation. We report it in as a percentage.
- **Algorithmic complexity.** It is defined as the total number of CPU cycles consumed, for processing control packets and forwarding data packets by a protocol during an experiment per 1,000 bits of data delivered at the destination (cycles/kbits).
- **Control efficiency.** It is defined as the number of control bits in the network per 1,000 bits of delivered data at the destination (control bits/kbits).
- **Control complexity.** It is defined as the total number of CPU cycles consumed in processing of control traffic per 1,000 bits of delivered data (CPU cycles/kbits).

8.11 Results

Packet delivery ratio and delay. Figure 8.15 shows the packet delivery ratio and delay of the protocols. *FP-Ant* has the highest packet delivery ratio, followed by *BeeSensor*. Target tracking, as expected, is a difficult scenario for *BeeSensor* and *AODV*. But the results demonstrate that the packet delivery ratio of *BeeSensor* is significantly higher than that of *EEABR* and *AODV*. The obvious reason for this superior performance is that it reduces the amount of control traffic in the network which results in quick convergence of the protocol. We also want to highlight an important point here: 100% packet delivery is neither expected nor desired in the majority of the applications of wireless sensor networks because different nodes generate redundant information.

It is worth mentioning that *EEABR* does not provide consistent performance. We investigated this issue, and the reason behind such a poor performance is that it launches the *forward ants* at regular intervals from all nodes in the network. As a result, the complete network becomes unstable due to a large amount of control traffic, and this behavior is more pronounced in larger topologies. Figure 8.15(b) shows the delay values of the protocols. Note that *BeeSensor* has the largest delay. The higher delay is due to the dynamic nature of an application, which frequently initiates a new route discovery process. Due to these regular discoveries of new routes, the probability that a packet may have to wait in an event cache (because of unavailability of the route) increases. Furthermore, *BeeSensor* is a multi-path routing protocol and tries to route packets through high-energy nodes instead of routing through shortest paths. This also results in increased delay of packets in *BeeSensor*. Delay of *EEABR* is slightly better compared to that *AODV* and *FP-Ant*.

Total energy and energy efficiency. Energy consumption is an important performance metric in wireless sensor networks. Figure 8.16 shows the total energy consumption in the network along with energy efficiency (note that in Figure 8.16, the y axes have a logarithmic scale). It is clear from the Figure 8.16(a) that *BeeSensor* not only consumes the least amount of total energy but its energy efficiency (see Figure 8.16(b)) is also better than the *AODV*, which we consider as a benchmark algorithm for energy consumption. It is interesting to note that *BeeSensor* consumes the least amount of total energy even though it delivers 20–25% additional data packets, which of course results in greater energy efficiency. The best energy consumption of *BeeSensor* is due to its low control efficiency (Figure 8.17(b)). High loss ratio (or low packet delivery ratio), as in case of AODV, results in more route discovery messages, which ultimately contributes to higher energy consumption. *FP-Ant*, as expected, is the worst protocol in terms of total energy consumption. It is also important to note that *FP-Ant* is not a pure flooding protocol but performs restrictive flooding of *forward ants* that carry the *data ants* as described in Section 8.8.1. One can easily extrapolate the energy consumption of a pure flooding algorithm. *EEABR* consumes two to three times less energy than *FP-Ant* but, due to its low packet delivery ratio, its energy efficiency is significantly poor. We would like to mention here that the simulator used for analysis is especially designed for wireless sensor networks and its radio and MAC models are realistic. If a node injects a large amount of traffic into

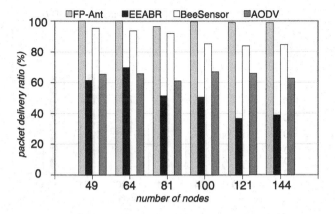

(a) Packet delivery ratio: Target tracking

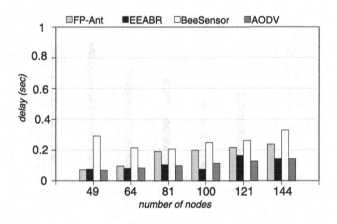

(b) Delay: Target tracking

Fig. 8.15. Packet delivery ratios and latencies for target tracking scenario

the network that exceeds the capacity of the MAC protocol, then this results in loss of the packets. Protocols having this behavior suffer. A flooding protocol like *FP-Ant* is an exception because it creates a large number of redundant paths which ensure a high packet delivery ratio at the cost of higher energy consumption and more control bits.

Control complexity. We measured this parameter separately from the algorithmic complexity just to highlight the significant amount of the processing power wasted in handling the control traffic. We have listed all these values in Table 8.13. *BeeSensor*'s performance is again superior compared to the rest of the protocols fol-

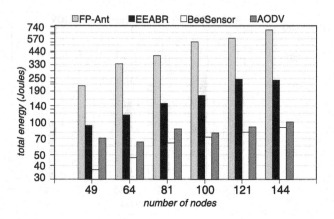

(a) Total energy consumption (J): Target tracking

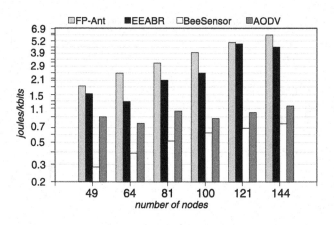

(b) Energy efficiency (J/kbits): Target tracking

Fig. 8.16. Total energy consumption and energy efficiency for target tracking scenario

lowed by *AODV*'s. We also listed the difference from the lowest value to provide a quantitative insight into the merits and demerits of reactive and proactive protocols. Note that the control complexity of *EEABR* in an 81-node network is better than that of *AODV* (see Table 8.13) even though *AODV* has delivered more packets than *EEABR* (refer to Figure 8.15(a)). This behavior shows the significance of optimizing the route discovery process. The results clearly indicate that a simple route discovery mechanism, coupled with agents that have simple behavior, is the key to the best performing protocols.

Application scenario	Number of nodes	CPU cycles (in millions)/kbits of data				%age Diff. from the lowest value		
		FP-Ant	EEABR	BeeSensor	AODV	FP-Ant	EEABR	AODV
Target tracking	49	5801	2804	866	2984	570	224	245
	64	10663	2518	943	3686	1031	167	291
	81	15727	4216	1208	5634	1202	249	366
	100	26759	5670	1646	5485	1526	244	233
	121	34372	11456	1670	6809	1958	586	308
	144	40905	14034	2112	8307	1837	565	293

Table 8.13. Control complexity

Lifetime. Lifetime of the network at the end of each experiment is listed in Table 8.14. The results show that *BeeSensor* has lifetimes better than or equal to those of *AODV*. It is interesting that *BeeSensor*, and even *AODV* achieve values of lifetime the same as or better than those of *EEABR* (especially in large networks).

Application scenarios	Number of nodes	% age remaining lifetime			
		FP-Ant	EEABR	BeeSensor	AODV
Target tracking	49	90	95	97	89
	64	88	96	97	95
	81	88	91	97	91
	100	88	95	98	98
	121	89	86	98	98
	144	87	96	98	98

Table 8.14. Lifetime of the network

8.12 Summary

In this chapter we presented our two bee-inspired energy-efficient routing protocols, *BeeAdHoc* and *BeeSensor*, for MANETs and sensor networks respectively. Both algorithms are simple and they basically need two types of agents for routing: scouts, which on demand discover new routes to the destinations, and forgers, which transport data packets and simultaneously evaluate the quality of the discovered routes. This simplicity results in a substantially smaller number of control packets sent; as a result, the algorithms are energy-efficient. We have verified through extensive simulations, which represent a wide spectrum of network conditions, that *BeeAdHoc* and *BeeSensor* deliver performance the same as or better than that of the state-of-the-art algorithms with significantly less energy expenditure.

As a continuation of our *Natural Engineering* approach introduced in the beginning of this book, we realized *BeeAdHoc* in mobile laptops running the Linux operating system. We validated our algorithm in simulated reality, quasi-reality, and reality scenarios and compared its performance with that of *AODV* and *OLSR*. An important outcome of this research was that the results in NS-2 simulation, simulated reality, quasi-reality, and reality had no direct correlation as is the case with *BeeHive*. Nevertheless, a similar pattern was observed in all above-mentioned scenarios: BeeAdHoc *delivered performance the same as or better than that of state-of-the-art algorithms*

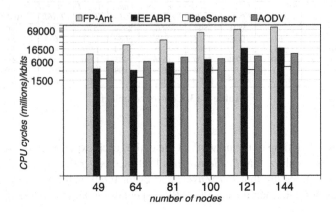

(a) Algorithmic complexity: Target tracking

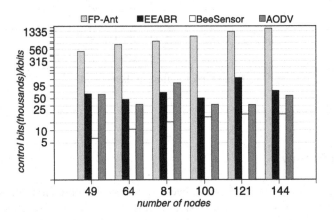

(b) Control efficiency: Target tracking

Fig. 8.17. Algorithmic complexity and control efficiency for target tracking scenario

with significantly less control overhead. The results reported in the chapter are intriguing enough to motivate researchers to develop self-organizing, simple, scalable, adaptive, and efficient routing protocols for emerging next-generation networks.

9

Conclusion and Future Work

This chapter concludes our book by emphasizing the scientific and technical contributions of our projects: BeeHive *and* BeeAdHoc. *We outline our vision of extensive future research that could follow the successful conclusion of our work. We place special emphasis on an intelligent and knowledgeable router, Nature-Inspired Decentralized and Autonomous Router (NIDAR), which can become a state-of-the-art router for networks of the new millennium. Finally, we suggest that the time has come to start a* Natural Engineering *program in our universities in order to successfully translate novel and cost-effective nature-/bio-inspired business solutions for highly competitive markets.*

9.1 Conclusion

We developed a dynamic, simple, efficient, robust, flexible, and scalable multi-path routing algorithm, *BeeHive*, for packet-switched fixed telecommunication networks. The algorithm is an important first step in endowing the network layer with intelligence and knowledge. An intelligent network layer is able to optimally manage its network resources. In this work, our focus was to provide a solution to the challenges in *traffic engineering* by designing a multi-path routing algorithm for IP networks. Such a paradigm will try to exploit the network bandwidth by utilizing the connectionless feature of IP. This is in contrast to existing approaches of either managing virtual circuits on top of the IP layer (MPLS) or administering the resources on a per-flow basis (RSVP). These approaches try to solve the QoS guarantee problems for streaming multimedia applications while utilizing *OSPF* at the IP layer. Our hypothesis is that simple and cost-effective solutions are possible for QoS routing if we reengineer the networking layer and the routing protocols.

An ideal or knowledge-empowered network layer should have the following features: *it should be able to discover and manage its routes in a decentralized and asynchronous fashion without the need to have access to the global view of the network topology; the diagnosis of a fault and its remedial management should be embedded*

*in the routing system; the network layer should be able to guarantee its resources to
streaming network traffic and should be able to enhance the network performance
for best-effort traffic; it should be able to do multiobjective optimization between
competing and conflicting demands according to the network state.*

Our observation is that different natural colony systems, for example a honeybee colony, are able to complete similar tasks through simple individuals that have
limited memory and processing abilities. However, they follow simple rules to coordinate their activities. As a result, an intelligent and coherent pattern emerges at
a colony level beyond the capabilities of any individual. According to Seeley [216],
*a honeybee colony can thoroughly monitor a vast region around the hive for rich
food sources, nimbly redistribute its foragers within an afternoon, fine-tune its nectar processing to match its nectar collecting, effect cross-inhibition between different
forager groups to boost its response differential between food sources, precisely regulate its pollen intake in relation to its ratio of internal supply and demand, and limit
the expensive process of comb building to times of critical need for additional storage space.* These observations motivated us to make the following hypotheses in the
beginning of our work, which we reproduce for the sake of clarity:

(a) **H1:** If a honeybee colony is able to adapt to countless changes inside the hive
or outside in the environment through simple individuals without any central
control, then an agent system based on similar principles should be able to adapt
itself to an ever-changing network environment in a decentralized fashion with
the help of simple agents who rely only on local information. This system should
be dynamic, simple, efficient, robust, flexible, reliable, and scalable because its
natural counterpart has all these features.

(b) **H2:** If designed with a careful engineering vision, nature-inspired solutions are
simple enough to be installed on real-world systems. Therefore, their benefit-to-
cost ratio should be better than that of existing real-world solutions.

In Chapter 3 and 8, we developed agent models inspired from the communication and evaluative features of a honeybee colony. The *bee agents* in our models
have simple behavior. As a result, the algorithms are able to take routing decisions in
a decentralized and asynchronous fashion. We have conducted extensive simulations
in OMNeT++ and NS-2 simulators to show the advantages of our algorithms over
existing state-of-the-art routing algorithms developed by the nature-inspired routing
community. Our algorithms are able to achieve better performance values but with
a simple agent model. Consequently, this resulted in not only low processing and
control overheads but also in low energy consumption in MANETs and WSNs. We
were able to collect a comprehensive set of performance values through our performance evaluation framework to study the behavior of routing protocols over a vast
operational landscape.

We defined a power metric for a routing protocol which models the performance
of a routing algorithm based on a number of parameters. We also defined a productivity metric that shows the benefit-to-cost ratio of a routing protocol in an unbiased
manner. We then developed a scalability model to study the scalability of a routing
protocol by conducting extensive experiments on six topologies varying in their size

and complexity. We concluded that *BeeHive* is scalable for most of the network configurations. The results discussed in Chapters 3, 4 and 8 are significant to confirm the validity of H1.

We then moved to a cardinal step in our *Natural Engineering* approach, in which we developed *engineering models* from the simulation models and implemented them in the network stack of the Linux kernel. In the case of *BeeHive*, we developed the same traffic generator in both the simulation and the application layer of the network stack of the Linux kernel. We then tested our algorithm in a real network of Linux routers and compared the performance values of our algorithm, both in simulation and in the real network, with *OSPF*. We can conclude from the results of the extensive experiments in Chapter 5 that the performance values of *BeeHive* were traceable from simulated to real networks. *BeeHive* is the first nature-inspired routing algorithm, according to our knowledge, that has been implemented and tested in real networks with the help of existing resources. Similarly, we tested our *BeeAdHoc* protocol in simulation, simulated reality, quasi-reality, and reality MANETs. Again we observed the same pattern of results in all scenarios: BeeAdHoc *delivered the same or better performance compared to existing routing protocols but with a significantly smaller number of control packets.* The success of our efforts lies in an engineering approach that we followed in our protocol development cycle. The success in this phase also validates H2.

BeeHive and *BeeAdHoc* discover and evaluate multiple paths in a deterministic fashion by utilizing a variant of breadth first search. It does not discover all possible multiple paths. Rather, only those multiple paths are utilized whose quality is above a certain threshold. *BeeHive* and *BeeAdHoc* then spread the data packets on multiple paths in a stochastic fashion in order to achieve better performance values. Our experience suggests that *BeeHive* and *BeeAdHoc* properly combine the deterministic elements with the stochastic elements in a routing algorithm. The proper classification of *BeeHive* and *BeeAdHoc* is shown in Figure 9.1. Note that *BeeSensor* deterministically discovers routes up to the first two hops and then switches to stochastic route discovery mode.

Concluding remark: We maintain the implicit assumption throughout our work that the resources of a network, including bandwidth, have to be utilized in an efficient manner. The existing fiber optic networks of terabyte capacities provide substantially greater bandwidth per user compared to the networks that existed only a couple of years ago. This might obviate the need for multi-path routing algorithms in the short term. However, the lesson of history is that whenever a resource is in abundance, new and powerful applications are developed that properly utilize that resource [260]. We strongly believe, therefore, that the need for a network layer that utilizes its resources in an efficient manner through intelligent routing algorithms cannot be underestimated in the long term.

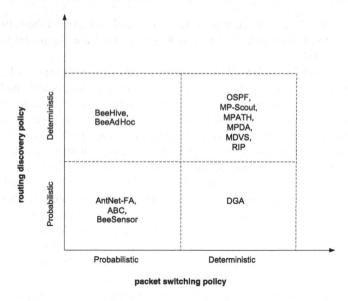

Fig. 9.1. Routing classification

9.2 Future Research

We believe that the objectives achieved in our projects *BeeHive* and *BeeAdHoc* are encouraging, especially for nature-inspired algorithms community, because the work presents simple nature-inspired multi-agent systems that are realized both in simulation and inside the network stack of the Linux operating system. After working for six years on these exciting projects, we honestly believe that the accomplishments have to be followed with rigorous design and development efforts in order to translate the prototype to an intelligent real-world routing system. We did not emphasize the related implementation details too much because we were interested in illustrating a "proof of concept" in the first phase of the projects. Therefore, details about how to realize *BeeHive* as an Interior Gateway Routing Protocol (IGRP) for intra-domain routing are never addressed. We believe that the successful completion of "proof of concept" phase is encouraging for undertaking this research. The objective of this research must be to replace existing *OSPF* type algorithms with the *BeeHive* type agent-based algorithms in a systematic fashion. We now highlight two shortcomings of the *BeeHive* protocol that make it unsuitable for QoS routing: (1) High jitter; and (2) Cyclic paths.

9.2.1 Quality of Service (QoS) Routing

Our results in Figure 9.2 demonstrate that under high network traffic load the jitter value of *BeeHive* is significantly higher compared with *OSPF*. The jitter values of

BeeHive are comparable to *OSPF* at MSIA = 4.6 sec (see Figure 9.2(a)). The jitter values of *BeeHive* are also comparable with *OSPF* at MSIA = 2.6 sec (see Figure 9.2(b)) with the exception of n57. However, the jitter values of *BeeHive* are significantly inferior to those of *OSPF* at MSIA = 1.6 sec (see Figure 9.2(c)). This is naturally due to the stochastic spreading of data packets on multiple paths as per their quality (remember that *BeeHive* delivers significantly more packets with smaller delays at higher loads). Consequently, any two subsequent packets from the same session may follow different paths and this leads to a higher jitter value. Therefore, we believe that future research has to address this issue. We propose three solutions to tackle this problem:

- *Time-based routing stability* approach will ensure that the routes, once selected, remain valid for a certain time t_{valid}. This way, the decision to stochastically route data packets does not happen at the arrival of each data packet. Rather it only happens at regular time intervals. We believe that by utilizing time-based stability of routing decisions, we can still achieve a better jitter value without compromising the performance of the algorithm. The disadvantage is that the stability interval is independent of the network traffic load.
- *Load-based routing stability* approach will ensure that the routes, once selected, remain valid for a certain number of data packets n_w. This means that the decision to stochastically route data packets is taken for every n_w packets. The advantage of this approach is that the stability interval adapts to the network traffic load: the stability interval will decrease with an increase in the traffic load and vice versa.
- *Priority routing* approach will treat the multimedia streaming packets in a priority manner in which the routes for these types of traffic are reserved at the start of an application session. We will still be able to distribute the traffic on multiple paths but only on a session basis rather than on a packet basis for these types of traffic. However, the normal network traffic will be handled in a best-effort manner.

9.2.2 Cyclic Paths

The data packets in the *BeeHive* algorithm can follow a cyclic path due to its stochastic routing property. The number of data packets that follow a cyclic path depend on the topology and the network traffic. Figure 9.3 shows the distribution of packets that followed cyclic paths in different topologies for different traffic patterns. We have given beneath each subfigure the topology name, the MSIA value, and the percentage of packets that followed cyclic paths. $P(1)$ is the percentage of packets that followed a cyclic path once, $P(2)$ is the percentage of packets that followed the cyclic paths twice, and $P(> 3)$ is the percentage of packets that followed the cyclic paths three times or more. We can safely conclude from our results that the cyclic paths are only short-lived and the probability that a packet might follow a cyclic path for the second time is approximately 1.0% (total probability) or less in almost all cases. The overall probability of following a cyclic path the second time in Figure 9.3(b) is 24% of 3.79%, which is 0.91%.

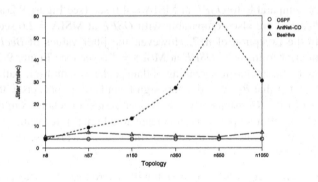

(a) MSIA = 4.6 sec

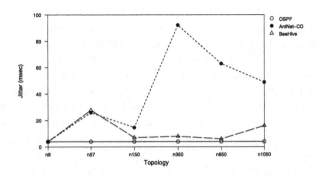

(b) MSIA = 2.6 sec

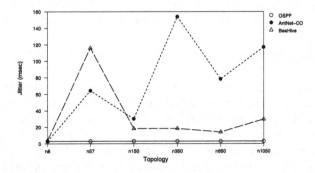

(c) MSIA = 1.6 sec

Fig. 9.2. Jitter (msec)

The results of our experiments suggest that even with the cyclic paths, *BeeHive* is able to outperform *OSPF*. Nevertheless, if we can make *BeeHive* a loop-free algorithm, then this will improve its acceptability, because the networking community puts significant emphasis on loop-free routing. However, the real challenge here is *to achieve loop freedom in a stochastic routing algorithm without utilizing agents that do not have a stack memory.* We now introduce our *BeeHivePlus* and *BeeHiveQos*, which will overcome the above-mentioned shortcomings of *BeeHive*. Consequently, *BeeHive* not only achieves loop-free routing but also small jitter without making any compromise on the performance under high network traffic loads.

BeeHivePlus and *BeeHiveQoS* for QoS Routing

The pilot studies reported in the previous sections clearly indicate that nature-inspired routing protocols like *BeeHive*, in spite of their superior performance, suffer from unwanted side effects such as short-lived loops and jitter fluctuations. These side effects are of course not desirable in QoS networks, which demand strict and predictable performance.

In [31] we enhanced *BeeHive* to enable it to do QoS-aware routing. *BeeHivePlus* overcomes these potential problems by utilizing the concept of *waterfall routing*, in which only those neighbors that are nearer, in terms of hops, to the destination than the current node are considered for selection as a next hop. As a result, a packet always moves in the direction of the destination, and loops are implicitly avoided. We use the concept of *time-based routing stability*, i.e., routing decisions for a destination remain fixed between the inter-arrival time period of two successive bee agents from the same destination. This feature ensures that packets for a short time period follow the same path. Consequently, it results in significant reduction in jitter fluctuations over that of *OSPF* (a singlepath algorithm). However, *BeeHivePlus*, in spite of having these desirable features of loop freedom and low jitter, provides similar performance to that of *BeeHive*. We skip the results for brevity, but interested readers are encouraged to consult [31].

We then derived from *BeeHivePlus* a novel algorithm, *BeeHiveQoS*, for QoS networks. The core mechanism in *BeeHiveQoS* is an intelligent hierarchical packet scheduler, which can be embedded in *BeeHivePlus* as well as in other schemes, and which provides soft guarantees to QoS-sensitive applications. The results of our experiments conducted on network topologies up to 150 nodes confirm that *BeeHiveQoS* is able to provide guarantees to QoS-sensitive applications. To conclude, our *BeeHive* protocol is not only efficient but also secure and QoS-aware. Recall that *BeeAdHoc* and *BeeSensor* protocols are host-intelligent; therefore, they do not suffer from the looping problem anyway. It will be interesting work to embed our intelligent hierarchical packet scheduler into *BeeAdHoc* and *BeeSensor* protocols for providing QoS-aware routing in MANETs and WSNs respectively.

9.2.3 Intelligent and Knowledgeable Network Engineering

We believe that this task can only be achieved by launching an interdisciplinary effort that requires cross-fertilization among different areas of engineering and science

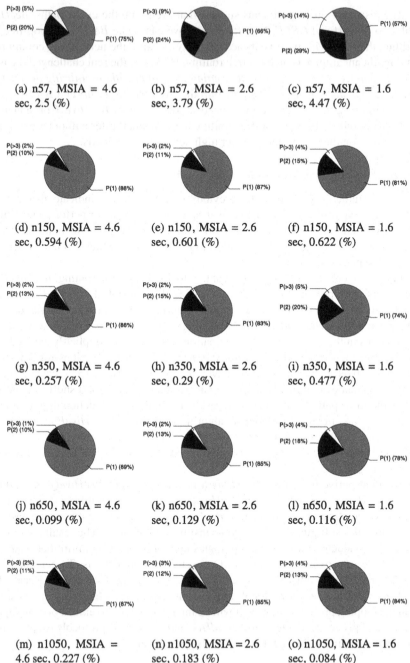

(a) n57, MSIA = 4.6 sec, 2.5 (%)

(b) n57, MSIA = 2.6 sec, 3.79 (%)

(c) n57, MSIA = 1.6 sec, 4.47 (%)

(d) n150, MSIA = 4.6 sec, 0.594 (%)

(e) n150, MSIA = 2.6 sec, 0.601 (%)

(f) n150, MSIA = 1.6 sec, 0.622 (%)

(g) n350, MSIA = 4.6 sec, 0.257 (%)

(h) n350, MSIA = 2.6 sec, 0.29 (%)

(i) n350, MSIA = 1.6 sec, 0.477 (%)

(j) n650, MSIA = 4.6 sec, 0.099 (%)

(k) n650, MSIA = 2.6 sec, 0.129 (%)

(l) n650, MSIA = 1.6 sec, 0.116 (%)

(m) n1050, MSIA = 4.6 sec, 0.227 (%)

(n) n1050, MSIA = 2.6 sec, 0.183 (%)

(o) n1050, MSIA = 1.6 sec, 0.084 (%)

Fig. 9.3. Distribution of packets that follow cyclic paths

on one the hand, and close cooperation between academia and industry on the other hand. The final aim of the research should be to design and develop a dedicated Nature-Inspired Decentralized and Autonomous Router (NIDAR) that can seamlessly replace existing routers in the Internet. NIDAR is to be designed from scratch with *Intelligence* and *Knowledge* as the key principles. The conceptual block-level diagram of NIDAR is shown in Figure 9.4. It shows the overwhelming complexity

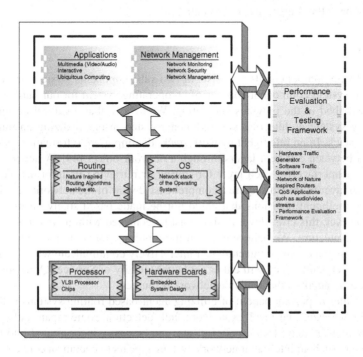

Fig. 9.4. Nature-inspired distributed and autonomous router

of achieving the task. Our vision is that the task can be achieved in an efficient and systematic fashion if we divide our objective into three important building blocks: hardware, software, and performance evaluation.

Hardware

NIDAR is to be designed with an engineering vision that simplifies its installation in real networks. As the routers are required to perform packet switching in real time, it is not advisable to host any additional computationally intensive tasks on such a system. This approach is in line with the recommendations made by Yu [295]. Our suggestion is to develop NIDAR hardware consisting of two subsystems: one subsystem solely responsible for the packet switching, and the second subsystem performing the agent processing tasks. We would like to realize the packet-switching component of our algorithm in a chip in order to achieve maximum efficiency.

A very attractive option can be the possibility of remote and on-the-fly upgrading of firmware software on the agent processing subsystem. This feature will help in downloading the new updates of the algorithm and in potential error fixes without bringing back the routers in service centers.

Software

The software component also consists of two subcomponents: system and application. The system component consists of an operating system for our hardware platform described in the previous subsection. We believe that a careful design of an operating system architecture is a rudimentary step in our realizing an intelligent network stack. In an intelligent network stack, the important information can be exchanged between different layers through a network object [49]. A network object defines an interface for each layer for storing or retrieving its information. In this way, each layer is able to acquire a comprehensive view of the network state in order to make intelligent decisions that could not be made based solely on its information. However, this cross-layer design must be achieved without violating the open system architecture. We believe that real-time multimedia audio and video applications utilizing such an architecture can interactively communicate with the network layer. Consequently, they will either get better Quality of Service (QoS) guarantees or gracefully adapt to the available network resources.

The other important benefit of utilizing agent-based routing systems is that *network management* that involves network fault detection, route repair and recovery, resource management, traffic policing, etc. can be easily incorporated into the network layer. Such an intelligent network system appears to become a core component of next-generation networks.

Performance evaluation

The performance evaluation of NIDAR in hardware and software with the help of a network traffic generator will be an important validation step. We believe that performance must be evaluated both in hardware and software to comprehensively analyze the true benefits of nature-inspired routing algorithms. The testing and evaluation should be done in real network environments with a set of different applications during an early stage of design and development in order to continuously improve the design of NIDAR through feedback channels. The outcome of this approach should report the quantitative gains over the existing routing systems. The evaluation and validation of a routing protocol is an important step of *protocol engineering* [140].

9.2.4 Bee Colony Metaheuristic

We should emphasize that the foraging model described in [217, 242] can be used for addressing any multiobjective optimization problems for dynamic environments. The important concepts of *BeeHive*, such as the *bee agent propagation* algorithm and *bee agent communication* paradigm, can be applied to any optimization problem that can be represented in a graph. However, the real strength of the *BeeHive* concept is its ability to function in dynamic environments, where scarce resources have to be managed in an optimum fashion, or in multiobjective optimization, in which competing and at times conflicting requirements have to be incorporated into the objective function. We believe that it is possible to generalize the bee behavior to develop a new metaheuristic for optimization problems that can be applied to different continuous and discrete optimization problems. We call it *Bee Hive Optimization* (BHO).

9.3 Natural Engineering: The Need for a Distinct Discipline

We believe now is a suitable time to seriously evaluate the feasibility of starting a distinct discipline of *Natural Engineering* in our universities. In this discipline, the novel algorithms and systems inspired by natural systems and developed by the natural computing community can be engineered for real-world applications. However, the major emphasis should be on cost-effective design that can bring substantial savings in costs of designing, manufacturing, installing, and maintaining such systems. We believe that such an engineering discipline requires a strong interdisciplinary effort, resulting in revolutionary design and development paradigms, paving the way to the development of products that are impossible to achieve with classical engineering principles.

References

1. Netfilter: firewalling, NAT, and packet mangling for Linux. http://www.netfilter.org.
2. ITU G. 711: Pulse code modulation (PCM) of voice frequencies, November 1988.
3. In H.-P. Schwefel, I. Wegener, and K. Weinert, editors, *Advances in Computational Intelligence: Theory and Practice*, Natural Computing Series. Springer-Verlag, 2003.
4. *VoIP Standards and Protocols*. Faulkner Information Services, 2003.
5. R. Ahuja, S. Keshav, and H. Saran. Design, implementation, and performance measurement of a native-mode ATM transport layer (extended version). *IEEE/ACM Transaction on Networks*, 4(4):502–515, 1996.
6. U. Aickelin, P. Bentley, S. Cayzer, J. Kim, and J. McLeod. Danger theory: the link between AIS and IDS. In *Proceedings of the 2nd International Conference on Artificial Immune Systems*, volume 2728 of *LNCS*, pages 147–155. Springer, 2003.
7. U. Aickelin and S. Cayzer. The danger theory and its applications to artificial immune systems. In *Proceedings of the 1st International Conference on Artificial Immune Systems*, volume 2728 of *LNCS*, pages 141–148. Springer, 2002.
8. J.N. Al-Karaki and A.E. Kamal. Routing techniques in wireless sensor networks: a survey. *IEEE Wireless Communications*, 11(6):6–28, Dec. 2004.
9. S. Appleby and S. Steward. Mobile software agents for control in telecommunications networks. *BT Technology Journal*, 12(2):104–113, Apr. 1994.
10. A.F. Atlasis, M.P. Saltouros, and A.V. Vasilakos. On the use of a stochastic estimator learning algorithm to the ATM routing problem: a methodology. *Computer Communications*, 21:538–546, 1998.
11. S. Avallone, M. D'Arienzo, M. Esposito, A. Pescapé, S.P. Romano, and G. Ventre. Mtools. *IEEE Networks Magazine: Software Tools for Networking*, 16(5):3, October 2002.
12. S. Avallone, D. Emma, A. Pescapé, and G. Ventre. A distributed multiplatform architecture for traffic generation. In *Proceedings of International Symposium on Performance Evaluation of Computer and Telecommunication Systems (SPECTS)*, San Jose, California (USA), July 2004.
13. S. Avallone, A. Pescapé, and G. Ventre. Distributed Internet Traffic Generator (D-ITG): analysis and experimentation over heterogeneous networks. In *ICNP 2003 poster Proceedings, International Conference on Network Protocols 2003*, Atlanta, Georgia (USA), November 2003.
14. S. Avallone, A. Pescapé, and G. Ventre. Analysis and experimentation of Internet Traffic Generator. In *New2an'04, Next Generation Teletraffic and Wired/Wireless Advanced Networking*, pages 70–75, 2004.

15. D. Awduche, A. Chiu, A. Elwalid, I. Widjaja, and X. Xiao. RFC 3272: Overview and principles of internet traffic engineering, May 2002.

16. P. Baldi, P. Frasconi, and P. Smyth. *Modeling the Internet and the Web: Probabilistic Methods and Algorithms*. Wiley, 2003.

17. B. Baran and R. Sosa. A new approach for AntNet routing. In *Ninth International Conference on Computer Communications and Networks*, pages 303–308, Las Vegas, NV, USA, 2000.

18. N. Bean and A. Costa. An analytic modelling approach for network routing algorithms that use "ant-like" mobile agents. *Computer Networks*, 49(2):243–268, 2005.

19. D. Bertsekas and R. Gallager. *Data Networks*. Prentice-Hall, Englewood Cliffs, NJ, USA, 1992.

20. H.G. Beyer and H.P. Schwefel. Evolution strategies – a comprehensive introduction. *Natural Computing*, 1(1):3–52, 2002.

21. E. Bonabeau, M. Dorigo, and G. Theraulaz. *Swarm Intelligence: From Natural to Artificial Systems*. Oxford University Press, 1999.

22. E. Bonabeau, M. Dorigo, and G. Theraulaz. Inspiration for optimization from social insect behaviour. *Nature*, 406:39–42, 2000.

23. E. Bonabeau, F. Hénaux, S. Guérin, D. Snyers, P. Kuntz, and G. Theraulaz. Routing in telecommunications networks with ant-like agents. In *Intelligent Agents for Telecommunication Applications, Second International Workshop, IATA '98, Paris, France, July 1998, Proceedings*, volume 1437 of *Lecture Notes in Computer Science*, pages 60–71. Springer, 1998.

24. E. Bonabeau, G. Theraulaz, J. Deneubourg, S. Aron, and S. Camazine. Self-organization in social insects. *Trends in Ecology and Evolution*, 12(5):188–193, May 1997.

25. G. Booch. *Object-Oriented Analysis and Design with Applications*. Addison Wesley Longman Inc, 1994.

26. A. Botta, D. Emma, S. Guadagno, and A. Pescapé. Performance evaluation of heterogeneous network scenarios. Technical report, Dipartimento di Informatica e Sistemistica, Universita' di Napoli Federico II, 2004.

27. J.A. Boyan and M.L. Littman. Packet routing in dynamically changing networks: a reinforcement learning approach. *Advances in Neural Information Processing Systems*, 6:671–678, 1993.

28. R. Braden, D. Clark, and S. Shenker. RFC 1633: Integrated services in the internet architecture: an overview, June 1994.

29. F.M.T. Brazier, B. Dunin-Keplicz, N.R. Jennings, and J. Treur. DESIRE: Modelling multi-agent systems in a compositional formal framework. *Int. J. Cooperative Inf. Syst.*, 6(1):67–94, 1997.

30. J. Broch, D.A. Maltz, D.B. Johnson, Y.C. Hu, and J. Jetcheva. A performance comparison of multi-hop wireless ad hoc network routing protocols. In *Proceedings of Fourth ACM/IEEE Conference on Mobile Computing and Networking (MobiCom)*, pages 85–97, 1998.

31. R. Brüntrup. Quality of service in von der natur inspirierten routing-algorithmen (in German). Master's thesis, LSIII, University of Dortmund, Germany, August 2006.

32. L.T. Bui, J. Branke, and A. Hussein. Multiobjective optimization for dynamic environments. In *Congress on Evolutionary Computation*. IEEE, 2006.

33. S. Camazine, J. Deneubourg, N. Franks, J. Sneyd, G. Theraulaz, and E. Bonabeau. *Self-Organization in Biological Systems*. Princeton University Press, Princeton, NJ, second edition, 2003.

34. T. Camilo, C. Carreto, J.S. Silva, and F. Boavida. An energy-efficient ant-based routing for wireless sensor networks. In *Int. Workshop on Ant Colony Optimization and Swarm Intelligence, ANTS 2006, Sep. 2006*.

35. K. Camp. *IP Telephony Demystified*. McGraw-Hill Companies Inc, 2003.

36. D. Cantor and M. Gerla. Optimal routing in a packet-switched network. *IEEE Transactions on Computers*, 23:1062–1068, 1974.

37. L. Carrillo, J.L. Marzo, L. Fàbrega, P. Vilà, and C. Guadall. Ant colony behaviour as routing mechanism to provide quality of service. In *Ant Colony Optimization and Swarm Intelligence, 4th International Workshop, ANTS 2004, Brussels, Belgium, September 5 – 8, 2004, Proceedings*, volume 3172 of *Lecture Notes in Computer Science*, pages 418–419. Springer, 2004.

38. L. Carrillo, J.L. Marzoa, D. Harle, and P. Vila. A review of scalability and its application in the evaluation of the scalability measure of AntNet routing. In Palau Salvador, editor, *IASTED Communication Systems and Networks CSN 2003*, pages 317–323, Benalmadena, Spain, 2003.

39. W. Chainbi, M. Jmaiel, and A.B. Hamadou. Conception, behavioral semantics and formal specification of multi-agent systems. In *Multi-Agent Systems: Theories, Languages, and Applications, 4th Australian Workshop on Distributed Artificial Intelligence, Brisbane, Queensland, Australia, July 13, 1998, Selected Papers*, volume 1544 of *Lecture Notes in Computer Science*, pages 16–28. Springer, 1998.

40. D.D. Champeaux, D. Lea, and P. Faure. *Object-Oriented System Development*. Addison Wesley, 1993.

41. J. Chen, P. Druschel, and D. Subramanian. An efficient multipath forwarding method. In *INFOCOM 98*, pages 1418–1425, 1998.

42. J. Chen, P. Druschel, and D. Subramanian. A new approach to routing with dynamic metrics. In *INFOCOM 99*, pages 661–670, 1999.

43. J. Chen, P. Druschel, and D. Subramanian. A simple, practical distributed multipath routing algorithm. TR98-320, Department of Computer Science, Rice University, July 1998.

44. P. Choi and D. Yeung. Predictive q-routing: A memory-based reinforcement learning approach to adaptive traffic control. *Advances in Neural Information Processing Systems*, 8:945–951, 1996.

45. Cisco Systems, Inc. Internetworking technology handbook, 2002.

46. D. Clark. RFC 817: Modularity and efficiency in protocol implementation. Technical report, MIT, July 1982.

47. D. Clark. The structuring of systems using upcalls. *Operating Systems Review*, 19(5):171–180, 1985.

48. D. Clark and D.L. Tennenhouse. Architectural considerations for a new generation of protocols. In *Proceedings of the ACM SIGCOMM'90*, pages 200–208, Philadelphia, Sep. 1990. ACM.

49. M. Conti, G. Maselli, G. Turi, and S. Giordano. Cross-layering in mobile ad hoc network design. *IEEE Computer*, 37(2):48–51, 2004.

50. Intel Corporation. Using the RDTSC instruction for performance monitoring. Application notes, Pentium II processor, Intel Corporation, 1997.

51. A. Costa. Analytic modelling of agent-based network routing algorithms. Ph.D. thesis, University of Adelaide, Australia, 2002.

52. L.H.M.K. Costa, S. Fdida, and O.C.M.B. Duarte. Developing scalable protocols for three-metric QoS routing. *Computer Networks: The International Journal of Computer and Telecommunications Networking*, 39(6):713–727, 2002.

53. M. Cox, R. Engelschall, S. Henson, and B. Laurie. The OpenSSL project. http://www.openssl.org.

54. Dipankar Dasgupta, editor. *Artificial Immune Systems and their Applications*. Springer-Verlag, 1998.

55. Leandro N. de Castro and Jonathan Timmis. *Artificial Immune Systems: A new computational intelligence approach*. Springer-Verlag, 2002.

56. K. Deb. *Multi-Objective Optimization using Evolutionary Algorithms*. John Wiley & Sons, UK, 2001.

57. J. Deneubourg, S. Aron, S. Goss, and J. Pasteels. The self-organizing exploratory pattern of the Argentine ant. *Journal of Insect Behavior*, 3:159–168, 1990.

58. G. Di Caro. Ant colony optimization and its application to adaptive routing in telecommunication networks. Ph.D. thesis, Université Libre de Bruxelles, Belgium, 2004.

59. G. Di Caro and M. Dorigo. AntNet: A mobile agents approach to adaptive routing. Technical Report IRIDIA/97-12, Université Libre de Bruxelles, Belgium, 1997.

60. G. Di Caro and M. Dorigo. An adaptive multi-agent routing algorithm inspired by ants behavior. In *Proceedings of 5th Annual Australasian Conference on Parallel Real Time Systems*, pages 261–272, 1998.

61. G. Di Caro and M. Dorigo. Ant colonies for adaptive routing in packet-switched communications networks. In *Parallel Problem Solving from Nature – PPSN V, LNCS 1498*, pages 673–682, Sept 1998.

62. G. Di Caro and M. Dorigo. AntNet: Distributed stigmergetic control for communication networks. *Journal of Artificial Intelligence Research*, 9:317–365, December 1998.

63. G. Di Caro and M. Dorigo. Extending AntNet for best effort quality-of-service routing. In *First International Workshop on Ant Colony Optimization*, Brussels, Belgium, October, 15-16 1998.

64. G. Di Caro and M. Dorigo. Mobile agents for adaptive routing. In *31st Hawaii International Conference on System Science*, pages 74–83, Big Island of Hawaii, 1998. IEEE Computer Society Press.

65. G. Di Caro and M. Dorigo. Two ant colony algorithms for best-effort routing in datagram networks. In *Proceedings of the Tenth IASTED International Conference on Parallel and Distributed Computing and Systems (PDCS'98)*, pages 541–546. IASTED/ACTA Press, 1998.

66. G. Di Caro, F. Ducatelle, and L.M. Gambardella. AntHocNet: an adaptive nature-inspired algorithm for routing in mobile ad hoc networks. *European Transactions on Telecommunications*, 16(5):443–455, 2005.

67. G. Di Caro and T. Vasilakos. Ant-SELA: ant-agent and stochastic automata learn adaptive routing tables for QoS routing in ATM networks. In *ANTS 2000 – From Ant Colonies to Artificial Ants: Second International Workshop on Ant Colony Optimization*, Sept. 8–9 2000.

68. E. Dijkstra. Self-stabilization in spite of distributed control. *Communications of the ACM*, pages 643–644, 1974.

69. E.W. Dijkstra. A note on two problems in connection with graphs. *Numerical Mathematics*, 1:269–271, 1959.

70. E.W. Dijkstra and C.S. Scholten. Termination detection for diffusing computations. *Information Processing Letters*, 11(1):1–4, 1980.

71. M. d'Inverno, D. Kinny, M. Luck, and M. Wooldridge. A formal specification of dMARS. In *Intelligent Agents IV, Agent Theories, Architectures, and Languages, 4th International Workshop, ATAL '97, Providence, Rhode Island, USA, July 24-26, 1997, Proceedings*, volume 1365 of *Lecture Notes in Computer Science*, pages 155–176. Springer, 1998.

72. S. Doi and M. Yamamura. BntNetL: Evaluation of its performance under congestion. *IEICE B*, pages 1702–1711, 2000.
73. S. Doi and M. Yamamura. BntNetL and its evaluation on a situation of congestion. *Electronics and Communications in Japan*, 85:31–41, 2002.
74. M. Dorigo, E. Bonabeau, and G. Theraulaz. Ant algorithms and stigmergy. *Future Generation Computer Systems*, 16(8):851–871, June 2000.
75. M. Dorigo and G. Di Caro. The ant colony optimization meta-heuristic. In D. Corne, M. Dorigo, and F. Glover, editors, *New Ideas in Optimization*, pages 11–32. McGraw-Hill, London, 1999.
76. M. Dorigo, G. Di Caro, and L.M. Gambardella. Ant algorithms for discrete optimization. *Artificial Life*, 5(2):137–172, 1999.
77. M. Dorigo, V. Maniezzo, and A. Colorni. Positive feedback as a search strategy, 1991.
78. M. Dorigo, V. Maniezzo, and A. Colorni. The ant system: optimization by a colony of cooperating agents. *IEEE Transactions on Systems, Man and Cybernetics–Part B*, 26(1):29–41, 1996.
79. M. Dorigo and T. Stützle. The ant colony optimization metaheuristic: algorithms, applications and advances. In F. Glover and G. Kochenberger, editors, *Handbook of Metaheuristics*, volume 57 of *International Series in Operations Research & Management Science*, pages 251–285. Kluwer Academic Publishers, Norwell, MA, 2003.
80. S.N. Dorogovtsev and J.F.F. Mendes. *Evolution of Networks: From Biological Nets to the Internet and WWW*. Oxford University Press, 2004.
81. P. Druschel, Larry L. Peterson, and Bruce S. Davie. Experiences with a high-speed network adaptor: a software perspective. In *Proceedings of the ACM SIGCOMM'94*, pages 2–13, 1994.
82. F. Ducatelle, G. Di Caro, and L.M. Gambardella. Using ant agents to combine reactive and proactive strategies for routing in mobile ad hoc networks. *International Journal of Computational Intelligence and Applications, Special Issue on Nature-Inspired Approaches to Networks and Telecommunications*, 5(2):169–184, 2005.
83. J. Durkin. *Voice-Enabling the Data Network*. CISCO Press, 2003.
84. A. Edwards and S. Muir. Experiences implementing a high performance TCP in userspace. In *SIGCOMM '95: Proceedings of the Conference on Applications, Technologies, Architectures, and Protocols for Computer Communication*, pages 196–205, New York, NY, USA, 1995. ACM Press.
85. A. Edwards, G. Watson, J. Lumley, D. Banks, C. Calamvokis, and C. Dalton. Userspace protocols deliver high performance to applications on a low-cost Gb/s LAN. In *SIGCOMM '94: Proceedings of the Conference on Communications Architectures, Protocols and Applications*, pages 14–23, New York, NY, USA, 1994. ACM Press.
86. L.M. Feeney. A taxonomy for routing protocols in mobile ad hoc networks. Technical Report ISRN:SICS-T-99/07-SE, Swedish Institute of Computer Science, Kista, Sweden, 1999.
87. L.M. Feeney. An energy consumption model for performance analysis of routing protocols for mobile ad hoc networks. *Mobile Networks and Applications*, 6(3):239–249, 2001.
88. L.M. Feeney and M. Nilsson. Investigating the energy consumption of a wireless network interface in an ad hoc networking environment. In *Proceedings of IEEE INFOCOM*, 2001.
89. S. Fenet and S. Hassas. ANT: a distributed network control framework based on mobile agents. In *International ICSC Congress on Intelligent Systems And Applications*, pages 831–837. ICSC Academic Press Editor, 2000.

90. S. Fenet and S. Hassas. ANT: a distributed problem-solving framework based on mobile agents. In *Mobile Agents Applications 2000 (12th International Conference on Systems Research, Informatics & Cybernetics)*, pages 39–44, 2000.

91. J.E. Flood (Ed.). *Telecommunication Networks*. Publishing & Inspec, 1997.

92. S. Forrest, A.S. Perelson, L. Allen, and R. Cherukuri. Self-nonself discrimination in a computer. In *Proceedings of the IEEE Symposium on Research in Security and Privacy*, pages 202–212. IEEE Computer Society Press, 1994.

93. R.L. Freeman. *Telecommunication System Engineering*. John Wiley & Sons, Inc, 2004.

94. E. Gafni and D. Bertsekas. Distributed algorithms for generating loopfree routes in networks with frequently changing topology. *IEEE Transactions on Communications*, 29:11–18, 1981.

95. R. Gallagher. A minimum delay routing algorithm using distributed computation. *IEEE Transactions on Communications*, 25:73–85, 1979.

96. M. Gallego-Schmid. Modified AntNet: software application in the evaluation and management of a telecommunication network. In Una-May O'Reilly, editor, *Graduate Student Workshop*, pages 353–354, Orlando, Florida, USA, 13 July 1999.

97. A. Giessler, J.D. Haenle, A. König, and E. Pade. Free buffer allocation – an investigation by simulation. *Computer Networks*, 2:191–208, 1978.

98. D.E. Goldberg. *Genetic Algorithms in Search, Optimization and Machine Learning*. Addison Wesley, Reading, MA, 1989.

99. A.Y. Grama, A. Gupta, and V. Kumar. Isoefficiency: measuring the scalability of parallel algorithms and architectures. *IEEE parallel and distributed technology: systems and applications*, 1(3):12–21, 1993.

100. P.P. Grassé. La reconstruction du nid et les coordinations interindividuelles chez bellicositermes natalensis et cubitermes sp. la théorie de la stigmergie: essai d'interprétation du comportement des termites constructeurs. *Insectes Sociaux*, 6:41–81, 1959.

101. R.S. Gray, G. Cybenko, D. Kotz, and D. Rus. Mobile agents: motivations and state of the art. In Jeffrey Bradshaw, editor, *Handbook of Agent Technology*. AAAI/MIT Press, 2002. Accepted for publication. Draft available as Technical Report TR2000-365, Department of Computer Science, Dartmouth College.

102. R. Guerin, S. Blake, and S. Herzog. Aggregating RSVP-based QoS requests, November 1997. http://tools.ietf.org/html/draft-guerin-aggreg-rsvp-00.

103. M. Günes, U. Sorges, and I. Bouazizi. ARA – the ant-colony based routing algorithm for MANETs. In *ICPPW '02: Proceedings of the 2002 International Conference on Parallel Processing Workshops*, pages 79–85, 2002.

104. A. Gupta and V. Kumar. Performance properties of large scale parallel systems. *Journal of Parallel and Distributed Computing*, 19(3):234–244, 1993.

105. A. Harsch. Design and development of a network infrastructure for swarm routing protocols inside Linux. Master's thesis, LSIII, University of Dortmund, Germany, July 2005.

106. R. Hauser, T. Przygienda, and G. Tsudik. Reducing the cost of security in link-state routing. In *SNDSS '97: Proceedings of the 1997 Symposium on Network and Distributed System Security*, page 93, Washington, DC, USA, 1997. IEEE Computer Society.

107. A.L.G. Hayzelden and J. Bigham. Agent technology in communications systems: an overview. *Knowledge Engineering Review*, 1999.

108. C.L. Hedrick. RFC 1058: Routing information protocol, June 1998.

109. W. Heinzelman, A. Chandrakasan, and H. Balakrishnan. Energy-efficient communication protocol for wireless microsensor networks. In *Proceeding of 33rd Hawaii Int'l. Conf. Sys. Sci., Jan. 2000*.

110. W. Heinzelman, J. Kulik, and H. Balakrishnan. Adaptive protocols for information dissemination in wireless sensor networks. In *MobiCom 99: Proceedings of the 5th Annual ACM/IEEE International Conference on Mobile Computing and Networking (MobiCom'99), Seattle, WA, August 1999*, 1999.

111. T. Hendtlass and M. Ali, editors. *Collective Intelligence and Priority Routing in Networks*. Springer Verlag, 2002.

112. J.L. Hennessey and D.A. Patterson. *Computer Organization and Design: The Hardware/Software Interface*. Morgan Kaufmann, 1995.

113. M. Heusse, D. Syners, S. Guerin, and P. Kuntz. Adaptive agent-driven routing and load balancing in communication networks. *Advances in Complex Systems*, 1(2-3):237–254, 1998.

114. G.N. Higginbottom. *Performace Evaluation of Communication Networks*. Artech House Inc, Norwood, MA, 1998.

115. V. Hilaire, A. Koukam, P. Gruer, and J.P. Müller. Formal specification and prototyping of multi-agent systems. In *Engineering Societies in the Agent World, First International Workshop, ESAW 2000, Berlin, Germany, August 21, 2000, Revised Papers*, volume 1972 of *Lecture Notes in Computer Science*, pages 114–127. Springer, 2000.

116. J. Holland. *Adaptation in Natural and Artificial Systems*. The University of Michigan Press, 1975.

117. Y. Hu, A. Perrig, and D.B. Johnson. Rushing attacks and defense in wireless ad hoc network routing protocols. Technical Report TR01-385, Department of Computer Science, Rice University, 1992.

118. Y.C. Hu, A. Perrig, and D.B. Johnson. Efficient security mechanisms for routing protocol. In *Proceedings of the Network and Distributed System Security Symposium, NDSS 2003, San Diego, California, USA*. The Internet Society, 2003.

119. Y.C. Hu, A. Perrig, and D.B. Johnson. Ariadne: a secure on-demand routing protocol for ad hoc networks. *Wireless Networks*, 11(1-2):21–38, 2005.

120. N.C. Hutchinson and L.L. Peterson. The x-kernel: an architecture for implementing network protocols. *IEEE Transactions on Software Engineering*, 17(1):64–76, 1991.

121. C.A. Iglesias, M. Garijo, J. Centeno-González, and Juan R. Velasco. Analysis and design of multiagent systems using MAS-common KADS. In *Intelligent Agents IV, Agent Theories, Architectures, and Languages, 4th International Workshop, ATAL '97, Providence, Rhode Island, USA, July 24-26, 1997, Proceedings*, volume 1365 of *Lecture Notes in Computer Science*, pages 313–327. Springer, 1997.

122. C. Intanagonwiwat, R. Govindan, and D. Estrin. Directed diffusion: a scalable and robust communication paradigm for sensor networks. In *Proc. ACM Mobi-Com 2000, Boston, MA, 2000*, 2000.

123. P. Jain. Validation of AntNet as a superior single path, single constrained algorithm. Master's thesis, Department of Computer Science and Engineering, University of Minnesota, USA, 2002.

124. N.R. Jennings. Agent-oriented software engineering in multi-agent system engineering. In *MultiAgent System Engineering, 9th European Workshop on Modelling Autonomous Agents in a Multi-Agent World, MAAMAW '99, Valencia, Spain, June 30 - July 2, 1999, Proceedings*, volume 1647 of *Lecture Notes in Computer Science*, pages 1–7. Springer, 1999.

125. P. Jogalekar and C.M. Woodside. A scalability metric for distributed computing applications in telecommunications. In *Proceedings of the 15th International Teletraffic Congress*, pages 101–110, June 1997.

126. P. Jogalekar and M. Woodside. Evaluating the scalability of distributed systems. In *HICSS '98: Proceedings of the Thirty-First Annual Hawaii International Conference on System Sciences–Volume 7*, page 524, Washington, DC, USA, 1998. IEEE Computer Society.

127. P. Jogalekar and M. Woodside. Evaluating the scalability of distributed systems. *IEEE Transactions on Parallel Distributed Systems*, 11(6):589–603, 2000.

128. D.B. Johnson and D.A. Maltz. Dynamic source routing in ad hoc wireless networks. In T. Imielinski and H.F. Korth, editors, *Mobile Computing*, pages 153–181. Kluwer Academic Publishers, 1996.

129. Christine E. Jones, Krishna M. Sivalingam, Prathima Agrawal, and Jyh-Cheng Chen. A survey of energy efficient network protocols for wireless networks. *Wireless Networks*, 7(4):343–358, 2001.

130. L.P. Kaelbling, M.L. Littman, and A.W. Moore. Reinforcement learning: a survey. *Journal of Artificial Intelligence Research*, 4:237–285, May 1996.

131. I. Kassabalidis, M.A. El-Sharkawi, and R.J. Marks. Adaptive-SDR: adaptive swarm-based distributed routing. In *Proceedings of the 2002 International Joint Conference on Neural Networks, 2002 IEEE World Congress on Computational Intelligence*, pages 2878–2883, 2002.

132. I. Kassabalidis, M.A. El-Sharkawi, R.J. Marks, P. Arabshahi, and A.A. Gray. Swarm intelligence for routing in communication networks. In *Global Telecommunications Conference GLOBECOM*, pages 3613–3617. IEEE, 2001.

133. J. Katz and M. Yung. Scalable protocols for authenticated group key exchange. In *Advances in Cryptology – CRYPTO 2003, 23rd Annual International Cryptology Conference, Santa Barbara, California, USA, August 17–21, 2003, Proceedings*, volume 2729 of *Lecture Notes in Computer Science*, pages 110–125. Springer, 2003.

134. J. Kay and J. Pasquale. The importance of non-data touching processing overheads in TCP/IP. In *SIGCOMM'93: Conference proceedings on Communications Architectures, Protocols and Applications*, pages 259–268, New York, NY, USA, 1993. ACM Press.

135. F. Kelly. Network routing. *Philosophical Transactions of the Royal Society*, 337:343–367, 1991.

136. F. Kelly, S. Zachary, and I. Zeidins. *Stochastic Networks Theory and Applications*. Oxford Science Publications, 1996.

137. S. Keshav. *An Engineering Approach to Computer Networking: ATM Networks, the Internet, and the Telephone Network*. Addison-Wesley Longman Publishing Co., Inc., Boston, MA, USA, 1997.

138. L. Kleinrock. On flow control in computer networks. In *Proceedings of the International Conference on Communications*, volume 2, pages 27.2.1–27.2.5, Toronto, Canada, June 1978. IEEE.

139. L. Kleinrock. Power and deterministic rules of thumb for probabilistic problems in computer communications. In IEEE, editor, *Proceedings of the International Conference on Communications*, ICC, pages 335–347, France, 1979.

140. Hartmut König. *Protocol Engineering* (in German). Teubner, 2003.

141. D. Kotz and R.S. Gray. Mobile agents and the future of the internet. *Operating Systems Review*, 33(3):7–13, 1999.

142. V. Kumar and A. Gupta. Analyzing scalability of parallel algorithms and architectures. *Journal of Parallel and Distributed Computing*, 22(3):379–391, 1994.

143. K. Kümmerle and H. Rudin. Packet and circuit switching: Cost/performance boundaries. *Computer Networks*, 2:3–17, 1978.

144. L. Lamport, R. Shostak, and M. Pease. The Byzantine generals problem. *ACM Transaction on Programming Languages and Systems*, 4(3):382–401, 1982.

145. G.M. Lee and J.S. Choi. A survey of multipath routing for traffic engineering. Term paper, Informations and Communications University, Korea, 2002.

146. S. Liang, A.N. Zincir-Heywood, and M.I. Heywood. The effect of routing under local information using a social insect metaphor. In *Proceedings of IEEE Congress on Evolutionary Computing*, May 2002.

147. S. Liang, A.N. Zincir-Heywood, and M.I. Heywood. Intelligent packets for dynamic network routing using distributed genetic algorithm. In *Proceedings of Genetic and Evolutionary Computation Conference*. GECCO, July 2002.

148. S. Lindsey and C. Raghavendra. Pegasis: Power-efficient gathering in sensor information systems. In *IEEE Aerospace Conf. Proc., 2002, vol. 3, 9–16, pp. 1125–30*, 2002.

149. R. Love. *Linux Kernel Development*. Novel Press, second edition, 2005.

150. C. Madukife. Development of a formal framework to analyze the behavior of swarm routing protocols. Master's thesis, LSIII, University of Dortmund, Germany, August 2005.

151. G. Malkin. RFC 2453: RIP version 2, November 1998.

152. V. Maniezzo and A. Carbonaro. Ant colony optimization: an overview. In C. Ribeiro, editor, *Essays and Surveys in Metaheuristics*, pages 21–44. Kluwer, 2001.

153. N. Mazhar and M. Farooq. BeeAIS: artificial immune system security for nature inspired, MANET routing protocol, BeeAdHoc. In *Proceedings of the 6th International Conference on Artificial Immune Systems*, volume 4628 of *LNCS*, pages 370–381. Springer, 2007.

154. N. Mazhar and M. Farooq. Vulnerability analysis and secuirty framework (BeeSec) for nature inspired MANET routing protocols. In *Proceedings of GECCO*, pages 102–109. ACM Press, 2007.

155. N. Mazhar and M. Farooq. A sense of danger: dendritic cells inspired artificial immune system for MANET security. In *Proceedings of GECCO*. ACM Press, 2008.

156. A. Medina, A. Lakhina, I. Matta, and J. Byers. BRITE: universal topology generation from a user's perspective. Technical Report BU-CS-TR-2001-003, Boston University, 1 2001.

157. A. Medina, A. Lakhina, I. Matta, and J. Byers. BRITE: universal topology generation from a user's perspective. In *Proceedings of the International Workshop on Modeling, Analysis and Simulation of Computer and Telecommunications Systems–MASCOTS '01*. Cincinnati, Ohio, August 2001.

158. A. Medina, I. Matta, and J. Byers. On the origin of power laws in internet topologies. *ACM Computer Communication Review*, 30(2):18–28, April 2000.

159. J. Mehdi. *Stochastic Processes*. John Wiley Publication, 1983.

160. D. Merkle, M. Middendorf, and A. Scheidler. Dynamic decentralized packet clustering in networks. In *Applications of Evolutionary Computing*, pages 574–583. Springer Verlag, April 2005.

161. T. Michalareas and L. Sacks. Link-state and ant-like algorithm behaviour for single-constrained routing. In *IEEE Workshop on High Performance Switching and Routing, HPSR 2001*, pages 302–305, May 2001.

162. T. Michalareas and L. Sacks. Stigmergic techniques for solving multi-constraint routing for packet networks. In Pascal Lorenz, editor, *Networking – ICN 2001, First International Conference, Colmar, France, July 9-13, 2001 Proceedings, Part 1 LNCS 2093*, pages 687–697. Springer-Verlag, 2001.

163. D.L. Mills. RFC 958: Network time protocol (NTP), September 1985.

164. N. Minar, K.H. Kramer, and P. Maes. *Cooperating Mobile Agents for Dynamic Network Routing*, chapter 12, pages 287–304. Springer-Verlag, 1999.

165. J. Mogul. IP network performance. In D.C. Lynch and M.T. Rose, editors, *Internet System Handbook*. Addison Wesley, 1993.

166. J. Mogul, R. Rashid, and M. Accetta. The packer filter: an efficient mechanism for user-level network code. In *SOSP '87: Proceedings of the Eleventh ACM Symposium on Operating Systems Principles*, pages 39–51, New York, NY, USA, 1987. ACM Press.

167. R. Mortier. Internet traffic engineering. Technical Report UCAM-CL-TR-532, University of Cambridge, Computer Laboratory, April 2002.

168. C. Moschovitis, H. Poole, T. Schuyler, and T. Senft. *History of the Internet: A Chronology, 1843 to the Present*. ABC-CLIO, 1999.

169. J.T. Moy. *OSPF Anatomy of an Internet Routing Protocol*. Addison-Wesley, 1998.

170. J.T. Moy. *OSPF Complete Implementation*. Addison-Wesley, 2000.

171. M. Munetomo. Designing genetic algorithms for adaptive routing algorithms in the internet. In *Proceedings of GECCO'99 Workshop on Evolutionary Telecommunications: Past, Present and Future*. Orlando, Florida, July 1999.

172. M. Munetomo. *Network Routing with the Use of Evolutionary Methods, in Computational Intelligence in Telecommunication Networks*. CRC Press, 2000.

173. M. Munetomo, Y. Takai, and Y. Sato. An adaptive network routing algorithm employing path genetic operators. In *Proceedings of the Seventh International Conference on Genetic Algorithms*, pages 643–649. Morgan Kaufmann Publishers, 1997.

174. S.L. Murphy and M.R. Badger. Digital signature protection of the OSPF routing protocol. In *SNDSS '96: Proceedings of the 1996 Symposium on Network and Distributed System Security (SNDSS '96)*, page 93, Washington, DC, USA, 1996. IEEE Computer Society.

175. W. Nachtigall. *Bionik; Grundlagen und Beispiele für Ingenieure und Naturwissenschaftler* (in German). Springer-Verlag, 2002.

176. A. Newell. Physical symbol systems. *Cognitive Science*, 4:135–183, 1980.

177. P. Nii. The blackboard model of problem solving. *AI Mag*, 7(2):38–53, 1986.

178. K. Oida and A. Kataoka. Lock-free AntNet and its evaluation adaptiveness. *Journal of IEICE B (in Japanese)*, J82-B(7):1309–1319, 1999.

179. K. Oida and M. Sekido. An agent-based routing system for QoS guarantees. In *IEEE International Conference on Systems, Man, and Cybernetics*, pages 833–838, 1999.

180. K. Oida and M. Sekido. ARS: An efficient agent-based routing system for QoS guarantees. *Computer Communications*, 23:1437–1447, 2000.

181. E. Osborne and A. Simha. *Traffic Engineering with MPLS*. Cisco Press, 2002.

182. S. Ossowski and A. García-Serrano. Social structure in artificial agent societies: Implications for autonomous problem-solving agents. In *Intelligent Agents V, Agent Theories, Architectures, and Languages, 5th International Workshop, ATAL '98, Paris, France, July 4-7, 1998, Proceedings*, volume 1555 of *Lecture Notes in Computer Science*, pages 133–148. Springer, 1995.

183. G.I. Papadimitriou. A new approach to the design of reinforcement schemes for learning automata: stochastic estimator learning algorithms. *IEEE Trans. Knowl. Data Eng.*, 6(4):649–654, 1994.

184. R. Pastor-Satorras and A. Vespignani. *Evolution and Structure of the Internet: A Statistical Physics Approach*. Cambridge University Press, 2004.

185. A.S. Perelson and G.F. Oster. Theoretical studies of clonal selection: minimal antibody repertoire size and reliability of self-nonself discrimination. *Journal of Theoretical Biology*, 81:645–670, 1993.

186. C. Perkins and P. Bhagwat. Highly dynamic destination-sequenced distance-vector routing (DSDV) for mobile computers. In *Proceedings of ACM SIGCOMM'94 Conference on Communications Architectures, Protocols and Applications*, pages 234–244, 1994.

187. C. Perkins and E. Royer. Ad-hoc on-demand distance vector routing. In *Proceedings of Second IEEE Workshop on Mobile Computing Systems and Applications*, pages 90–100, 1999.

188. R. Perlman. *Network Layer Protocols with Byzantine Robustness*. PhD thesis, Department of Electrical Engineering and Computer Science, Massachusetts Institute of Technology, 1988. Report MIT/LCS/TR 429.

189. L.L. Peterson and B.S. Davie. *Computer Networks: A Systems Approach*. Morgan Kaufmann Publishers, 2000.

190. J. Postel. RFC 768: User datagram protocol, August 1980.

191. J. Postel. RFC 793: Transmission control protocol, September 1981.

192. J. Postel and J.K. Reynolds. RFC 959: File transfer protocol, October 1985.

193. G.N. Purdy. *Linux iptables Pocket Reference*. O'Reilly & Associates, 2004.

194. C.S. Raghavendra, K.M.S. Krishna, and T. Znati, editors. *Wireless Sensor Networks*. Springer-Verlag, 2004.

195. Ronald L. Rivest, Adi Shamir, and Leonard M. Adleman. A method for obtaining digital signatures and public-key cryptosystems. *Commun. ACM*, 21(2):120–126, 1978.

196. J.A. Robinson. A machine-oriented logic based on the resolution principle. *Journal of ACM*, 12(1):23–41, January 1965.

197. R.D. Rosner. Circuit and packet switching. *Computer Networks*, 1:7–26, 1976.

198. S.M. Ross. *Stochastic Processes*. John Wiley Publication, 1983.

199. T. Rossmann and C. Tropea. *Bionik; Aktuelle Forschungsergebnisse in Natur-, Ingenieur- und Geisteswissenschaft* (in German). Springer-Verlag, 2004.

200. M. Roth and S. Wicker. Termite: ad-hoc networking with stigmergy. In *Proceedings of Golbecom*, volume 22, pages 2937–2941, 2003.

201. M. Roth and S. Wicker. Termite: emergent ad-hoc networking. In *Proceedings of the Second Mediterranean Workshop on Ad-Hoc Networks*, 2003.

202. M. Roth and S. Wicker. Asymptotic pheromone behavior in swarm intelligent MANETs: An analytical analysis of routing behavior. In *Sixth IFIP IEEE International Conference on Mobile and Wireless Communications Networks (MWCN)*, October 2004.

203. E. Royer and C. Toh. A review of current routing protocols for ad-hoc mobile wireless networks. *IEEE Personal Communications*, 1999.

204. S. Russell and P. Norvig. *Artificial Intelligence: A Modern Approach*. Prentice Hall, second edition, 2002.

205. T.N. Saadawi and M.H. Ammar. *Fundamentals of Telecommunication Networks*. John Wiley & Sons, Inc, 1994.

206. M. Saleem and M. Farooq. Beesensor: a bee-inspired power aware routing protocol for wireless sensor networks. In *M. Giacobini et al. (Eds.), Lecture Notes in Computer Science, LNCS 4449*, pages 81–90. Springer Verlag, 2007.

207. M. Saleem and M. Farooq. A framework for empirical evaluation of nature inspired routing protocols for wireless sensor networks. In *Proceedings of Congress on Evolutionary Computing (CEC)*, pages 751–758. IEEE, 2007.

208. H.G. Sandalidis, C.X. Mavromoustakis, and P. Stavroulakis. Ant based probabilistic routing with pheromone and antipheromone mechanisms. *Communication Systems*, 17:55–62, 2004.

209. H.G. Sandalidis, C.X. Mavromoustakis, and P.P. Stavroulakis. Performance measures of an ant based decentralised routing scheme for circuit switching communication networks. *Soft Comput.*, 5(4):313–317, 2001.

210. C. Santivanez, B. McDonald, I. Stavrakakis, and R. Ramanathan. On the scalability of ad hoc routing protocols. In *Proceedings of IEEE INFOCOM 2002*. IEEE, June 2002.

294   References

211. S.R. Sarukkai, P. Mehta, and R.J. Block. Automated scalability anaylsis of message-passing parallel programs. *IEEE Parallel and Distributed Technology: Systems and Applications*, 3(4):21–32, 1995.

212. R. Schoonderwoerd and O. Holland. Minimal agents for communications network routing: the social insect paradigm. *Software Agents for Future Communication Systems*, (1), 1999.

213. R. Schoonderwoerd, O. Holland, and J. Bruten. Ant-like agents for load balancing in telecommunications networks. In *Agents*, pages 209–216, 1997.

214. R. Schoonderwoerd, O.E. Holland, J.L. Bruten, and L.J.M. Rothkrantz. Ant-based load balancing in telecommunications networks. *Adaptive Behavior*, 5(2):169–207, 1996.

215. R.H. Schwartz and J. Banchereau. Immune tolerance. *The Immunologist 4*, pages 211–218, 1996.

216. T.D. Seeley. *The Wisdom of the Hive*. Harvard University Press, London, 1995.

217. T.D. Seeley and W.F. Towne. Collective decision making in honey bees: how colonies choose among nectar sources. *Behavior Ecology and Sociobiology*, 12:277–290, 1991.

218. R. Serfozo. *Introduction to Stochastic Networks*. Springer-Verlag, 1999.

219. R.W. Shirey. Security architecture for internet protocols. a guide for protocol design and standards, 1994. Internet Draft: draft-irtf-psrg-secarch-sect-100.txt.

220. A. Silberschatz and P.B. Galvin. *Operating System Concepts (4th Edition)*. Addison-Wesley, 1994.

221. K.M. Sim and W.H. Sun. Ant colony optimization for routing and load-balancing: survey and new directions. *IEEE Transactions on Systems, Man and Cybernetics–Part A*, 33(5):560–572, 2003.

222. G. Simon. Probabilistic wireless network simulator. http://www.isis.vanderbilt.edu/projects/nest/prowler/.

223. H.A. Simon. *Administrative Behavior: A Study of Decision-Making Processes in Administrative Organization*. Free Press, New York, 1976.

224. M.C. Sinclair. Evolutionary telecommunications: a summary. In *Proceedings of GECCO'99 Workshop on Evolutionary Telecommunications: Past, Present and Future*. Orlando, Florida, July 1999.

225. Munindar P. Singh. Agent communication languages: rethinking the principles. *IEEE Computer*, 31(12):40–47, 1998.

226. S. Singh, M. Woo, and C.S. Raghavendra. Power-aware routing in mobile ad hoc networks. In *Proceedings of Fourth ACM/IEEE Conference on Mobile Computing and Networking (MobiCom)*, pages 181–190, 1998.

227. A. Sivasubramaniam, U. Ramachandran, and H. Venkateswaran. A comparative evaluation of techniques for studying parallel system performance. Technical report, College of Computing, Georgia Institute of Technology, Sept. 1994. Technical Report GIT-CC-94/38.

228. B.R. Smith, S. Murthy, and J.J. Garcia-Luna-Aceves. Securing distance-vector routing protocols. In *Proceedings of the Symposium on Network and Distributed Systems Security*, pages 85–92, 1997.

229. C.U. Smith. Designing high-performance distributed applications using software performance engineering: a tutorial. In *22nd International Computer Measurement Group Conference, December 10-13, 1996, San Diego, CA, USA, Proceedings*, pages 498–507. Computer Measurement Group, 1996.

230. C.U. Smith and L.G. Williams. Building responsive and scalable Web applications. In *26th International Computer Measurement Group Conference, December 10-15, 2000, Orlando, FL, USA, Proceedings*, pages 127–138. Computer Measurement Group, 2000.

231. C.U. Smith and L.G. Williams. Performance and scalability of distributed software architectures: an SPE approach. *Parallel and Distributed Computing Practices*, 3(4), 2000.

232. P.G. Spirakis and C.D. Zaroliagis. *Distributed Algorithm Engineering*, pages 197–228. Springer-Verlag New York, Inc., New York, NY, USA, 2002.

233. R.E. Steuer. *Multiple Criteria Optimization: Theory, Computation and Application*. Wiley, New York, 1986.

234. W.R. Stevens. *TCP/IP Illustrated: The Protocols*, volume 1. Addison Wesley, 1994.

235. W.R. Stevens. *TCP/IP Illustrated: TCP for Transactions, HTTP, NTTP, and the UNIX Domain Protocols*, volume 3. Addison Wesley, 1996.

236. W.R. Stevens. *UNIX Network Programming: Networking APIs – Sockets and XTI*, volume 1. Addison-Wesley, first edition, 1997.

237. W.R. Stevens. *UNIX Network Programming, Interprocess Communication*, volume 2. Addison Wesley, 2 edition, 1999.

238. P. Stone and M.M. Veloso. Multiagent systems: a survey from a machine learning perspective. *Auton. Robots*, 8(3):345–383, 2000.

239. T. Stützle and H.H. Hoos. MAX-MIN ant system. *Future Generation Computer Systems*, 16(8):889–914, 2000.

240. Z. Subing and L. Zemin. A QoS routing algorithm based on ant algorithm. In *Proceedings of IEEE International Conference on Communications (ICC'01)*, pages 1587–1591, 2001.

241. D. Subramanian, P. Druschel, and J. Chen. Ants and reinforcement learning: a case study in routing in dynamic networks. In *Proceedings of 15th Joint Conference on Artificial Intelligence (IJCAI 97)*, pages 832–839. Morgan Kaufmann, San Francisco, CA, 1997.

242. D.J.T. Sumpter. From bee to society: an agent-based investigation of honey bee colonies. Ph.D. thesis, The University of Manchester, UK, 2000.

243. X.H. Sun and L.M. Ni. Scalable problems and memory-bounded speedup. *Journal of Parallel and Distributed Computing*, 19(1):27–37, 1993.

244. S. Tadrus and L. Bai. A QoS network routing algorithm using multiple pheromone tables. In *Web Intelligence*, pages 132–138, 2003.

245. H.A. Taha. *Operations Research*. John Wiley & Sons, 1982.

246. A.S. Tanenbaum. *Modern Operating Systems*. Internals and Design Principles. Prentice-Hall International, 1992.

247. P. Tarasewich and P.R. McMullen. Swarm intelligence: power in numbers. *Communications of ACM*, 45(8):62–67, 2002.

248. C.A. Thekkath, T.D. Nguyen, E. Moy, and E.D. Lazowska. Implementing network protocols at user level. In *SIGCOMM'93: Conference proceedings on Communications architectures, protocols and applications*, pages 64–73, New York, NY, USA, 1993. ACM Press.

249. M. Thirunavukkarasu. Reinforcing reachable routes. Master's thesis, Virginia Polytechnic Institute and State University, 2004.

250. R.A. Tintin and D.I. Lee. Intelligent and mobile agents over legacy, present and future telecommunication networks. In *First International Workshop on Mobile Agents for Telecommunication Applications (MATA'99)*, pages 109–126. World Scientific Publishing Ltd., 1999.

251. K.S. Trivedi. *Probability and Statistics with Reliability, Queuing and Computer Science Applications*. John Wiley – Interscience Publication, 2002.

252. R. van der Put. Routing in packet switched networks using agents. Master's thesis, KBS, Delft University of Technology, Netherlands, 1998.

253. R. van der Put. Routing in the fax factory using mobile agents. Technical report, KPN Research, 1998.

254. S. Varadarajan, N. Ramakrishnan, and M. Thirunavukkarasu. Reinforcing reachable routes. *Computer Networks*, 43(3):389–416, 2003.

255. A. Varga. OMNeT++: Discrete event simulation system: user manual. http://www.omnetpp.org.

256. B. Vetter, F. Wang, and S.F. Wu. An experimental study of insider attacks for OSPF routing protocol. In *ICNP '97: Proceedings of the 1997 International Conference on Network Protocols (ICNP '97)*, page 293, Washington, DC, USA, 1997. IEEE Computer Society.

257. C. Villamizar. OSPF optimized multipath (OSPF-OMP). In *Proceedings of the fourty-fourth Internet Engineering Task Force*, INTERNET DRAFT, draft-ietf-ospf-omp-02, Minneapolis, MN, USA, February 1999.

258. T. von Eicken, A. Basu, V. Buch, and W. Vogels. U-Net: a user-level network interface for parallel and distributed computing. In *Proceedings of ACM Symposium on Operating systems principles*, pages 40–53, 1995.

259. K. von Frisch. *The Dance Language and Orientation of Bees*. Harvard University Press, Cambridge, 1967.

260. S. Vutukury. Multipath routing mechanisms for traffic engineering and quality of service in the internet. Ph.D. thesis, University of California, Santa Cruz, 2001.

261. S. Vutukury and J. Garcia-Luna-Aceves. An algorithm for multipath computation using distance-vectors with predecessor information. In *Proceedings of 8th International Conference of IEEE Computer Communications and Networks*, pages 534–539. IEEE Press, 1999.

262. S. Vutukury and J. Garcia-Luna-Aceves. A distributed algorithm for multipath computation. In *Proceedings of GLOBECOM*, pages 1689–1693, 1999.

263. S. Vutukury and J.J. Garcia-Luna-Aceves. A simple approximation to minimum-delay routing. *ACM SIGCOMM Computer Communication Review*, 29(4):227–238, 1999.

264. S. Vutukury and J.J. Garcia-Luna-Aceves. MDVA: a distance-vector multipath routing protocol. In *INFOCOM*, pages 557–564, 2001.

265. C.J. Watkins. Learning from delayed rewards. Ph.D. thesis, Psychology Department, University of Cambridge, UK, 1989.

266. C.J. Watkins and P. Dayan. Q-learning. *Machine Learning*, 1:279–292, 1992.

267. R.W. Watson and S.A. Mamrak. Gaining efficiency in transport services by appropriate design and implementation choices. *ACM Transactions on Computer Systems*, 5(2):97–120, 1987.

268. H.F. Wedde and M. Farooq. A performance evaluation framework for nature inspired routing algorithms. In *Applications of Evolutionary Computing, LNCS 3449*, pages 136–146. Springer Verlag, March 2005.

269. H.F. Wedde and M. Farooq. The wisdom of the hive applied to mobile ad-hoc networks. In *Proceedings of the IEEE Swarm Intelligence Symposium*, pages 341–348, 2005.

270. H.F. Wedde, M. Farooq, T. Pannenbaecker, B. Vogel, C. Mueller, J. Meth, and R. Jeruschkat. BeeAdHoc: an energy efficient routing algorithm for mobile ad-hoc networks inspired by bee behavior. In *Proceedings of ACM GECCO*, pages 153–160, 2005.

271. H.F. Wedde, M. Farooq, T. Pannenbaecker, B. Vogel, C. Mueller, J. Meth, R. Jeruschkat, M. Duhm, L. Bensmann, G. Kathagen, K. Moritz, R. Zeglin, and T. Büning. BeeHive: an energy-aware scheduling and routing framework. Technical report–PG439, LSIII, School of Computer Science, University of Dortmund, 2004.

272. H.F. Wedde, M. Farooq, C. Timm, J. Fischer, M. Kowalski, M. Langhans, N. Range, C. Schletter, R. Tarak, M. Tchatcheu, F. Volmering, S. Werner, and K. Wang. BeeAdHoc: an efficient, secure, scalable routing framework for mobile adhoc networks. Technical report–PG460, LSIII, School of Computer Science, University of Dortmund, 2005.

273. H.F. Wedde, M. Farooq, and Y. Zhang. BeeHive: an efficient fault-tolerant routing algorithm inspired by honey bee behavior. In *Ant Colony Optimization and Swarm Intelligence, LNCS 3172*, pages 83–94. Springer Verlag, Sept 2004.

274. H.F. Wedde, C. Timm, and M. Farooq. BeeHiveAIS: a simple, efficient, scalable and secure routing framework inspired by artificial immune systems. In *Proceedings of the PPSN IX*, volume 4193 of *Lecture Notes in Computer Science*, pages 623–632. Springer Verlag, September 2006.

275. H.F. Wedde, C. Timm, and M. Farooq. BeeHiveGuard: a step towards secure nature inspired routing algorithms. In *Applications of Evolutionary Computing*, volume 3907 of *Lecture Notes in Computer Science*, pages 243–254. Springer Verlag, April 2006.

276. M. Weiser. The computer for the 21st century. *Scientific American*, pages 933–940, 1991.

277. M. Weiser. Hot topics: ubiquitous computing. *IEEE Computer*, 1993.

278. M. Weiser. Some computer science issues in ubiquitous computing. *Commun. ACM*, 36(7):74–84, 1993.

279. M. Weiser. The world is not a desktop. *Interactions*, 1(1):7–8, 1994.

280. G. Weiss, editor. *Multiagent Systems: A Modern Approach to Distributed Artificial Intelligence*. The MIT Press, San Francisco, CA, 1999.

281. A.R.P. White. *SynthECA: A Synthetic Ecology of Chemical Agents*. PhD thesis, Department of Systems and Computer Engineering, Carleton University, August 2000.

282. T. White. Routing with swarm intelligence. Technical report SCE-97-15, Systems and Computer Engineering Department, Carleton University, Canada, 1997.

283. T. White and B. Pagurek. Towards multi-swarm problem solving in networks. In *3rd International Conference on Multi-Agent Systems (ICMAS 1998), 3–7 July 1998, Paris, France*, pages 333–340. IEEE Computer Society, 1998.

284. T. White and B. Pagurek. Application oriented routing with biologically-inspired agents. In *Proceedings of Genetic Evolutionary Computation Conference (GECCO)*, pages 1453–1454, San Francisco, CA, USA, July 1999. Orlando, Florida, Morgan Kaufmann Publishers Inc.

285. T. White and B. Pagurek. Emergent behaviour and mobile agents. In *Proceedings of the workshop on Mobile Agents Coordination Cooperation Autonomous Agents*, Washington, Seattle, May 1-5 1999.

286. T. White, B. Pagurek, and D. Deugo. Biologically-inspired agents for priority routing in networks. In S.M. Haller and G. Simmons, editors, *Proceedings of the Fifteenth International Florida Artificial Intelligence Research Society Conference, May 14-16, 2002, Pensacola Beach, Florida, USA*, pages 282–287. AAAI Press, 2002.

287. T. White, B. Pagurek, and F. Oppacher. ASGA: Improving the ant system by integration with genetic algorithms. In J.R. Koza, W. Banzhaf, K. Chellapilla, K. Deb, M. Dorigo, D.B. Fogel, M.H. Garzon, D.E. Goldberg, H. Iba, and R. Riolo, editors, *Genetic Programming 1998: Proceedings of the Third Annual Conference*, pages 610–617, University of Wisconsin, Madison, Wisconsin, USA, 22-25 1998. Morgan Kaufmann.

288. T. White, B. Pagurek, and F. Oppacher. Connection management using adaptive agents. In *Proceedings of 1998 International Conference on Parallel and Distributed Processing Techniques and Applications (PDPTA'98)*, pages 802–809. CSREA Press, 1998.

289. L.G. Williams and C.U. Smith. PASA(SM): An architectural approach to fixing software performance problems. In *28th International Computer Measurement Group Conference, December 8-13, 2002, Reno, Nevada, USA, Proceedings*, pages 307–320. Computer Measurement Group, 2002.

290. M. Woodside. Scalability metrics and analysis of mobile agent systems. In *Revised Papers from the International Workshop on Infrastructure for Multi-Agent Systems*, pages 234–245, London, UK, 2001. Springer-Verlag.

291. M. Wooldridge and N.R. Jennings. Agent theories, architectures, and languages: A survey. In *Intelligent Agents, ECAI-94 Workshop on Agent Theories, Architectures, and Languages, Amsterdam, The Netherlands, August 8-9, 1994, Proceedings*, volume 890 of *Lecture Notes in Computer Science*, pages 1–39. Springer, 1995.

292. M. Wooldridge, N.R. Jennings, and D. Kinny. A methodology for agent-oriented analysis and design. In *ACM Third International Conference on Autonomous Agents*, pages 69–76, 1999.

293. G.R. Wright and W.R. Stevens. *TCP/IP Illustrated: The Implementation*, volume 2. Addison Wesley, 1995.

294. Y. Yang, A.N. Zincir-Heywood, M.I. Heywood, and S. Srinivas. Agent-based Routing Algorithms on a LAN. In *IEEE Canadian Conference on Electrical and Computer Engineering*, 1442-1447 2002.

295. J. Yu. RFC 2791: Scalable routing design principles, July 2000.

296. Y. Yu, D. Estrin, and R. Govindan. Geographical and energy-aware routing: a recursive data dissemination protocol for wireless sensor networks. UCLA Computer Science Department Technical Report, UCLA-CSD TR-01-0023, May, 2001.

297. L. Zhang, S. Deering, and D. Estrin. RSVP: a new resource reservation protocol. *IEEE Communications Magazine*, 31(9):8–18, 1993.

298. Y. Zhang. Routing modeling application simulation environment. http://www2.parc.com/isl/groups/era/nest/Rmase/.

299. Y. Zhang. Design and implementation of bee agents based algorithm for routing in high speed, adaptive and fault-tolerant networks. Master's thesis, LSIII, University of Dortmund, Germany, 2004.

300. Y. Zhang, L.D. Kuhn, and M.P.J. Fromherz. Improvements on ant routing for sensor networks. In *Int. Workshop on Ant Colony Optimization and Swarm Intelligence, ANTS. 2004, Sep. 2004*.

301. W. Zhong and D. Evans. When ants attack: security issues for stigmergic systems. UVA CS Technical Report, CS-2002-23, Department of Computer Science, University of Virginia, April 2002.

302. H. Zhu. Formal specification of agent behaviour through environment scenarios. In *Formal Approaches to Agent-Based Systems, First International Workshop, FAABS 2000 Greenbelt, MD, USA, April 5-7, 2000, Revised Papers*, volume 1871 of *Lecture Notes in Computer Science*, pages 263–277. Springer, 2000.

303. H. Zhu. SLABS: A formal specification language for agent-based systems. *International Journal of Software Engineering and Knowledge Engineering*, 11(5):529–558, 2001.

Index

Natural Computing Series

H.-P. Schwefel, I. Wegener, K. Weinert (Eds.): **Advances in Computational Intelligence. Theory and Practice.** VIII, 325 pages. 2003

A. Ghosh, S. Tsutsui (Eds.): **Advances in Evolutionary Computing. Theory and Applications.** XVI, 1006 pages. 2003

L.F. Landweber, E. Winfree (Eds.): **Evolution as Computation.** DIMACS Workshop, Princeton, January 1999. XV, 332 pages. 2002

M. Hirvensalo: **Quantum Computing.** 2nd ed., XI, 214 pages. 2004 (first edition published in the series)

A.E. Eiben, J.E. Smith: **Introduction to Evolutionary Computing.** XV, 299 pages. 2003

A. Ehrenfeucht, T. Harju, I. Petre, D.M. Prescott, G. Rozenberg: **Computation in Living Cells. Gene Assembly in Ciliates.** XIV, 202 pages. 2004

L. Sekanina: **Evolvable Components. From Theory to Hardware Implementations.** XVI, 194 pages. 2004

G. Ciobanu, G. Rozenberg (Eds.): **Modelling in Molecular Biology.** X, 310 pages. 2004

R.W. Morrison: **Designing Evolutionary Algorithms for Dynamic Environments.** XII, 148 pages, 78 figs. 2004

R. Paton†, H. Bolouri, M. Holcombe, J.H. Parish, R. Tateson (Eds.): **Computation in Cells and Tissues. Perspectives and Tools of Thought.** XIV, 358 pages, 134 figs. 2004

M. Amos: **Theoretical and Experimental DNA Computation.** XIV, 170 pages, 78 figs. 2005

M. Tomassini: **Spatially Structured Evolutionary Algorithms.** XIV, 192 pages, 91 figs., 21 tables. 2005

G. Ciobanu, G. Păun, M.J. Pérez-Jiménez (Eds.): **Applications of Membrane Computing.** X, 441 pages, 99 figs., 24 tables. 2006

K.V. Price, R.M. Storn, J.A. Lampinen: **Differential Evolution.** XX, 538 pages, 292 figs., 48 tables and CD-ROM. 2006

J. Chen, N. Jonoska, G. Rozenberg: **Nanotechnology: Science and Computation.** XII, 385 pages, 126 figs., 10 tables. 2006

A. Brabazon, M. O'Neill: **Biologically Inspired Algorithms for Financial Modelling.** XVI, 275 pages, 92 figs., 39 tables. 2006

T. Bartz-Beielstein: **Experimental Research in Evolutionary Computation.** XIV, 214 pages, 66 figs., 36 tables. 2006

S. Bandyopadhyay, S.K. Pal: **Classification and Learning Using Genetic Algorithms.** XVI, 314 pages, 87 figs., 43 tables. 2007

H.-J. Böckenhauer, D. Bongartz: **Algorithmic Aspects of Bioinformatics.** X, 396 pages, 118 figs., 9 tables. 2007

P. Siarry, Z. Michalewicz (Eds.): **Advances in Metaheuristics for Hard Optimization.** XVI, 481 pages, 66 figs., 83 tables. 2008

J. Knowles, D. Corne, K. Deb (Eds.): **Multiobjective Problem Solving from Nature. From Concepts to Applications.** XVI, 412 pages, 178 figs., 53 tables. 2008

P.F. Hingston, L.C. Barone, Z. Michalewicz (Eds.): **Design by Evolution.** XII, 362 pages, 143 figs., 20 tables. 2008

C. Blum, D. Merkle (Eds.): **Swarm Intelligence: Introduction and Applications.** X, 284 pages, 70 figs., 14 tables. 2008

M. Farooq: **Bee-Inspired Protocol Engineering.** XX, 306 pages, 128 figs., 61 tables. 2009